Protein Interactions: Research and Development

Protein Interactions: Research and Development

Edited by **Anton Torres**

New York

Published by Callisto Reference,
106 Park Avenue, Suite 200,
New York, NY 10016, USA
www.callistoreference.com

Protein Interactions: Research and Development
Edited by Anton Torres

International Standard Book Number: 978-1-63239-520-7 (Hardback)

Printed in the United States of America.

Contents

Preface

This book, written by renowned experts, provides advanced researches and concepts in the field of protein interactions. The book presents the interaction between proteins and other biomolecules, which is crucial to all features of biological procedures such as cell growth and differentiation. Hence, examination and modulation of protein interactions are of vital importance, as it not only discloses the mechanism governing cellular movement, but also leads to possible agents for the treatment of a broad range of disorders. The purpose of this book is to emphasize some of the most recent advances in the study of protein interactions, along with modulation of protein interactions, improvements in systematic methods, etc. It provides various examples of protein interaction as the book demonstrates the significance and the opportunities for further exploration of protein interactions.

This book is a result of research of several months to collate the most relevant data in the field.

When I was approached with the idea of this book and the proposal to edit it, I was overwhelmed. It gave me an opportunity to reach out to all those who share a common interest with me in this field. I had 3 main parameters for editing this text:

1. Accuracy – The data and information provided in this book should be up-to-date and valuable to the readers.
2. Structure – The data must be presented in a structured format for easy understanding and better grasping of the readers.
3. Universal Approach – This book not only targets students but also experts and innovators in the field, thus my aim was to present topics which are of use to all.

Thus, it took me a couple of months to finish the editing of this book.

I would like to make a special mention of my publisher who considered me worthy of this opportunity and also supported me throughout the editing process. I would also like to thank the editing team at the back-end who extended their help whenever required.

Editor

Examples of Protein Interactions

Autophagy-Mediated Defense Response of Mouse Mesenchymal Stromal Cells (MSCs) to Challenge with *Escherichia coli*

N.V. Gorbunov[1,*], B.R. Garrison[1], M. Zhai[1], D.P. McDaniel[2],
G.D. Ledney[3], T.B. Elliott[3] and J.G. Kiang[3,*]

[1]The Henry M. Jackson Foundation for the Advancement of Military Medicine, Inc.
[2]The Department of Microbiology and Immunology, School of Medicine,
[3]Radiation Combined Injury Program, Armed Forces Radiobiology Research Institute,
Uniformed Services University of the Health Sciences, Bethesda, Maryland,
USA

1. Introduction

Symbiotic microorganisms are spatially separated from their animal host, e.g., in the intestine and skin, in a manner enabling nutrient metabolism as well as evolutionary development of protective physiologic features in the host such as innate and adaptive immunity, immune tolerance, and function of tissue barriers (1,2). The major interface barrier between the microbiota and host tissue is constituted by epithelium, reticuloendothelial tissue, and mucosa-associated lymphoid tissue (MALT) (2,3).

Traumatic damage to skin and the internal epithelium in soft tissues can cause infections that account for 7% to 10% of hospitalizations in the United States (4). Moreover, wound infections and sepsis are an increasing cause of death in severely ill patients, especially those with immunosupression due to exposure to cytotoxic agents and chronic inflammation (4). It is well accepted that breakdown of the host-bacterial symbiotic homeostasis and associated infections are the major consequences of impairment of the "first line" of anti-microbial defense barriers such as the mucosal layers, MALT and reticuloendothelium (1-3). Under these impairment conditions of particular interest then is the role of sub-mucosal structures, such as connective tissue stroma, in the innate defense compensatory responses to infections.

The mesenchymal connective tissue of different origins is a major source of multipotent mesenchymal stromal cells (i.e., colony-forming-unit fibroblasts) (5, 6). Recent discovery of immunomodulatory function of mesenchymal stromal cells (MSCs) suggests that they are essential constituents that control inflammatory responses (6-7).

Recent *in vivo* experiments demonstrate promising results of MSC transfusion for treatment of acute sepsis and penetrating wounds (7-9). The molecular mechanisms underlying MSC

* Corresponding Authors

action in septic conditions are currently under investigation. It is known to date that (i) Gram-negative bacteria can induce an inflammatory response in MSCs *via* cascades of Toll-like receptor (type 4) and the nucleotide-binding oligomerization domain-containing protein 2 (NOD2) complexes recognizing the conserved pathogen-associated molecular patterns; (ii) activated MSCs can modulate the septic response of resident myeloid cells; and (iii) activated MSCs can directly suppress bacterial proliferation by releasing antimicrobial factors (10, 11).

Considering all of the above factors including the fact that MSCs are ubiquitously present in the sub-mucosal structures and conjunctive tissue, one would expect involvement of these cells in formation of antibacterial barriers and host-microbiota homeostasis. From this perspective our attention was attracted by the phagocytic properties of mesenchymal fibroblastic stromal cells documented in an early period of their investigation (5, 12). The phagocytosis mechanism is closely and synchronously connected with the cellular mechanisms of biodegradation mediated by the macroautophagy-lysosomal (autolysosomal) system (13-15). The last one decomposes proteins and organelles as well as bacteria and viruses inside cells and, therefore, is considered as a part of the innate defense mechanism (13- 15).

Macroautophagy (hereafter referred to as autophagy) is a catabolic process of bulk lysosomal degradation of cell constituents and phagocytized particles (16). Autophagy dynamics in mammalian cells are well described in recent reviews (14, 17-20). Thus, it was proposed that autophagy is initiated by the formation of the phagophore, followed by a series of steps, including the elongation and expansion of the phagophore, closure and completion of a double-membrane autophagosome (which surrounds a portion of the cytoplasm), autophagosome maturation through docking and fusion with an endosome (the product of fusion is defined as an amphisome) and/or lysosome (the product of fusion is defined as an autolysosome), breakdown and degradation of the autophagosome inner membrane and cargo through acid hydrolases inside the autolysosome, and recycling of the resulting macromolecules through permeases (14). These processes, along with the drastic membrane traffic, are mediated by factors known as autophagy-related proteins (i.e., ATG-proteins) and the lysosome-associated membrane proteins (LAMPs) that are conserved in evolution (21). The autophagic pathway is complex. To date there are over 30 ATG genes identified in mammalian cells as regulators of various steps of autophagy, e.g., cargo recognition, autophagosome formation, etc. (14, 22). The core molecular machinery is comprised of (i) components of signaling cascades, such as the ULK1 and ULK2 complexes and class III PtdIns3K complexes, (ii) autophagy membrane processing components, such as mammalian Atg9 (mAtg9) that contributes to the delivery of membrane to the autophagosome as it forms, and two conjugation systems: the microtubule-associated protein 1 (MAP1) light chain 3 (i.e., LC3) and the Atg12–Atg5–Atg16L complex. The two conjugation systems are proposed to function during elongation and expansion of the phagophore membrane (14, 19, 22, 23). A conservative estimate of the autophagy network counts over 400 proteins, which, besides the ATG-proteins, also include stress-response factors, cargo adaptors, and chaperones such as p62/SQSTM1 and heat shock protein 70 (HSP70) (15, 19, 22, 24, 26-28).

Autophagy is considered as a cytoprotective process leading to tissue remodeling, recovery, and rejuvenation. However, under circumstances leading to mis-regulation of the autolysosomal pathway, autophagy can eventually cause cell death, either as a precursor of apoptosis in apoptosis-sensitive cells or as a result of destructive self-digestion (29).

Based on this information we hypothesized that challenge of MSCs with *Escherichia coli* (*E. coli*) can induce a complex process where bacterial phagocytosis is accompanied by activation of autolysosomal pathway and stress-adaptive responses in MSCs. The objective of this current chapter is to provide evidence of this hypothesis.

2. Hypothesis test: Experimental procedures and technical approach

2.1 Bone marrow stromal cells

Bone marrow stromal cells were obtained from 3- to 4-month-old B6D2F1/J female mice using a protocol adapted from STEMCELL Technologies, Inc., and were expanded and cultivated in hypoxic conditions (5% O_2, 10% CO_2, 85% N_2) for approximately 30 days in MESENCULT medium (STEMCELL Technologies, Inc.) in the presence of antibiotics. Phenotype, proliferative activity, and colony-forming ability of the cells were analyzed by flow cytometry and immunofluorescence imaging using positive markers for mesenchymal stromal cells: CD44, CD105, and Sca1. The results of these analyses showed that the cultivated cells displayed properties of mesenchymal stromal clonogenic fibroblasts.

The experiments were performed in a facility accredited by the Association for the Assessment and Accreditation of Laboratory Animal Care-International (AAALAC-I). All animals used in this study received humane care in compliance with the Animal Welfare Act and other federal statutes and regulations relating to animals and experiments involving animals and adhered to principles stated in the Guide for the Care and Use of Laboratory Animals, NRC Publication, 1996 edition.

2.2 Challenge of MSCs with *Escherichia coli* bacteria

MSC cultures of approximate 80% confluency were challenged with proliferating *E. coli* (1×10^7 microorganisms/ml) for 1-5 h in antibiotic-free media. For assessment of the cellular alteration ≥ 5 h the incubation medium was replaced with fresh medium containing penicillin and streptavidin antibiotics. Bacteria-cell interaction was monitored with time-lapse microscopy using DIC imaging of MSCs and fluorescence imaging of *E. coli* labeled with PSVue® 480, a fluorescent cell tracking reagent (www.mtarget.com). At the end of the experiments the cells were either (i) harvested, washed, and lysed for qRT-PCR and immunoblot analyses, (ii) fixed for transmission electron microscopy and fluorescence confocal imaging, or (iii) used live for imaging of Annexin V reactivity, dihydrorhodamine 123, a sensitive indicator of peroxynitrite reactivity, and colony formation. With this protocol the cells were tested for (i) phagocytic activity; (ii) autolysosomal activity; (iii) production of reactive oxygen (ROS) and nitrogen species, (iii) stress responses to *E. coli*; (iv) genomic DNA damage and pro-apoptotic alterations; and (v) colony-forming ability. The results of observations indicated that challenge with *E. coli* did not diminish viability and colony forming ability of the cells under the selected conditions (Fig.1). Stimulation of MSCs with *E. coli* resulted in expression of the proinflammatory genes, IL-1α, IL-1β, IL-6, and iNOS, as determined with qRT-PCR analysis.

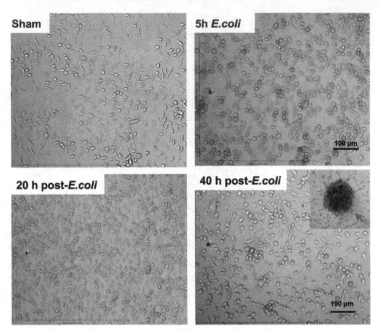

Conditions: MSCs were incubated with ~1x10⁷/ml *E. coli* for 5 h in medium (without antibiotics). After 5 h the medium was replaced with fresh medium (with antibiotics) and MSCs were incubated for another 40 h. Inset: formation of colonies (red arrowhead) occurred at 72 h post-exposure to *E. coli*.

Fig. 1. Bright field microscopy of MSCs challenged with *E. coli*. Images presented in the panels are MSCs at different time-points following exposure of MSCs to *E. coli*.

2.3 Analysis of the cell proteins

Proteins from MSCs were extracted in accordance with the protocol described previously (30). The aliquoted proteins (20 µg total protein per gel well) were separated on SDS-polyacrylamide slab gels (NuPAGE 4-12% Bis-Tris; Invitrogen, Carlsbad, CA). After electrophoresis, proteins were blotted onto a PDVF membrane and the blots were incubated with antibodies (1 µg/ml) raised against MAP LC3, Lamp-1, p62/SQSTM1, p65(NFκB), Nrf2, HSP70, iNOS, and actin (Abcam, Santa Cruz Biotechnology Inc., LifeSpan Biosciences, Inc., eBiosciences) followed by incubation with species-specific IgG peroxidase conjugate. IgG amounts did not alter after radiation. IgG, therefore, was used as a control for protein loading.

2.4 Immunofluorescent staining and image analysis

MSCs (5 specimens per group) were fixed in 2% paraformaldehyde and analyzed with fluorescence confocal microscopy following labeling (30). Normal donkey serum and antibody were diluted in phosphate-buffered saline (PBS) containing 0.5% BSA and 0.15% glycine. Any nonspecific binding was blocked by incubating the samples with purified normal donkey serum (Santa Cruz Biotechnology, Inc., Santa Cruz, CA) diluted 1:20. Primary antibodies were raised against MAP LC3, Lamp-1, p62/SQSTM1, p65(NFκB),

Nrf2, Tom 20, and iNOS. That was followed by incubation with secondary fluorochrome-conjugated antibody and/or streptavidin-AlexaFluor 610 conjugate (Molecular Probes, Inc., Eugene OR), and with Hoechst 33342 (Molecular Probes, Inc., Eugene OR) diluted 1:3000. Secondary antibodies used were AlexaFluor 488 and AlexaFluor 594 conjugated donkey IgG (Molecular Probes Inc., Eugene OR). Negative controls for nonspecific binding included normal goat serum without primary antibody or with secondary antibody alone. Five confocal fluorescence and DIC images of crypts (per specimen) were captured with a Zeiss LSM 7100 confocal microscope. The immunofluorescence image analysis was conducted as described previously (30).

2.5 Transmission Electron Microscopy (TEM)

MSCs in cultures were fixed in 4% formaldehyde and 4% glutaraldehyde in PBS overnight, post-fixed in 2% osmium tetroxide in PBS, dehydrated in a graduated series of ethanol solutions, and embedded in Spurr's epoxy resin. Blocks were processed as described previously (30). The sections of embedded specimens were analyzed with a Philips CM100 electron microscope.

2.6 RNA isolation and qRT-PCR

Total cellular RNA was isolated from MSC pellets using the Qiagen RNeasy miniprep kit, quantified by measuring the absorbance at 260nm on a Nanodrop, and qualified by electrophoresis on a 1.2% agarose gel. cDNA was synthesized using Superscript II (Invitrogen) and qRT-PCR was performed using SYBR Green iQ Supermix (Bio-Rad), each according to the manufacturers' instructions. The quality of qRT-PCR data were verified by melt curve analysis, efficiency determination, agarose gel electrophoresis, and sequencing. Relative gene expression was calculated by the method of Pfaffl using the formula $2^{-\Delta\Delta Ct}$(31).

2.7 Statistical analysis

Statistical significance was determined using one-way ANOVA followed by post-hoc analysis with pair-wise comparison by Tukey-Kramer test. Significance is reported at a level of $p < 0.05$.

3. Response of MSCs to challenge with *E. coli*

3.1 Phagocytosis and autolysosomal degradation of *E. coli* bacteria by MSCs

TEM images presented in Fig. 2 show different stages of cell-bacterium interaction. The uptake of microorganisms occurred in at least two independent events. The first event encompassed engulfing and taking in particles by the cell membrane extrusions (Fig. 2A1). The second event was tethering and "zipping" of adhered particles by the cell plasma membrane (Fig. 2A2 – 2A5). The time–lapse fluorescence microscopy observation indicated that these events proceeded quickly and the uptake process required a few minutes (not shown). Thereafter, a significant amount of bacteria in MSCs was observed within 1 h of co-incubation of the cells. The phagocytized bacteria were subjected to autolysosomal degradation (Fig 2B). Formation of the double-membrane autophagosomes, which incorporated bacteria, was observable in MSCs at 3 h of co-incubation and during a further period of observation. Fusion of autophagosomes with lysosomes also occurred at this period. Fragmentation of bacterial constituents was observed at 5 h of co-incubation and appearance of bacterial "ghosts" at 24 h (Fig. 2B).

Various cells eliminate bacterial microorganisms by autophagy, and this elimination is in many cases crucial for host resistance to bacterial translocation. Although autophagy is a non-selective degradation process, autophagosomes do not form randomly in the cytoplasm, but rather sequester the bacteria selectively (32, 33). Therefore, autophagosomes that engulf microbes are sometimes much larger than those formed during degradation of cellular organelles, suggesting that the elongation step of the autophagosome membrane is involved in bacteria-surrounding autophagy (33). The mechanism underlying selective induction of autophagy at the site of microbe phagocytosis remains unknown. However, it is likely mediated by pattern recognition receptors, stress-response elements, and adaptor proteins, e.g., p62/SQSTM1, which target bacteria and ultimately recruit factors essential for formation of autophagosomes (13,14, 33, 34).

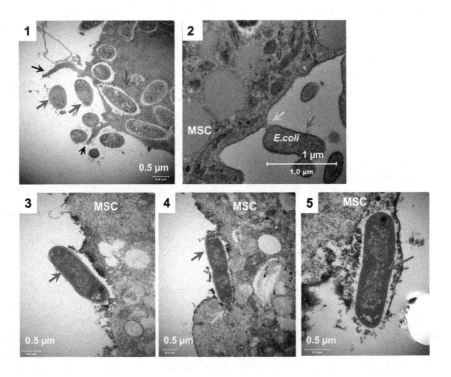

A

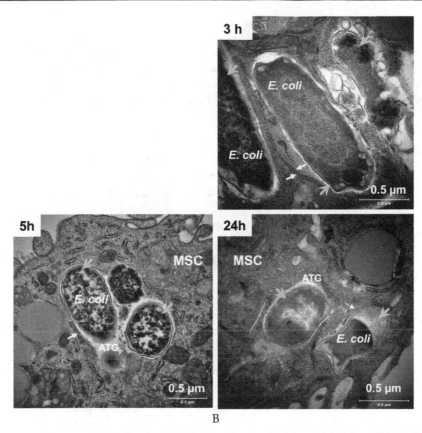

B

Conditions: MSCs were incubated with ~1x10⁷/ml *E. coli* either for 3 h or 5 h in MesenCult Medium (without antibiotics). After 5 h the medium was replaced with fresh medium (with antibiotics) and MSCs were incubated for another 19 h.

Fig. 2. Transmission electron micrographs (TEM) of *E. coli* phagocytosis by MSCs and autolysosomal degradation of phagocytized bacteria.
A) Panel A1: Engulfing and up-take of bacteria (red arrows) by the cell plasma membrane extrusions (black arrows). Panels A2-A5: Tethering and zipping (green arrows) and up-take of bacteria (red arrows) by the cell plasma membrane. Specimens were fixed at 3 h co-incubation of MSCs with bacteria.
B) Autolysosomal degradation of phagocytized bacteria at different time-points after exposure of MSCs to *E. coli* (green arrows). Autophagosome (ATG) membranes are indicated with yellow arrows. Lysosome fusion with autophagosomes is indicated with red arrows.

The results of TEM were corroborated by the data obtained with immunoblotting and immunofluorescence confocal imaging of autophagy MAP (LC3) protein, lysosomal LAMP1 and the ubiquitin-associated target adaptor p62. A key step in the autophagosome biogenesis is the conversion of light-chain protein 3 type I (LC3-I, also known as ubiqitin-like protein, Atg8) to type II (LC3-II). The conversion occurs via the cleavage of the LC3-I carboxyl terminus by a redox-sensitive Atg4 cysteine protease. The subsequent binding of the modified LC3-I to phosphatidylethanolamine, i.e., process of lipidation of LC3-I, on the isolation membrane, as it forms, is mediated by E-1- and E-2-like enzymes Atg7 and Atg3 (14). Therefore, conversion of LC3-I to LC3-II and formation of LC3-positive vesicles are considered to be a marker of activation of autophagy (14). A growing body of evidence suggests involvement of chaperone HSP70 in regulation of LC3-translocation. The results of immunoblot analysis of the proteins indicated an increase in the LC3-I to LC3-II – transition in the *E. coli* –challenged MSCs (Fig. 3).

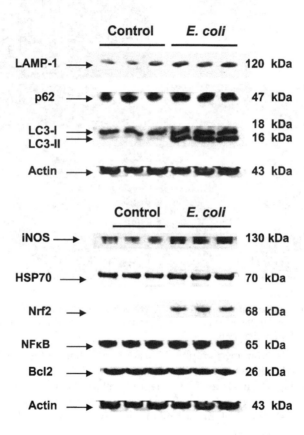

Conditions: MSCs were incubated with ~1x10⁷/ml *E. coli* for 3 h in MesenCult Medium (without antibiotics). After 3 h the medium was replaced with fresh medium (with antibiotics) and MSCs were further incubated for another 21 h.

Fig. 3. Immunoblotting analysis of LC3, LAMP1 autolysosomal proteins, p62 adaptor protein, and stress-response elements: NF-κB(p65), Nrf2, HSP70 in MSCs challenged with *E. coli*.

The images presented in Fig. 4A indicate an increase of formation of LC3-positive vesicles in MSCs challenged with *E. coli*. The LC3 immunoreactivity co-localized with immunoreactivity to LAMP1, a marker of lysosomes, indicating presence of fusion of autophagosomes with lysosomes, i.e., formation of autolysosomes (Fig. 4A). This effect

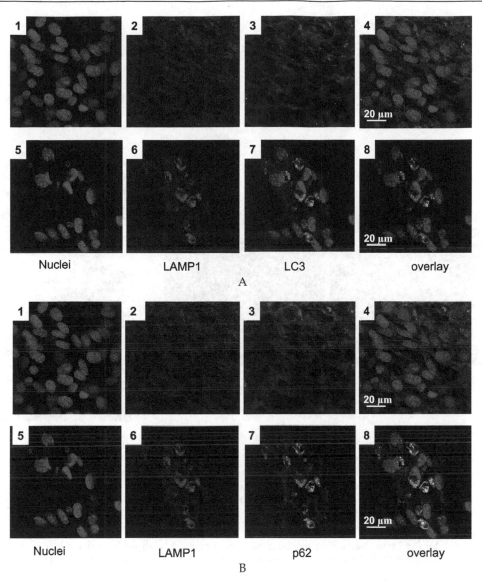

Conditions: MSCs were incubated with ~1x10⁷/ml *E. coli* for 3 h in MesenCult Medium (without antibiotics). After 3 h the medium was replaced with fresh medium (with antibiotics) and MSCs were further incubated for 21 h. Projections of LAMP1 protein (red channel) are shown in panels A2, A6, B2, and B6. Projections of LC3 protein (green channel) are shown in panels A3 and A7. Projections of p62 protein (green channel) are shown in panels B3 and B7. Counterstaining of nuclei was with Hoechst 33342 (blue channel). Panels A4, A8, B4, and B8 are overlay of signals acquired in the red, green, and blue channels. The confocal images were taken with pinhole setup to obtain 0.5 μm Z-sections.

Fig. 4. Immunofluorescence confocal imaging of the LC3, LAMP1, and p62 protein in MSCs challenged with *E. coli*. Panels A1-A4 and B1-B4 are control specimens. Panels A5-A8 and B5-B8 are challenged with *E. coli*.

was accompanied by the presence of immunoreactivity to p62, a marker of ubiquitin-dependent target transport, in autolysosomes that was associated with autophagy of *E. coli* (Fig. 4B, Fig. 5). The image analysis of autophagy was supported by results of immunoblotting of the proteins (Fig. 3). It should be noted that pre-incubation of cell cultures with wortmannin, an autophagy inhibitor, resulted in apoptotic transformations and ultimately loss of confluency approximately 3 h after challenge with *E. coli* (not shown).

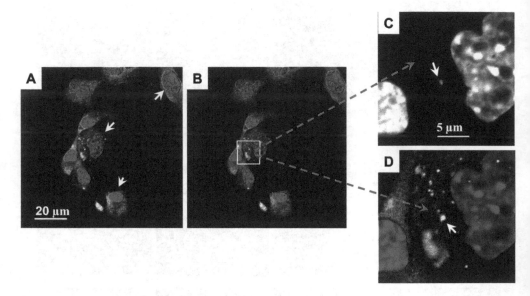

Panel A: Projection of FOXO3a (red channel; nuclear FOXO3a is indicated with yellow arrows) and p62 (green channel). Panel B: Projection of LC3 (red channel) and p62 (green channel). Counterstaining of nuclei with Hoechst 33342 appears in blue color. Panels C and D – selected area indicated in panel B.
Panel C: Signal acquired in the blue channel; bacterial DNA is indicated with white arrow.
Panel D: Signals acquired in the blue, red and green channels; co-localization of bacterial nucleus with p62 and LC3 proteins is indicated with white arrow.

Conditions: MSCs were incubated with ~1×10^7/ml *E. coli* for 3 h in MesenCult Medium (without antibiotics). After 3 h the medium was replaced with fresh medium (with antibiotics) and MSCs were incubated for further 21 h. The confocal images were taken with pinhole setup to obtain 0.5 μm Z-sections.

Fig. 5. Immunofluorescence confocal imaging of LC3, p62, phagocytized bacteria, and nuclear fraction of FOXO3a in MSCs challenged by *E. coli*.

Autolysosomal degradation of phagocytized bacteria can involve reactive oxygen and nitrogen species ultimately leading to up-regulation of stress-adaptive elements (13). Confocal fluorescence imaging of formation of reactive nitrogen species in autolysosomes was conducted using dihydrorhodamine 123, a sensitive indicator of peroxynitrite. The results of assessment of oxidative environment in the MSC autolysosomes containing *E. coli* are presented in Fig. 6. The appearance of reactivity to dihydrorhodamine 123 was likely

due to up-regulation of nitric oxide synthase induced in MSCs in response to challenge with
E. coli. It was hypothesized this increase in redox events in MSCs could at least in part
contribute to degradation of the phagocytized bacteria. Indeed, as shown in Fig. 7 bacterial
nuclei present in autolysosomes were positive to terminal deoxynucleotidyl transferase
dUTP nick-end labeling (TUNEL).

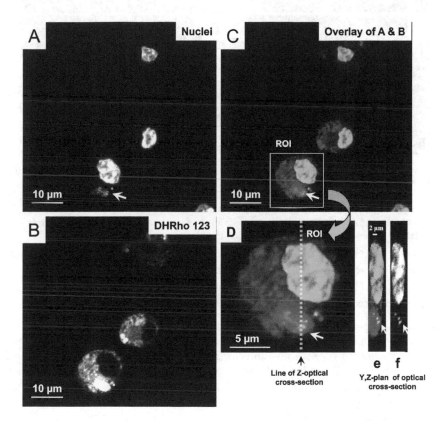

Panels A-C are projections of nuclei and oxidized fluorescent product of dihydrorhodamine 123. The
images acquired in the blue (panel A) and green (panel B) are shown in grayscale; then, the images were
overlaid in panel C in pseudo-colors that are "red" and "green", respectively. Panel D is the selected
area indicated in panel C, where nuclei are green, oxidized dihydrorhodamine 123 (DHRho 123) is red,
and co-localization of nuclei and DHRho 123 is in yellow colors. The presence of bacterial genomic
DNA in the autolysosome appears in yellow as result of interference of red and green colors.
Experimental conditions were the same as indicated in Fig. 5.

Fig. 6. Assessment of production of peroxynitrite in *E. coli*-challenged MSCs using
dihydrorhodamine 123 probe.

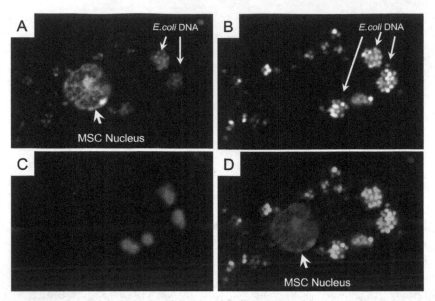

Panel A: Projection the nuclear DNA is indicated with yellow arrows (blue channel, counterstaining of nuclei with Hoechst 33342). Panel B: Projection of TUNEL-positive DNA (green channel).
Panel C: Projection of tyrosine-phosphorylated caveolin-1 (red channel). Panel D: Overlay of the images presented in panels A, B, and C. TUNEL-positive bacterial nuclei appear in yellow as result of interference of blue and green. TUNEL – positive staining of bacterial DNA occurred in autolysosomes. Experimental conditions were the same as indicated in Fig. 5.

Fig. 7. Assessment of bacterial DNA damage in *E. coli*-challenged MSCs using terminal deoxynucleotidyl transferase dUTP nick-end labeling (TUNEL).

3.2 Stress-response of MSCs following challenge with *E. coli* bacteria

General stress responses are characterized by conserved signaling modules that are interconnected to the cellular adaptive mechanisms. It is proposed that stress induced by inflammatory factors, microorganisms, and oxidants triggers a cascade of responses attributed to specific sensitive transcriptional and post-transcriptional mechanisms mediating inflammation, antioxidant response, adaptation, and remodeling (36-42). The components of the oxidative stress response employ a battery of redox-sensitive thiol-containing molecules, such as glutathione (GSH), thioredoxin 1 (TRX1)/thioredoxin reductase, apurinic/apyrimidinic endonuclease/redox effector factor-1 (APE/Ref-1), and transcription factors (such as nuclear factor-kappa B (NF-κB) and nuclear factor (erythroid-derived 2)-like 2 (Nrf2). Overall, these effector proteins play a major role in maintaining the steady-state intracellular balance between pro-oxidant production, antioxidant capacity, and repair of oxidative damage (39, 43). While NF-κB and Nrf2 are normally sequestered in the cytoplasm bound to their native inhibitors, i.e., IκB and Keap-1 respectively, bacterial products, pro-inflammatory factors, and oxidative stress can stimulate their translocation to the nucleus (38, 41, 44). NF-κB and Nrf2 are known to regulate numerous genes that play a crucial role in the host response to sepsis (40, 45) and therefore, have relevance to the current study. Regulation of Nrf2 function is controlled by numerous factors among which Nrf2 conjugates with Keap-1.

Dissociation of the Nrf2/Keap-1 complex results from a modification of cysteine residues in Keap-1 through either their conjugation or oxidation (40, 43, 45).

Two major redox systems, the GSH and TRX1 systems, control intracellular thiol/disulfide redox environments. While the GSH/GSSG couple provides a major cellular redox buffer, TRXs serve a more specific function in regulating redox-sensitive proteins (46). These two redox systems function at different sites in the Nrf2 signaling pathway: first, the cytoplasmic dissociation of Nrf2 is primarily regulated by cytoplasmic GSH concentrations, and second, the nuclear reduction of Nrf2 cysteine 506 (required for Nrf2 binding of DNA) is primarily regulated by TRX1 (45). Redox dependence of DNA-binding activity of NF-κB has been broadly discussed (39, 47). DNA-binding activity of NF-κB can drastically increase in the presence of the reduced form of the redox factor-1 (Ref-1) redox-converted by TRX (39, 47). It should be noted that up-regulation of Nrf2 and NF-κB via autophagy-dependent mechanisms can also occur *via* lysosomal degradation of IκB and Keap-1, (48). Therefore, we do not exclude autophagy-dependent activation of these transcriptional factors in *E. coli*-treated cells. Taking into consideration all of the above, one would assume that a battery of stress-sensitive mechanisms mediated by survival transcription factors such as NF-κB, Nrf2, and FOXO3a are involved in adaptive response of MSCs challenged with *E. coli*.

Immunoblot analysis of stress-response proteins indicated that control MSCs had relatively high amounts of constitutively present NF-κB. Challenge of cells with *E. coli* resulted in prompt (within 1 h) increases in the nuclear fraction of NF-κB as determined with confocal immunofluorescence imaging (not shown). But, we did not observe a similar pattern when we assessed nuclear Nrf2. That could be due to an extremely low level of constitutive Nrf2 in the cells (Fig. 3). A drastic increase in the nuclear fraction of NF-κB occurred during the period of the observation, i.e., 24 h post-exposure (Fig. 8). This effect was accompanied

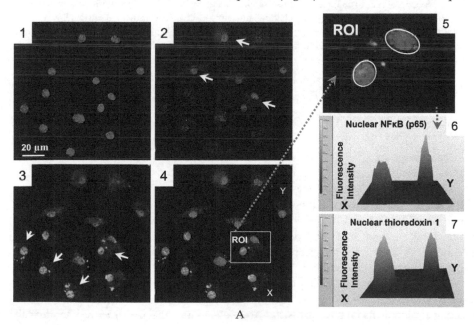

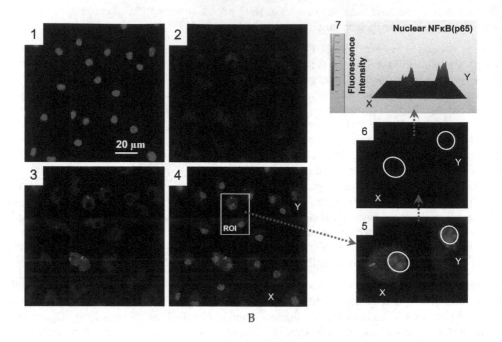

Panel 1: Projection of the nuclear DNA (blue channel, counterstaining of nuclei with Hoechst 33342).
Panel 2: Projection of NFκB(p65) (red channel, nuclear localization is indicated with yellow arrows).
Panel 3: Projection of thioredoxin 1 (green channel, nuclear localization is indicated with yellow arrows).
Panel 4: Overlay of the images presented in panels 1, 2, and 3. Panels 5-7: analysis of nuclear fractions of NFκB(p65) and thioredoxin 1 in ROI indicated in panel 4. Experimental conditions were the same as indicated in Fig. 5.

Fig. 8. Assessment of nuclear fractions of NF-κB(p65) and thioredoxin 1 in MSCs challenged with *E. coli*. (A) Challenge with *E. coli*.; (B) Control.

by transactivation of NF-κB-dependent proinflammatory factors such as IL-1α, IL-1β, IL-6, and iNOS (Fig. 9). Interestingly, pre-incubation of the cells with pyrrolidine dithiocarbamate, an inhibitor of NF-κB translocation, resulted in development of pro-apoptotic alterations and loss of confluency in *E. coli*-treated MSCs (not shown). The response to *E. coli*–induced stress was also associated with increases in nuclear fractions of Ref-1 and TRX-1 (Figs. 8 and 10); these reducing agents appeared in close proximity with the nuclear NF-κB (Figs. 8 and 10). Moreover, the MSC stress-response at 24 h was characterized by significant expression of Nrf2 protein (Fig. 3) that accumulated in cell nuclei (Fig. 11). Based on these observations we concluded that the MSC response to challenge with *E. coli* activates complex molecular machinery designed to eliminate environmental microorganisms and increase adaptive capacity to stress. That conclusion contributes to a broad perspective on the role of stromal cells in the host innate defense and on the cell molecular mechanisms mediating resistance of cells to damage. Considering that the cell can employ a battery of stress-response factors operating synchronously, we focused our attention on other cellular components that are crucial for cell survival, e.g., mitochondria, the caveolae vesicular system, and signaling cascades mediated by transcriptional factor FOXO3a.

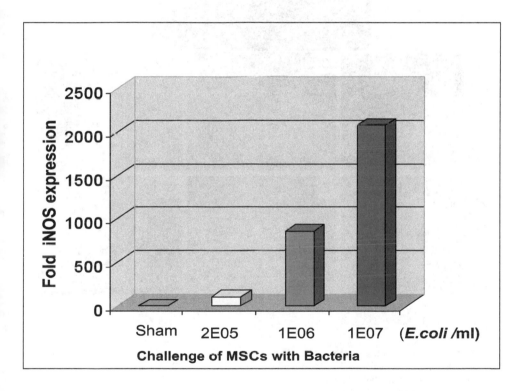

Fig. 9. qRT-PCR assessment of iNOS transactivation in MSCs challenged with *E. coli*. Conditions: MSCs were incubated with bacteria for 3 h in MesenCult Medium (without antibiotics). After 3 h the cells were harvested and lysed for extraction of RNA.

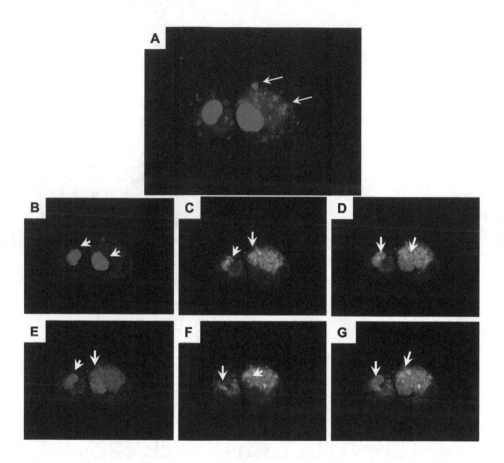

Panel A: Projection of the nuclear DNA (blue channel, high intensity; counterstaining of nuclei with Hoechst 33342). Bacterial nuclei are indicated with yellow arrow. Panel B: Projection of NFκB(p65) (red channel) and nuclear DNA (blue channel); nuclear co-localization of NFκB(p65) is indicated with white arrows. Panel C: Projection of Ref1 protein (green channel, nuclear localization is indicated with white arrows). Panel D: Overlay of the images presented in panels B and C. Nuclear co-localization of Ref1 and NFκB(p65) is indicated with white arrows. Panel E: Projection of Ref1 (red channel) and nuclear DNA (blue channel); nuclear co-localization of Ref1 is indicated with white arrows.
Panel F: Projection of thioredoxin 1 protein (green channel, nuclear localization is indicated with white arrows). Panel G: Overlay of the images presented in panels E and F. Nuclear co-localization of Ref1 and thioredoxin 1 is indicated with white arrows. Experimental conditions were the same as indicated in Fig. 5.

Fig. 10. Assessment of nuclear co-localization of NF-κB, thioredoxin 1, and Ref1 in MSCs challenged with *E. coli*.

FOXO3a, a member of a family of mammalian forkhead transcription factors of the class O, was recently proposed as mediator of diverse physiologic processes, including regulation of stress resistance and survival (49, 50). Thus, it is shown in our study that in response to oxidative stress, FOXO3a along with Nrf2 can promote cell survival by inducing the expression of antioxidant enzymes and factors involved in cell cycle withdrawal, such as the cyclin-dependent kinase inhibitor (CKI) p27 (50). We analyzed FOXO3a transcriptional factor in MSCs responding to *E. coli* challenge. Fig. 12 shows that the presence of *E. coli* increased FOXO3a protein in MSCs. The data suggest that, indeed, this FOXO3a transcriptional factor is also implicated in the stress-response to *E. coli* challenge.

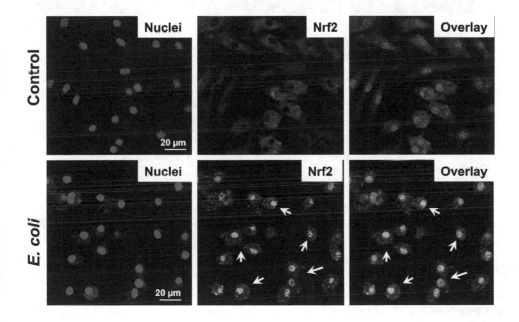

Fig. 11. Assessment of nuclear fractions of Nrf2 in MSCs challenged with *E. coli*. Counterstaining of nuclear DNA was with Hoechst 33342 (blue channel). Nrf2 staining is in green. Nrf2 localized in nuclei appears in turquoise/green color due to interference of "green" and "blue" (indicated with arrows). Experimental conditions were the same as indicated in Fig. 5.

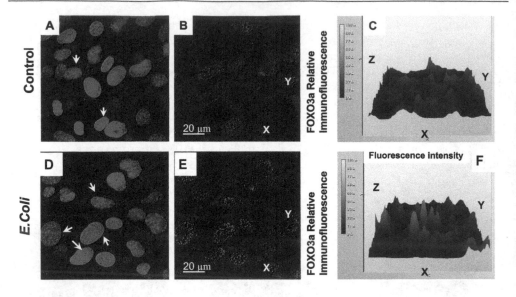

Control: Panel A-C. Challenged with *E. coli*.: Panels D-F.
Panels A and D: Projection of FOXO3a (red channel) and nuclear DNA (blue channel); nuclear localization of FOXO3a is indicated with white arrows. Panels B and E: Projection of FOXO3a protein (red channel only). Panel C and F: Relative intensity of the FOXO3a immunofluorescence shown in panels B and E, respectively. Experimental conditions were the same as indicated in Fig. 5.

Fig. 12. Immunofluorescence assessment of nuclear fraction of FOXO3a in MSCs challenged with *E. coli*.

4. Conclusion

Multipotent fibroblast-type mesenchymal cells are the essential components of the stroma, which supports tissue barriers and integrity (51). Disturbance in the stroma, composed of endothelial, fibroblastic and myofibroblastic cells as well as macrophages and other inflammatory cells - can be a critical step triggering bacterial translocation and sepsis - exacerbating a variety of injury types. This chapter aims to define whether MSCs can contribute to antibacterial innate defense mechanisms.

The antibacterial defense response of MSCs was characterized by extensive phagocytosis and inactivation of *E. coli* mediated by autolysosome mechanisms. *E. coli*-challenged MSCs showed increased transactivation of NF-κB, Nrf2, and FOXO3a stress-response transcriptional factors and associated expression of proinflammatory mediators. These observations were accompanied by a compensatory antioxidant response of MSCs mediated by nuclear translocation of Nrf2, Ref-1 and thioredoxin 1.

Taken together our data support the hypothesis that (i) MSCs contribute to the innate defense response to bacterial infection; (ii) the mechanism of MSC responses involves specific macroautophagy and nitroxidation mediated by iNOS; and (iii) MSCs are armed against self-injury by the mechanisms degrading phagocytized *E. coli*.

5. Acknowledgements

The authors thank HM1 Neil Agravante and Ms. Dilber Nurmemet for their technical support.

5.1 Grants

This work was supported by AFRRI Intramural RAB2CF (to JGK) and NIAID YI-AI-5045-04 (To JGK).

There are no ethical and financial conflicts in the presented work.

5.2 Disclaimer

The opinions or assertions contained herein are the authors' private views and are not to be construed as official or reflecting the views of the Uniformed Services University of the Health Sciences, AFRRI, the United States Department of Defense, or the National Institutes of Health.

6. References

[1] Turner HL, Turner JR. Good fences make good neighbors: Gastrointestinal mucosal structure. Gut Microbes. 2010; 1(1):22-9.

[2] Hooper LV, Macpherson AJ. Immune adaptations that maintain homeostasis with the intestinal microbiota. Nat Rev Immunol. 2010; 10(3):159-69.

[3] Wershil BK, Furuta GT. Gastrointestinal mucosal immunity. J Allergy Clin Immunol. 2008; 121(2 Suppl):S380-3.

[4] Vinh DC, Embil JM. Rapidly progressive soft tissue infections. Lancet Infect Dis. 2005; 5(8):501-13.

[5] Friedenstein A. Stromal-hematopoietic interrelationships: Maximov's ideas and modern models. Haematol Blood Transfus. 1989; 32:159-67.

[6] Sethe S, Scutt A, Stolzing A. Aging of mesenchymal stem cells. Ageing Res Rev. 2006; 5(1):91-116.

[7] Mezey E, Mayer B, Németh K. Unexpected roles for bone marrow stromal cells (or MSCs): a real promise for cellular, but not replacement, therapy. Oral Dis. 2010; 16(2):129-35.

[8] English K, French A, Wood KJ. Mesenchymal stromal cells: facilitators of successful transplantation? Cell Stem Cell. 2010; 7(4):431-42.

[9] McFarlin K, Gao X, Liu YB, Dulchavsky DS, Kwon D, Arbab AS, Bansal M, Li Y, Chopp M, Dulchavsky SA, Gautam SC. Bone marrow-derived mesenchymal stromal cells accelerate wound healing in the rat. Wound Repair Regen. 2006; 14(4):471-8.

[10] Tyndall A and Pistoia V. Mesenchymal stem cells combat sepsis. Nat Med. 2009; 15: 18 – 20.

[11] Krasnodembskaya A, Song Y, Fang X, Gupta N, Serikov V, Lee JW, Matthay MA. Antibacterial effect of human mesenchymal stem cells is mediated in part from secretion of the antimicrobial peptide LL-37. Stem Cells. 2010; 28(12):2229-38.

[12] Hall SE, Savill JS, Henson PM, Haslett C. Apoptotic neutrophils are phagocytosed by fibroblasts with participation of the fibroblast vitronectin receptor and involvement of a mannose/fucose-specific lectin. J Immunol. 1994; 153(7):3218-27.

[13] Levine B, Mizushima N, Virgin HW. Autophagy in immunity and inflammation. Nature. 2011; 469(7330):323-35.

[14] Yang Z, Klionsky DJ. Eaten alive: a history of macroautophagy. Nat Cell Biol. 2010; 12(9):814-22;

[15] Yano T, Kurata S. Intracellular recognition of pathogens and autophagy as an innate immune host defence. J Biochem. 2011; 150(2):143-9.

[16] Mizushima N, Levine B, Cuervo AM, Klionsky DJ. Nature. Autophagy fights disease through cellular self-digestion. 2008; 451:1069-75.

[17] Klionsky DJ. The Autophagy Connection. Dev Cell. Author manuscript; available in PMC 2011 July 20.

[18] Tooze SA, Yoshimori T. The origin of the autophagosomal membrane. Nat Cell Biol. 2010; 12(9):831-5.

[19] Weidberg H, Shvets E, Elazar Z. Biogenesis and cargo selectivity of autophagosomes. Annu Rev Biochem. 2011; 80:125-56.

[20] Eskelinen EL. New insights into the mechanisms of macroautophagy in mammalian cells. Int Rev Cell Mol Biol. 2008; 266:207-47.

[21] Eskelinen EL, Saftig P. Autophagy: a lysosomal degradation pathway with a central role in health and disease. Biochim Biophys Acta. 2009; 1793(4):664-73.

[22] Mizushima N, Levine B. Autophagy in mammalian development and differentiation. Nat Cell Biol. 2010; 12(9):823-30.

[23] Kabeya Y, Mizushima N, Yamamoto A, Oshitani-Okamoto S, Ohsumi Y, Yoshimori T. LC3, GABARAP and GATE16 localize to autophagosomal membrane depending on form-II formation. J Cell Sci. 2004; 117(Pt 13):2805-12.

[24] Behrends C, Sowa ME, Gygi SP, Harper JW. Network organization of the human autophagy system. Nature. 2010; 466(7302):68-76.

[25] Lipinski MM, Hoffman G, Ng A, Zhou W, Py BF, Hsu E, Liu X, Eisenberg J, Liu J, Blenis J, Xavier RJ, Yuan J. A genome-wide siRNA screen reveals multiple mTORC1 independent signaling pathways regulating autophagy under normal nutritional conditions. Dev Cell. 2010; 18(6):1041-52.

[26] Viiri J, Hyttinen JM, Ryhänen T, Rilla K, Paimela T, Kuusisto E, Siitonen A, Urtti A, Salminen A, Kaarniranta K. p62/sequestosome 1 as a regulator of proteasome inhibitor-induced autophagy in human retinal pigment epithelial cells. Mol Vis. 2010; 16:1399-414.

[27] Ryhänen T, Hyttinen JM, Kopitz J, Rilla K, Kuusisto E, Mannermaa E, Viiri J, Holmberg CI, Immonen I, Meri S, Parkkinen J, Eskelinen EL, Uusitalo H, Salminen A, Kaarniranta K. Crosstalk between Hsp70 molecular chaperone, lysosomes and proteasomes in autophagy-mediated proteolysis in human retinal pigment epithelial cells. J Cell Mol Med. 2009; 13(9B):3616-31.

[28] Behl C. BAG3 and friends: co-chaperones in selective autophagy during aging and disease. Autophagy. 2011; 7(7):795-8.

[29] Sridhar S, Botbol Y, Macian F, Cuervo AM. Autophagy and Disease: always two sides to a problem. J Pathol. 2011 Oct 12. doi: 10.1002/path.3025. [Epub ahead of print];

[30] Gorbunov NV, Garrison BR, Kiang JG. Response of crypt Paneth cells in the small intestine following total-body gamma-irradiation. Int J Immunopathol Pharmacol. 2010; 23(4):1111-23.

[31] Pfaffl MW. A new mathematical model for relative quantification in real-time RT-PCR. Nucleic Acids Res. 2001 29(9):2002-7.

[32] Nakagawa I, Amano A, Mizushima N, Yamamoto A, Yamaguxhi H, Kamimoto T, Nara A, Funao J, Nakata M, Tsuda K, Hamada S, Yoshimori T. Autophagy defends cells against invading Group A Streptococcus. Science. 2004; 306:1037-40.

[33] Yano T, Kurata S. Intracellular recognition of pathogens and autophagy as an innate immune host defence. J Biochem. 2011; 150(2):143-9.

[34] Xiao G. Autophagy and NF-kappaB: fight for fate. Cytokine Growth Factor Rev. 2007; 18(3-4):233-43.

[35] Ichimura Y, Komatsu M. Selective degradation of p62 by autophagy. Semin Immunopathol. 2010; 32(4):431-6.

[36] Kültz D. Molecular and evolutionary basis of the cellular stress response. Annu Rev Physiol. 2005; 67: 225-57.

[37] Burhans WC, Heintz NH. The cell cycle is a redox cycle: linking phase-specific targets to cell fate. Free Radic Biol Med. 2009; 47(9):1282-93

[38] Sebban H, Courtois G. NF-kappaB and inflammation in genetic disease. Biochem Pharmacol. 2006; 72(9):1153-60.

[39] Janssen-Heininger YM, Mossman BT, Heintz NH, Forman HJ, Kalyanaraman B, Finkel T, Stamler JS, Rhee SG, van der Vliet A. Redox-based regulation of signal transduction: principles, pitfalls, and promises. Free Radic Biol Med. 2008; 45(1):1-17.

[40] Nguyen T, Nioi P, Pickett CB. The Nrf2-Antioxidant Response Element Signaling Pathway and Its Activation by Oxidative Stress J Biol Chem. 2009; 284(20):13291-5.

[41] Thimmulappa RK, Lee H, Rangasamy T, Reddy SP, Yamamoto M, Kensler TW, Biswal S. Nrf2 is a critical regulator of the innate immune response and survival during experimental sepsis. J Clin Invest. 2006; 116(4):984-95.

[42] Nivon M, Richet E, Codogno P, Arrigo AP, Kretz-Remy C. Autophagy activation by NFkappaB is essential for cell survival after heat shock. Autophagy. 2009; 5(6):766-83.

[43] Maher J, Yamamoto M. The rise of antioxidant signaling--the evolution and hormetic actions of Nrf2. Toxicol Appl Pharmacol. 2010; 244(1):4-15.

[44] Cronin JG, Turner ML, Goetze L, Bryant CE, Sheldon IM. Toll-Like Receptor 4 and MYD88-Dependent Signaling Mechanisms of the Innate Immune System Are Essential for the Response to Lipopolysaccharide by Epithelial and Stromal Cells of the Bovine Endometrium. Biol Reprod. 2011 Nov 2. [Epub ahead of print].

[45] Hansen JM, Watson WH, Jones DP. Compartmentation of Nrf-2 redox control: regulation of cytoplasmic activation by glutathione and DNA binding by thioredoxin-1. Toxicol Sci. 2004; 82(1):308-17.

[46] Jones DP. Methods Enzymol. 2002; 348:93-112. Redox potential of GSH/GSSG couple: assay and biological significance.

[47] Nishi T, Shimizu N, Hiramoto M, Sato I, Yamaguchi Y, Hasegawa M, Aizawa S, Tanaka H, Kataoka K, Watanabe H, Handa H. Spatial redox regulation of a critical cysteine residue of NF-kappa B in vivo. J Biol Chem. 2002; 277(46):44548-56.

[48] Weidberg H, Shvets E, Elazar Z. Biogenesis and cargo selectivity of autophagosomes. Annu Rev Biochem. 2011; 80:125-56.

[49] Miyamoto K, Araki KY, Naka K, Arai F, Takubo K, Yamazaki S, Matsuoka S, Miyamoto T, Ito K, Ohmura M, Chen C, Hosokawa K, Nakauchi H, Nakayama K, Nakayama KI, Harada M, Motoyama N, Suda T, Hirao A. Foxo3a is essential for maintenance of the hematopoietic stem cell pool. Cell Stem Cell. 2007; 1(1):101-12.

[50] Burhans WC, Heintz NH. The cell cycle is a redox cycle: linking phase-specific targets to cell fate. Free Radic Biol Med. 2009; 47(9):1282-93.

[51] Powell DW, Pinchuk IV, Saada JI, Chen X, Mifflin RC. Mesenchymal cells of the intestinal lamina propria. Annu Rev Physiol. 2011; 73:213-37.

MOZ-TIF2 Fusion Protein Binds to Histone Chaperon Proteins CAF-1A and ASF1B Through Its MOZ Portion

Hong Yin[1], Jonathan Glass[1] and Kerry L. Blanchard[2]
[1]Department of Medicine and the Feist-Weiller Cancer Center,
LSU Health Sciences Center, Shreveport, LA
[2]Eli Lilly & Company, Indianapolis, IN
USA

1. Introduction

We previously identified a MOZ-TIF2 (transcriptional intermediary factor 2) fusion gene from a young female patient with acute myeloid leukemia (AML) (Liang et al., 1998). MOZ related chromosome translocations include MOZ-CREB-binding protein (MOZ-CBP, t(8;16)(p11;p13)), MOZ-P300(t(8;22)(p11;q13)), MOZ-TIF2(inv(8)(p11q13), and MOZ-NCOA3(t(8;20)(p11;q13)) (Esteyries et al., 2008; Troke et al., 2006). In an animal model, the MOZ-TIF2 fusion product successfully induced the occurrence of AML (Deguchi et al., 2003). Though the mechanisms for leukemogenesis of this fusion protein are poorly understood, analysis of functional domains in the MOZ-TIF2 fusion protein discloses at least two distinct functional domains: 1) the MYST domain containing the C2HC nucleosome recognition motif and the histone acetyltransferase motif in the MOZ portion and 2) the CID domain containing two CBP binding motifs in the TIF2 portion. Together these domains were responsible for AML in mice caused by injecting bone marrow cells transduced with retrovirus containing the MOZ-TIF2 fusion gene. Furthermore, MOZ-TIF2 conferred the properties of leukemic stem cells (Huntly et al., 2004). The MOZ-TIF2 transduced mouse common myeloid progenitors and granulocyte-monocyte progenitors exhibited the ability to serially replated *in vitro*. The cell line derived from transduced progenitors could induce AML in mice. Interestingly, the C543G mutation in C2HC nucleosome recognition motif or in the CBP binding motif (LXXLL) blocked the self-renewal function of MOZ-TIF2 transduced progenitors. More recently, a study using PU.1 deficient mice demonstrated that the interaction between MOZ-TIF2 and PU.1 promoted the expression of macrophage colony–stimulating factor receptor (CSF1R). Cells with high expression of CSF1R are potential leukemia initiating cells(Aikawa et al., 2010). Models suggesting that aberrant transcription by the interaction between MOZ fusion proteins and transcription factors, AML1, p53, PU1, or NF-kB have been well reviewed(Katsumoto et al., 2008).

MOZ as a fusion partner of MOZ-TIF2 is a member of MYST domain family (MOZ/YBF2/SAS2/TIP60) and acetylates histones H2A, H3 and H4 as a histone acetyltransferase (HAT) (Champagne et al., 2001; Kitabayashi et al., 2001). MOZ is a cofactor in the regulation of transcriptional activation of several target genes important to hematopoiesis, such as Runx1 and PU.1 (Bristow and Shore, 2003; Katsumoto et al., 2006; Kitabayashi et al., 2001). MOZ$^{-/-}$ mice died at embryonic day 15 and exhibited a significant decrease of mature erythrocytes (Katsumoto et al., 2006). The histone acetyltransferase activity of MOZ is required to maintain normal functions of hematopoietic stem cells (HSC) (Perez-Campo et al., 2009). Mice with mutation at HAT or MYST domain (G657E) showed a decreased population of HSC in fetal liver. The lineage-committed hematopoietic progenitors from fetal liver cells with HAT$^{-/-}$ mutant had reduced colony formation ability.

In our attempt to find proteins that interact with the fusion protein by using as bait a construct of the MOZ N-terminal fragment, encoding the first 759 amino acids of MOZ-TIF2 fusion gene and containing the H15, PHD, and MYST domains, we were able to isolate two proteins, the p150 subunit or subunit A of the human chromatin assembly factor-1 (p150/CAF-1A) and the human anti-silencing function protein 1 homolog B (ASF1B). Both of these proteins were verified to interact with the MOZ partner of MOZ-TIF2 fusion in the yeast two-hybrid system. The interaction has been further characterized by co-immunoprecipitation, protein pull-down assays, and co-localization by immunohistochemistry. The differences in the interactions of CAF-1A and ASF1B with wild type MOZ and the MOZ-TIF2 fusion proteins may contribute to leukemogenesis.

2.Materials and methods

2.1 The sources of cDNAs and plasmid constructions

The cDNA for MOZ was kindly provided by Julian Borrow (Center for Cancer Research, Massachusetts Institute of Technology, MA) and TIF2 was a kind gift from Hinrich Gronemeyer (Institut de Genetique et de Biologie Moleculaire et Cellulaire, France). A full length MOZ-TIF2 fusion was created by inserting a RT-PCR fragment crossing the MOZ–TIF2 fusion site into the Hind3 site of wild type of human MOZ and the Sac1 site of human TIF2 in pBluescript KS phagemid vector (pBlueKS). The cDNAs for CAF-1A and ASF1B were screened and rescued from Human Bone Marrow MATCHMAKER cDNA Library (BD Biosciences Clontech Palo Alto, CA) by the yeast two-hybrid system using the N-terminal fragment of the MOZ-TIF2 fusion as bait. The cDNAs from the positive clones, which were in the pACT2 vector, were switched into the pBlueKS vector at EcoRI and XhoI sites and sequenced with a T7 primer. The resulting sequences were identified in the NCBI GenBank as the subunit A (p150) of human chromatin assembly factor-1 (GenBank accession No. NM-005483) and human anti-silencing function protein 1 homolog B (GenBank accession No. AF279307). The full length of both cDNAs was confirmed by DNA sequencing with gene specific primers. For the visualization of the expression and localization in mammalian cells, the full length of MOZ, MOZ-TIF2, TIF2, CAF-1A, and ASF1B were subcloned in frame into the C-terminal fluorescent protein Vector, pEGFP or pDsRed2 (BD Biosciences Clontech, Palo Alto, CA) to generate fluorescent fusion proteins. For studies of protein-protein interaction *in vitro*, glutathione S-transferase (GST) fusions of MOZ fragments were constructed in the pGEX vector (Amersham Biosciences, Piscataway, NJ). Briefly, the full length MOZ cDNA was digested with Asp718/Bgl2 from pBlueKS-MOZ and was ligated into the pET-30a (EMD Biosciences, Inc. Novagen Madison, WI) plasmid at Asp718/BamH1 site to create the pET-30a-MOZ construct. A PET-30a-MOZ-1/759 (amino acids 1 to 759) construct was generated by removing a Hind3/Hind3 fragment from pET-30a-MOZ and then re-ligating. This fragment was then switched from pET-30a vector to pGEX-4T at a Not1/Xho1 site to construct the pGEX-4T-MOZ-1/759. The pGEX-4T-MOZ-1/313 (amino acids 1 to 313) containing H15 and the PHD domain was generated by the deletion of a 1515 base pair fragment from pGEX-4T-MOZ-1/759 with Hind3 /Blin1 followed by re-ligation. The pGEX-6P-MOZ-488/703 plasmid was constructed by inserting an EcoRV to Eag1 fragment of MOZ (amino acids 488 to 703) containing the C2HC motif and acetyl-CoA binding region to pGEX-6P-2 vector at Sma1/Eag1 sites. To create pET-30a-CAF-1A, the pBlueKS-CAF-1A was first digested with XhoI and then digested partially with NcoI. A 3.1 kb fragment was recovered by agarose electrophoresis and was ligated to NcoI/XhoI sites of pET-30a vector. The pET-30c-ASF1B was constructed by inserting the 1 kilobase EcoR1/Hind3 fragment of pBLueKS-ASF1B into the pET-30c vector at EcoR1 /Hind3 sites.

2.2 Yeast two-hybrid screen

pGBD-MOZ-MYST, a bait plasmid with a fusion of the N-terminal fragment of MOZ-TIF2 to the GAL4 DNA binding domain was constructed by inserting a 2.3 kb fragment encoding amino acids 1 to 759 of human MOZ to BamH1/blunted Bgl 2 sites in the pGBD-C3 vector (James et al., 1996). The bait plasmid was transformed into the yeast host PJ69-2A and mated with pre-transformed Human Bone Marrow MATCHMAKER cDNA Library according to the manufacturer's instruction. The mating culture was plated on 25 x 150 mm triple dropout (TDO) dishes (SD/-His/-Leu/-Trp) and 25 x 150 mm quadruple dropout (QDO) dishes (SD/-Ade/-His/-Leu/-Trp). After incubation for 7 and 14 days, the more than 100 colonies which grew on TDO and QDO dishes were picked for re-screening on SD/-His, SD/-Ade/ and QDO dishes. A total of five colonies were grown from the second screening. The plasmids from each colony were rescued and transformed into KC8 cells. All of the plasmids were re-transformed into the yeast host PJ69-2A and Y187; no auto-transcription activation of any reporter was seen. The pVA3.1 plasmids containing either the murine p53 in PJ69-2A or the PTD1-1 with SV 40 large T antigen in Y187 were used as controls for DNA binding domain and activation domain fusions. The plasmids from positive clones were subjected to restriction enzyme mapping which showed two potential interacting genes which were subsequently sequenced and identified with the NCBI database.

2.3 Co-localization of MOZ or MOZ-TIF2 and CAF-1A or ASF1B

To identify the co-localization of expressed fluorescent fusion proteins, HEK293 cells were grown in DMEM (Mediatech Cellgro, VA) containing 10% fetal bovine serum (FBS) and co-transfected by pEGFP-MOZ or pEGFP-MOZ-TIF2 and pDsRed2-CAF-1A or pDsRed2-ASF1B with Lipofectamine 2000 (Invitrogen, Carlsbad, CA). Briefly, cells were grown on a coverslip in a 12-well plate a day before the transfection in the antibiotic-free medium to reach 80-90% confluence on the next day. 1.6 µg of DNA in 100 µl of Opti-MEM I Reduced Serum Medium (Invitrogen, Carlsbad, CA) was mixed with 100 µl of diluted Lipofectamine 2000 reagent. After incubation for 20 min. at room temperature, the DNA-Lipofectamine 2000 complex was added to the cells and 48 hours later, subcellular location of expressed fluorescent fusion proteins was examined with a Zeiss fluorescent microscope equipped with Axiocam system and by a laser scanning confocal microscope (Bio-Rad Laser Scanning

System Radiance 2000/Nikon Eclipse TE300 microscope). To examine the subcellelular localization of endogenously expressed MOZ and CAF-1A, HEK293 and Hela cells were fixed with 4% paraformaldehyde and then blocked with Ultra V block (Lab Vision Co.CA). For some experiments pre-extraction with 0.3%Triton-X100 was conducted. The fixed cells were then incubated with antibody against MOZ (N-19, Santa Cruz Biotechnology, Inc, Santa Cruz, CA) at 1:100 and /or antibody against CAF-1A (a kind gift from Dr. Bruce Stillman, Cold Spring Harbor, NY). In some experiments, the antibody against CAF-1A and ASF1B were purchased from Cell Signaling Technology, MA. The immunofluorescence of MOZ, CAF-1A, or ASF1B was observed as described above for examination of expressed EGFP fusion proteins.

2.4 Co-immunoprecipitation and immunoblotting

HEK293 cells were transfected with EGFP fusions of MOZ, MOZ-TIF2, or TIF2. After 48 hours of transfection, whole cell lysates was prepared with plastic individual homogenizers in the lysis buffer [50 mM NaCL, 5mM KCL, 1mM EDTA, 20 mM HEPES, pH 7.6, 10% glycerol, 0.5% NP-40, and protease inhibitor cocktails (Roche Applied Science, IN)]. Immunoprecipitation was conducted with an antibody against EGFP (BD Biosciences, Palo Alto, CA). Briefly, 2 µg of anti-EGFP antibody and protein A/G-agarose (Santa Cruz Biotechnology, Santa Cruz, CA) were added to 0.8 ml of cell lysate (about 500 µg protein) and incubated overnight at 4°C with rotation. The precipitate was collected by centrifugation, extensively washed, subjected to SDS-PAGE, transferred onto Hybond-ECL nitrocellulose membrane (Amersham Pharmacia Biotech, Piscataway, NJ), and examined by immunoblotting with the antibody against CAF-1A.

2.5 Expression of GST fusion proteins and GST pull down assay

E. *coli* BL21-CodonPlus®(DE3)-RIL Competent Cells (Stratagene, La Jolla, CA) were transformed with pGEX vectors containing cDNA fragments MOZ-1/759, MOZ-1/313, or MOZ-488/703 and grown in LB medium. To induce protein expression isopropyl β-D-thiogalactopyranoside (IPTG) was added at final concentration of 1mM when the A_{600} of the cultures reached 0.6 to 0.8. After three more hours of growth at 28° C, cells were collected by centrifugation and resuspended in cold PBS containing 1% Triton X-100 and protease inhibitor cocktail and kept on ice for 30 minutes. Cell lysates were prepared by ultrasonication followed by centrifugation at 15,000 rpm for 30 minutes at 4°C. GST fusion proteins were purified with the GST Purification Module (Amersham Pharmacia Biotech, Piscataway, NJ). Purified GST fusion proteins were examined with SDS-PAGE followed by Coomassie Blue staining. To perform GST pull down affinity assays [35S]Methionine-labeled proteins were first produced with Single Tube Protein® System 3 or EcoPro™ T7 system (EMD Biosciences, Inc. Novagen, Madison, WI) from pET 30 vectors carrying full length of CAF-1A or ASF1B. The binding reaction was conducted with 5µl of *in vitro*-translated protein and 3-5 µg of GST alone or GST fusion protein attached to Sepharose 4B beads in 200 µl binding buffer (50mM Tris-HCI , pH 8.0, 100 mM NaCl, 0.3 mM DTT, 10mM MgCl2, 10% glycerol, 0.1% NP40). The reaction was conducted at 4 °C for 1 hour followed by five washes with 400 µl of binding buffer. The final pellet was separated by SDS-PAGE, autoradiography performed, and radioactivity detected with a phosphorimager.

2.6 *In Vitro* protein binding assay with S-tagged fusion protein

The S-tagged fusion of ASF1B was expressed from pET-30c-ASF1B in E. *coli* BL21-CodonPlus® (DE3)-RIL cells after induction with 0.8 mM of IPTG and purification with S-tagged agarose beads. The fusion protein on agarose beads was incubated with 150µl (about 600 µg of protein) of cell extract from HEK293 cells transfected with pEGFP fusion protein. The beads were pelleted, washed, and the "pull-down" proteins examined as described above with the anti-EGFP antibody.

2.7 RNA isolation and microarray analysis

RNA was isolated from stably transfected U937 cells with TRI Reagent® (Molecular Research Center, Inc., Cincinati, OH). The analysis of gene expression profile was conducted on the Human Genome U95A Array (Affymetrix, Inc., Santa Clara, CA). The cRNA was synthesized from 10µg of total RNA. The hybridization and signal detection was completed in the Core Facility at LSUHSC-Shreveport according to the standard Affymetrix protocol. The human U95A array represents 12,256 oligonucleotides of known genes or expression tags. The expression profile was analyzed with GeneSifter software. In pairwise analysis, the quality was set as 0.5 for at least one group in order to minimize the effect of low intensity or poor quality spots. Genes with a $\geq$ 2-fold change and with P<0.05 in a student T-test were considered as either significantly up or down regulated genes. To find genes either commonly or differentially expressed in the gene list, we set the quality as 1 to obtain positive expressed genes in pattern navigation analysis. The analysis results were exported for Venn Diagram analysis using the GeneSifter intersector tool.

3. Results

3.1 Screening for MOZ interacting proteins by the yeast two-hybrid system

A MOZ cDNA fragment encoding amino acids 1 to 759 cloned into pGBD was used as the bait in the yeast two-hybrid system in which the prey was a human cDNA bone marrow library. After a second screening five β-galactosidase positive clones grew on SD/-His plates. To eliminate any of these clones as representing false positive clones, plasmid DNA from each clone was rescued using KC8 cells and transformed into PJ69-2A cells carrying pGDB-MOZ-MYST. The transformants were then selected on five different media: –Trp/ -Leu, -His, -His+5mM 3-amino-1,2,4-triazole (3-AT), -His+10mM 3-AT, and –Ade and interaction with the MOZ fragment was verified in all five of the clones (Figure 1). Clone 3.1 grew on –His, -His+10mM 3-AT, and –Ade medium indicative of a strong physical interaction; the other clones only grew on –His and –His + 5 mM 3-AT, but not on –Ade, indicating a weaker interaction. DNA sequencing of the putatively strongly MOZ interacting protein demonstrated that the cDNA encoded the full length CAF-1A. The more weakly interacting cDNAs represented the entire coding region of ASF1B.

3.2 Identify the interaction between MOZ and CAF-1A in human cells

In yeast, the MYST family member Sas2 was found to interact with Cac1, the largest subunit of Saccharomyces cerevisiae chromatin assembly factor-I (CAF-1) (Meijsing and Ehrenhofer-Murray, 2001) but it is not known if the interaction between the homologous proteins in mammalian cells, MOZ and CAF-1A, takes place in human cells and if any interaction occurs between the MOZ-TIF2 fusion protein and CAF-1A. To address these areas we looked for interactions by co-immunoprecipitation using transfections with the

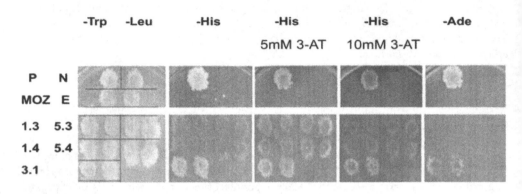

Fig. 1. Protein interaction between MOZ and CAF-1A or ASF1B in the yeast two-hybrid system. The yeast two-hybrid system was used with pretransformed Matchmaker libraries as detailed in the Methods. The bait was the fragment encoding amino acids 1 to 759 of the human MOZ gene in the pGAL 4 DNA-BD vector. In the upper panel controls are plated on 5 different selection media: **P**, positive control diploid with plasmid pDT1-1 encoding an AD/SV40 large T-antigen fusion protein and pVA3-1 carrying DNA-BD/murine P53 fusion protein. **N**, negative control diploid. MOZ, a diploid with GAL4 DNA-BD+ MOZ fragment of amino acids 1 to 759. **E**, a diploid with GAL4 DNA-BD vector only. In the lower panel the five clones (1.3, 1.4, 3.1, 5.3, and 5.4) that were positive after a second screening were plated in duplicate on the same media. Clones 1.3, 1.4, 5.3, and 5.4 show an interaction between MOZ and ASF1B; clone 3.1 shows an interaction between the MOZ and CAF-1A. Trp, tryptophan, Leu, leucine, His, histidine, Ade, adenine, 3-AT, 3-amino-1,2,4,triazole.

MOZ and MOZ-TIF2 fusion constructs into HEK293 cells which express CAF-1A (Figure 2). In these experiments the HEK293 cells were transfected with EGFP fusions of MOZ, MOZ-TIF2 and TIF2, the expressed fusion proteins precipitated with anti-EGFP antibody and the presence of co-precipitated CAF-1A assayed by western blot analysis.   Only with the product of the EGFP-MOZ construct was a significant amount of CAF-1A precipitated (Figure 2A); a far smaller amount was precipitated with MOZ-TIF2. By comparison to the intensity of the CAF-1A band in the input lane, which represents 10% of the amount of lysate subjected to immunoprecipitation, approximately 35-40% of the HEK293 cell CAF-1A was estimated to be co-precipitated with the transfected MOZ. In contrast, less than 10% of the CAF-1A co-precipitated with MOZ-TIF2 (Figure 2A). The differences in the amount of CAF-1A precipitated were not a result of altered expression of CAF-1A or of differences in expression levels of the transfectants as the expression of CAF-1A was not affected by any of the three transfectants (Figure 2B) and the EGFP-tagged MOZ and MOZ-TIF2 proteins showed similar levels of expression, while TIF2 showed a 2-3 fold higher expression than MOZ and MOZ-TIF2 (Figure 2C).

3.3 The MOZ portion of MOZ-TIF2 fusion interacts physically with CAF-1A through the N-terminal of MOZ

Using the yeast two-hybrid system we have shown that CAF-1A interacted with a MOZ fragment extending from amino acids 1 to 759. Within this region are PHD (amino acids 195-320) and MYST (amino acids 562-750) domains that are potential sites for the interaction with (Figure 3A) (Champagne et al., 1999).

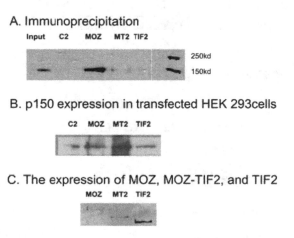

A. Immunoprecipitation

B. p150 expression in transfected HEK 293cells

C. The expression of MOZ, MOZ-TIF2, and TIF2

Fig. 2. **Co-precipitation of CAF-1A (p150) with EGFP-tagged MOZ, MOZ-TIF2, and TIF2.** The EGFP constructs of MOZ, MOZ-TIF2, and TIF2 were transfected into HEK293 cells. **Panel A.** After 48 hours, whole cell extracts were prepared in lysis buffer and subjected to immunoprecipitation with anti-EGFP antibody, followed by SDS-PAGE, and western blot analysis with anti-p150 antibodies. The input lane corresponds to 10% of the amount of lysate subjected to immunoprecipitation. Lane C2 represents the pEGFP-C2 vector alone and MT2 represents MOZ-TIF2. **Panel B.** The lysates of the various transfectants were subjected to SDS-PAGE followed by western blot analysis with anti- p150 antibody to demonstrate the expression level of p150 in the transfected cells. **Panel C.** The same lysates used in Panel B were subjected to a western blot analysis with anti-EGFP antibody to demonstrate the expression of EGFP-tagged MOZ, MOZ-TIF2 and TIF2.

To further define the region containing the binding domain, a pull down assay using GST fusion proteins was established. First, a GST-tagged MOZ fragment from amino acids 1 to 759 was used to pull down CAF-1A and to demonstrate that the GST did not interfere with the MOZ-CAF-1A interactions shown earlier (Figure 3B). We then generated two GST-tagged MOZ fragments, one encompassing amino acids 1-313 (MOZ-1/313) containing the H15 and PHD domains and the other from amino acids 488-703 (MOZ-488/703) including the C2HC motif and acetyl-CoA binding region (Figure 3 C, left panel). These peptides were used with [35S]methionine labeled CAF-1A synthesized in an *in vitro* translation system and interactions detected with a GST pull down assay (Figure 3C). For equivalent amounts of fusion peptides more MOZ-1/313 was bound to CAF-1A than MOZ-488/703 (Figure 3C). As a percentage of the input radioactivity, MOZ-1/313 pulled down about 30 % of the [35S]methionine labeled CAF-1A while MOZ-488/703 pulled down only 14%. Further analysis of domain interactions showed that strongest binding was seen between MOZ-1/313 and CAF-1A-176/327 among all peptides (Figure 3D). CAF-1A-176/327 pulled down about 328% of [^{35}S]methionine labeled MOZ-1/313 and pulled down only 76% of MOZ-488/703 while CAF-1A-620/938 pulled down 20% and 28% of MOZ-1/313 and MOZ-488/703, respectively.

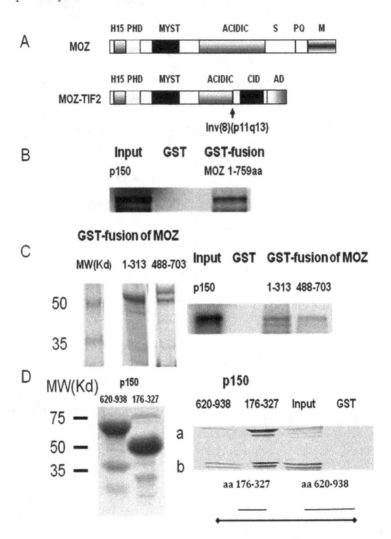

Fig. 3. The interaction between MOZ fragments and CAF-1A (p150). GST-tagged MOZ fragments were expressed and purified with glutathione Sepharose 4B as described in Materials and Methods. [^{35}S]-methionine labeled p150 protein was produced from a T7-driven pET-30 plasmid with an *in vitro* translation system. A, binding assay was conducted with [^{35}S]-methionine labeled p150 and the GST-tagged MOZ fragments. The input lane is 10% of the [^{35}S] methionine p150 protein added to the binding assay. A, schematic structure

of MOZ and MOZ-TIF2. B, interaction between p150 and the MOZ fragment from amino acids 1 to 759 using the binding assay as described in the Materials and Methods. C, left panel, SDS-PAGE of the purified GST-MOZ-1/313 and GST-MOZ-488/703 peptides to demonstrate that the peptides were of the expected molecular weights; right panel, as described in Materials and Methods [^{35}S]-methionine labeled p150 synthesized in a cell-free translation system was incubated *in vitro* with equivalent amounts of GST fusions with MOZ-1/313 or MOZ-488/703, the resulting complexes isolated by GST-pull down assay, and the amount of [^{35}S]-methionine labeled p150 detected by radioautography following SDS-PAGE. D, left panel, SDS-PAGE of the purified GST-p150-176/327 and GST-p150-620/938 fusion peptides; right panel, GST pulldown assays as described in C with [^{35}S]-methionine labeled MOZ-1/313 (a) and MOZ-488/703(b) peptides. The bottom line indicates the full length p150 protein.

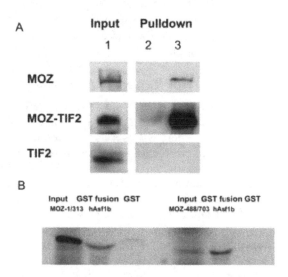

Fig. 4. ASF1B interacts with MOZ and MOZ-TIF2. **Panel A.** HEK293 cells were transfected with EGFP-MOZ, EGFP-MOZ-TIF2, and EGFP-TIF2 as detailed in the Materials and Methods. At 48 hours after transfection cell lysates were incubated with S–tagged ASF1B absorbed to S-tag agarose beads and after extensive washing the proteins bound to ASF1B were analyzed by SDS-PAGE with subsequent western blot analysis with anti-GFP antibody. Lane 1, 10% of input; lane 2, S-tag protein alone; lane 3, S-tagged ASF1B. **Panel B.** GST pull down assays were performed as detailed above incubating GST- ASF1B with [^{35}S]-methionine labeled MOZ-1/313 or MOZ-488/703 peptides synthesized in a cell-free translation system as described in the Material and Methods.

3.4 Confirmation of ASF1B as an interacting protein of MOZ and MOZ-TIF2

The yeast two-hybrid system also revealed a cDNA encoding another protein, ASF1B, which interacted with the MOZ-1/759 fragment. To verify the interaction between MOZ and ASF1B and to examine if the MOZ-TIF2 fusion protein also interacts with ASF1B, we conducted pull down assays and examined co-localization of proteins similar to the studies with CAF-1A. A S-tag fusion cDNA with ASF1B was created in the pET-30c vector and the fusion protein was labeled with [^{35}S]methionine by an in vitro transcription/translation system. The expressed fusion protein was purified with S-tag agarose beads and incubated with cell lysates containing expressed EGFP fusions of MOZ, MOZ-TIF2 and TIF2. Subsequently, EGFP proteins that interacted with ASF1B were identified by western blot analysis with an anti-EGFP antibody (Figure 4A). Both EGFP-MOZ and EGFP-MOZ-TIF2 could be demonstrated to interact with ASF1B. MOZ-TIF2 appeared to interact more strongly with the percentage of EGFP fusion protein bound to ASF1B approximately 240% over the input for MOZ-TIF2 and 70% for MOZ, respectively. TIF2 showed no binding to ASF1B. To further identify the ASF1B binding domain in MOZ, the GST-tagged ASF1B was incubated with [^{35}S]methionine labeled MOZ-1/313 and MOZ-488/703 (Figure 4B). The MOZ-488/703 fragment showed stronger binding to ASF1B than MOZ-1/313. The percentage of ASF1B bound to the MOZ-1/313 fragment represented about 25% of the input while the percentage of ASF1B bound to the MOZ-488/703 fragment was 150% of the input.

3.5 The co-localization of MOZ and MOZ-TIF2 with CAF-1A and ASF1B

To further verify the interaction of MOZ with CAF-1A, we first examined by indirect immunohistochemistry the localization of endogenous MOZ and CAF-1A in Hela cells to determine if the subcellular distribution was similar by confocal immunofluorescence microscopy (Figure 5A). In Hela cells both MOZ and CAF-1A were predominately localized in interphase nuclei (Figure 5A-a). As the chromatin condensed in metaphase MOZ distributed dominantly in cytoplasm and disassociated from the spindle-chromosome in some cells (Figure 5A-b and 5A-c). CAF-1A was observed either to disassociate from (Figure 5A-b) or bind to spindle-chromosomes (Figure 5A-c). However, cytoplasmic co-localization of MOZ and CAF-1A was still seen as detected by the persistence of yellow by confocal microscopy. In anaphase, with paired chromosome separation, CAF-1A was still bound to the spindle-chromosome but MOZ was fully dissociated (Figure 5A-d) but with persistent co-localization of both in the cytoplasm. To determine if the MOZ-TIF2 fusion protein has similar localization as MOZ and co-localized with CAF-1A, HEK293 cells were transfected with EGFP-MOZ or EGFP-MOZ-TIF2 and DsRed2-CAF-1A (Figure 5B). Both EGFP-MOZ and EGFP-MOZ-TIF2 showed a predominantly nuclear localization in HEK293 cells in interphase. However, the, EGFP-MOZ-TIF2 fusion protein appeared in larger aggregates compared to the more homogenously distributed MOZ. In the merged image the MOZ co-localization with CAF-1A appeared stronger than the MOZ-TIF2-CAF-1A co-localization (Figure 5B, top panel, merge). To examine the binding of MOZ, MOZ-TIF2, and CAF-1A to the interphase chromatin we conducted pre-extraction with Triton-X100 in EGFP-MOZ and EGFP-MOZ-TIF2 transfected HEK293 cells (Figure 5C). In the interphase, all three proteins, EGFP-MOZ, EGFP-MOZ-TIF2, and CAF-1A showed resistance to pre-extraction and the co-localization with DAPI-stained DNA. Similarly, the co-localization of EGFP-MOZ-TIF2 with ASF1B was shown in transfected HEK293 cells (Figure 6A). Interestingly, EGFP-MOZ-TIF2 exhibited stronger co-localization with DsRed2-ASF1B than EGFP-MOZ in pre-extracted HEK293 cells (Figure 6B, merge).

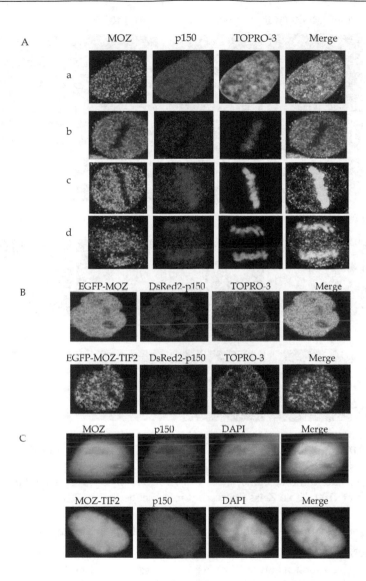

Fig. 5. Subcellular localization of MOZ, MOZ-TIF2, CAF-1A (p150). **A**. Indirect immunofluorescence of MOZ (green) and p150 (red) in HeLa cells at interphase and metaphase observed by confocal microscopy with the nuclei stained with Topro-3. **B**. Confocal microscope images were obtained of HEK293 cells co-transfected with EGFP-MOZ and DsRed2-p150 or EGFP-MOZ-TIF2 and DsRed2-p150 as detailed in the Materials and Methods and nuclei stained with Topro-3. **C**. HEK293 cells transfected with EGFP-MOZ (green) and EGFP-MOZ-TIF2 (green) and stained with anti-p150 antibody after pre-extraction with Triton-X100. The fluorescent images were obtained at x100 with a Zeiss fluorescent microscope.

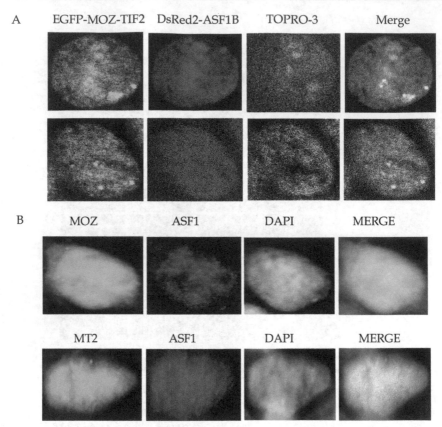

Fig. 6. **A**. Confocal microscope images were obtained of HEK293 cells co-transfected with EGFP-MOZ-TIF2 and DsRed2-ASF1B. The nuclei were stained with Topro-3. **B**. HEK293 cells were transfected with EGFP-MOZ or EGFP-MOZ-TIF2. 48 hours later, cells were pre-extracted, fixed, and immune-stained with anti-ASF1B antibody. Fluorescent images were photographed at x100 with a Zeiss fluorescent microscope.

3.6 Altered gene expression profile in U937 cells stably transfected with MOZ-TIF2

CAF-1 and ASF1, as histone chaperon proteins are essential in maintaining the nucleosome structure after DNA replica and in DNA repair. In yeast, CAF-1 and ASF1 are regulators of global gene expression (Zabaronick and Tyler, 2005). However, if MOZ and MOZ-TIF2, as proteins that associate with CAF-1 and ASF1, affect global gene expression is not known. We established stable transfection clones from U937 cells with forced expression of MOZ and MOZ-TIF2 and analyzed global gene expression of these cell clones. Compared to the expression profile of control cells stably transfected with pcDNA3 vector alone, MT2 caused a > 2-fold change in expression with 181 genes increasing and 106 genes decreasing expression ($p = 0.01$). Over expression of wild type MOZ also altered gene expression (>2-fold increase in 132 genes and >2-fold decrease in 88 genes, $p=0.01$). In addition, a differential gene expression signature was seen between MOZ and MOZ-TIF2 in a Venn diagram analysis (Figure 7). The signature-expressed genes are 189 with pcDNA3, 84 with MOZ, and 427 with MOZ-TIF2, respectively. Further pairwise analysis of differential expression of genes between MOZ and MOZ-TIF2 indicated that there 28 genes increasing over 2 fold (Table 1) and 34 genes decreasing over 2 fold (Table 2) in MOZ-TIF2 compared with that in MOZ. The altered genes between MOZ and MOZ-TIF2 are involved in multiple cell functions such as signal transduction, cell response to stimulus, cell cycle, chromosome structure, development, and tumor progression.

Ratio	p-value	Gene Name
6.41	0.002777	Transcribed locus, weakly similar to XP_537423.2 PREDICTED: similar to LINE-1 reverse transcriptase homolog [Canis familiaris
5.78	0.001076	Malic enzyme 3, NADP(+)-dependent, mitochondrial
4.8	0.042772	Testis derived transcript (3 LIM domains)
4.01	0.021676	Bone morphogenetic protein 1
3.71	0.031657	Interleukin 8 receptor, beta
3.63	0.012968	NADH dehydrogenase (ubiquinone) 1 beta subcomplex, 6, 17kDa
3.09	0.042154	calreticulin
2.73	0.047444	Actin binding LIM protein 1
2.55	0.040271	Ribosome binding protein 1 homolog 180kDa (dog)
2.43	0.016348	vesicle amine transport protein 1 homolog (T. californica)
2.41	0.004052	Calreticulin
2.39	0.039279	histone cluster 1, H2bi
2.37	0.039548	insulin-like growth factor binding protein 2, 36kDa
2.36	0.012714	Inversin
2.36	0.048093	PTK2B protein tyrosine kinase 2 beta
2.29	0.010907	RAP1 interacting factor homolog (yeast)
2.22	0.00445	CD160 molecule
2.13	0.042832	Carnitine palmitoyltransferase 1B (muscle)
2.13	0.010751	PCTAIRE protein kinase 1
2.12	0.006671	Neutrophil cytosolic factor 4, 40kDa
2.11	0.039481	neurogranin (protein kinase C substrate, RC3)
2.1	0.036043	Tyrosine 3-monooxygenase/tryptophan 5-monooxygenase activation protein, epsilon polypeptide
2.09	0.002837	Mediator complex subunit 21
2.06	0.008281	Insulin-like growth factor binding protein 2, 36kDa
2.04	0.032872	acylphosphatase 2, muscle type
2.03	0.006805	FK506 binding protein 1A, 12kDa
2.02	0.006539	Neurochondrin
2	0.001062	Syntaxin 5

Table 1. Up-regulated genes in MOZ-TIF2 vs MOZ.

Ratio	p-value	Gene Name
14.58	0.034066	Fibroblast growth factor receptor 2
10.51	0.048971	Transcribed locus
5.7	0.020122	Spectrin, beta, non-erythrocytic 1
5.3	0.007972	Sulfotransferase (Sulfokinase) like gene, a putative GS2 like gene
5.17	0.005226	Defensin, beta 1
3.92	0.00956	chorionic somatomammotropin hormone-like 1
3.68	0.002866	elongation factor, RNA polymerase II, 2
3.64	0.04087	RAP2A, member of RAS oncogene family
3.57	0.040904	CD2 molecule
3.57	0.039775	Met proto-oncogene (hepatocyte growth factor receptor)
3.34	6.34E-05	Adipose differentiation-related protein
2.99	0.049188	ATPase, Ca++ transporting, plasma membrane 4
2.61	0.013043	X-ray repair complementing defective repair in Chinese hamster cells 2
2.49	0.002229	regulatory solute carrier protein, family 1, member 1
2.47	0.018837	CMP-N-acetylneuraminate monooxygenase) pseudogene
2.45	0.049139	Angiogenic factor with G patch and FHA domains 1
2.45	0.026164	SCY1-like 3 (S. cerevisiae)
2.36	0.003507	spermidine/spermine N1-acetyltransferase 1
2.34	0.046867	ATPase, class VI, type 11A
2.27	0.027531	ecotropic viral integration site 2A
2.22	0.045457	Ubiquitin specific peptidase like 1
2.21	0.035349	Cyclin-dependent kinase 6
2.17	0.000434	CDC14 cell division cycle 14 homolog B (S. cerevisiae)
2.17	0.032088	Kruppel-like factor 10
2.17	0.049619	Starch binding domain 1
2.16	0.023854	Homeodomain interacting protein kinase 3
2.15	0.030337	Ectodermal-neural cortex (with BTB-like domain)
2.14	0.010869	Angiogenic factor with G patch and FHA domains 1
2.12	0.049079	Reversion-inducing-cysteine-rich protein with kazal motifs
2.11	0.042149	suppressor of Ty 3 homolog (S. cerevisiae)
2.08	0.037371	Nuclear receptor subfamily 1, group D, member 2
2.08	0.022954	cytochrome P450, family 1, subfamily A, polypeptide 1
2.06	0.004142	Peroxisomal biogenesis factor 5
2.06	0.033184	Fem-1 homolog c (C. elegans)

Table 2. Down-regulated genes in MOZ-TIF2 vs MOZ.

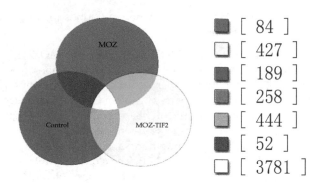

Fig. 7. The Venn diagram of signature gene expression among pcDNA3, MOZ, and MOZ-TIF2. The positive expressed genes were picked up as described in Materials and Methods. The number in brackets indicates the signature genes.

4. Discussion

In order to gain understanding of the function of the MOZ-TIF2 fusion protein we used the yeast two-hybrid system to screen a human bone marrow cDNA library and identified two proteins, CAF-1A and ASF1B, that interacted with the MOZ partner of MOZ-TIF2. The CAF-1A is the largest subunit of CAF-1 which is responsible for bringing histones H3 and H4 to newly synthesized DNA to constitute a nucleosome during DNA replication and DNA repair (Moggs et al., 2000; Shibahara and Stillman, 1999; Smith and Stillman, 1989). CAF-1 controls S-phase progression in euchromatic DNA replication (Klapholz et al., 2009). During chromatin assembly CAF-1 is localized at the replication loci through the association with the proliferation cell nuclear antigen (PCNA), interacting with the N-terminal PCNA binding motif in the CAF-1A. CAF-1 has also been shown to have a role in transcription regulation and epigenetic control of gene expression by interacting with methyl-CpG binding protein and by contributing non-methylation dependent gene silencing (Reese et al., 2003; Sarraf and Stancheva, 2004; Tchenio et al., 2001). A dominant-negative mutant of CAF-1A arrests cell cycle in S-phase (Ye et al., 2003). The loss of CAF-1 is lethal in human cells and increases the sensitivity of cells to UV and other DNA damaging reagents (Game and Kaufman, 1999; Nabatiyan and Krude, 2004). In addition, CAF-1 has been suggested as a clinical marker to distinguish quiescent from proliferating cells (Polo et al., 2004). ASF1B, the other MOZ-TIF2 interacting protein identified, is one of two human ASF1 proteins and participates in chromatin assembly by interacting with the p60 unit of CAF-1 (Mello et al., 2002). The function of ASF1 overlaps with CAF-1 but contributes mainly to chromatin-mediated gene silencing (Meijsing and Ehrenhofer-Murray, 2001; Mello et al., 2002; Osada et al., 2001). In the process of nucleosome formation during DNA replication, ASF1 synergizes functionally with CAF-1 by binding histone H3/H4 and delivers histone H3 and H4 dimers to CAF-1 (Tyler et al., 1999; Tyler et al., 2001). As with CAF-1 mutations, mutations in ASF1 raise the sensitivity of cells to DNA damage (Daganzo et al., 2003; Emili et al., 2001; Le et al., 1997). In yeast, the absence of ASF1 leads to enhanced genetic instability and sister chromatid exchange (Prado et al., 2004). Recent study revealed that the expression of ASF1B, like CAF-1A, was proliferation-dependent (Corpet et al., 2011). Both CAF-1 and ASF1 are important in maintaining genetic stability and hence mutations or aberrant expression in either may contribute to carcinogenesis.

Our initial results demonstrated that the MOZ portion of the MOZ-TIF2 fusion protein interacted with the human CAF-1A and ASF1B. These associations are consistent with previous findings that a MYST family member in yeast, SAS (something about silencing) protein, interacts with Cac1, a yeast homologue of human CAF-1A, and yeast ASF1 and that the interaction contributed to the silencing of the ribosomal DNA locus (Meijsing and Ehrenhofer-Murray, 2001). However, in our experiments with the yeast two-hybrid system, the association between the MOZ-1/759 fragment and CAF-1A was stronger than the interaction of the MOZ-1/759 fragment and ASF1B. The clones of MOZ-1/759 and CAF-1A grew in both –His and –Ade selection media while the clones of MOZ-1/759 and ASF1B grew only in the –His medium. These results suggest that the intensity of interaction of the MOZ fragment with each chaperone is different and the interactions may involve different domains of MOZ. With the GST pull-down assays, we were able to verify the physical interactions using purified proteins and to begin probing the regions of MOZ involved in the interactions. Our results demonstrated that CAF-1A bound primarily to the N-terminus of MOZ (MOZ-1/313) while ASF1B bound to the domain containing C2HC motif and acetyl-CoA binding region (MOZ-488/703). To exclude possible indirect interactions caused by using a mammalian transcription/ translation system, the pull-down assay was also conducted using an E. coli translation system (EcoPro™ T7 System, EMD Biosciences, Novagen, San Diego, CA) with the same interactions being seen again (data not shown). The binding of CAF-1A and ASF1B to two distinct regions within the MOZ fragment involved in the MOZ-TIF2 fusion protein suggests that MOZ-TIF2 positively influences participation in chromatin assembly.

The experiments reported here also begin to shed some light on aberrant function of the MOZ-TIF2 fusion protein by comparing semi-quantatively the strength of association of CAF-1A and ASF1B with MOZ and MOZ-TIF2. In the co-immunopreciptiation and S-tagged pull down experiments, CAF-1A appeared to interact more strongly with MOZ than MOZ-TIF2. These observations were confirmed by the increased co-localization seen in confocal microscopy of the co-transfected cells at interphase. The converse was seen in the interactions of ASF1B with an apparent greater intensity of interaction of ASF1B with MOZ-TIF2 than MOZ alone. Again, this interaction was confirmed in pre-extracted HEK293 cells. It seems that MOZ-TIF2 fusion protein changed the binding priorities of MOZ. These differences may occur because of the necessity of appropriate folding or other higher order structural changes in the full-length MOZ, which are obviated in the fusion protein. In addition, we noticed that the localization of MOZ and CAF-1A was altered in mitotic cells, suggesting that the function of interactions in chromatin assembly and modification depend on cell division cycle. Previously, CAF-1 has been observed to disassociate from chromosomes during the M phase and to be inactivated in mitosis (Marheineke and Krude, 1998). However, we have seen the binding of CAF-1A to the spindle-like chromosome during the metaphase and anaphase in immune-stained Hela cells. It is not clear if the altered association of CAF-1A with chromosome indicates a physiological process during the mitosis or is the artificial results either of fixation and stain process or the limitation of the antibody. A further investigation is necessary to determine the dynamic change of the association.

Using stably transfected U937 cells, we were able to find MOZ-TIF2-correlated changes in the global expression profile of genes and identify a signature-expression profile for MOZ-TIF2. However, as MOZ and TIF2 function as transcription co-factors and as CAF-1 and ASF1 are regulators of global transcription the altered gene expression by MOZ-TIF2 cannot be ascribed to the interaction of MOZ-TIF2 with CAF-1A and ASF1B alone. Interestingly, inspite of 427 expressed signature genes of MOZ-TIF2, only 62 genes were found with over two-fold significant change between MOZ-TIF2 and MOZ, suggesting that differences in expression level between MOZ and MOZ-TIF2 of most most signature genes signature genes could be relatively small.

We are currently examining the hypothesis that the association of MOZ-TIF2 with chromatin assembly factors affects the nucleosome structure and/or histone modification such that histone acetylation status would contribute to leukemogenesis. This hypothesis assumes that the MOZ-TIF2 fusion protein may alter constitution of the chromatin assembly factor complex and then change global gene expression. A possible target for this type of altered function would be that the fusion protein could alter the recruitment of CBP to the complex via LXXLL motifs in TIF2 portion (Voegel et al., 1998; Yin et al., 2007).

5. Conclusions

We demonstrate that both MOZ and MOZ-TIF2 interacts with ASF1B via its MYST domain and interacts with CAF-1A via its zinc finger domain. MOZ and MOZ-TIF2 co-localize with CAF-1A and ASF1B in interphase nuclei. MOZ-TIF2, compared to MOZ, preferentially binds to ASF1B rather than to CAF-1A. MOZ-TIF2 interferes with the function of wild type MOZ and alters global gene expression in U937 cells.

6. References

Aikawa, Y., Katsumoto, T., Zhang, P., Shima, H., Shino, M., Terui, K., Ito, E., Ohno, H., Stanley, E. R., Singh, H., et al. (2010). PU.1-mediated upregulation of CSF1R is crucial for leukemia stem cell potential induced by MOZ-TIF2. Nat Med 16, 580-585, 581p following 585.

Bristow, C. A., and Shore, P. (2003). Transcriptional regulation of the human MIP-1alpha promoter by RUNX1 and MOZ. Nucleic Acids Res 31, 2735-2744.

Champagne, N., Bertos, N. R., Pelletier, N., Wang, A. H., Vezmar, M., Yang, Y., Heng, H. H., and Yang, X. J. (1999). Identification of a human histone acetyltransferase related to monocytic leukemia zinc finger protein. J Biol Chem 274, 28528-28536.

Champagne, N., Pelletier, N., and Yang, X. J. (2001). The monocytic leukemia zinc finger protein MOZ is a histone acetyltransferase. Oncogene 20, 404-409.

Corpet, A., De Koning, L., Toedling, J., Savignoni, A., Berger, F., Lemaitre, C., O'Sullivan, R. J., Karlseder, J., Barillot, E., Asselain, B., et al. (2011). Asf1b, the necessary Asf1 isoform for proliferation, is predictive of outcome in breast cancer. Embo J 30, 480-493.

Daganzo, S. M., Erzberger, J. P., Lam, W. M., Skordalakes, E., Zhang, R., Franco, A. A., Brill, S. J., Adams, P. D., Berger, J. M., and Kaufman, P. D. (2003). Structure and function of the conserved core of histone deposition protein Asf1. Curr Biol 13, 2148-2158.

Deguchi, K., Ayton, P. M., Carapeti, M., Kutok, J. L., Snyder, C. S., Williams, I. R., Cross, N. C., Glass, C. K., Cleary, M. L., and Gilliland, D. G. (2003). MOZ-TIF2-induced acute myeloid leukemia requires the MOZ nucleosome binding motif and TIF2-mediated recruitment of CBP. Cancer Cell 3, 259-271.

Emili, A., Schieltz, D. M., Yates, J. R., 3rd, and Hartwell, L. H. (2001). Dynamic interaction of DNA damage checkpoint protein Rad53 with chromatin assembly factor Asf1. Mol Cell 7, 13-20.

Esteyries, S., Perot, C., Adelaide, J., Imbert, M., Lagarde, A., Pautas, C., Olschwang, S., Birnbaum, D., Chaffanet, M., and Mozziconacci, M. J. (2008). NCOA3, a new fusion partner for MOZ/MYST3 in M5 acute myeloid leukemia. Leukemia 22, 663-665.

Game, J. C., and Kaufman, P. D. (1999). Role of Saccharomyces cerevisiae chromatin assembly factor-I in repair of ultraviolet radiation damage in vivo. Genetics 151, 485-497.

Huntly, B. J., Shigematsu, H., Deguchi, K., Lee, B. H., Mizuno, S., Duclos, N., Rowan, R., Amaral, S., Curley, D., Williams, I. R., et al. (2004). MOZ-TIF2, but not BCR-ABL, confers properties of leukemic stem cells to committed murine hematopoietic progenitors. Cancer Cell 6, 587-596.

James, P., Halladay, J., and Craig, E. A. (1996). Genomic libraries and a host strain designed for highly efficient two-hybrid selection in yeast. Genetics 144, 1425-1436.

Katsumoto, T., Aikawa, Y., Iwama, A., Ueda, S., Ichikawa, H., Ochiya, T., and Kitabayashi, I. (2006). MOZ is essential for maintenance of hematopoietic stem cells. Genes Dev 20, 1321-1330.

Katsumoto, T., Yoshida, N., and Kitabayashi, I. (2008). Roles of the histone acetyltransferase monocytic leukemia zinc finger protein in normal and malignant hematopoiesis. Cancer Sci 99, 1523-1527.

Kitabayashi, I., Aikawa, Y., Nguyen, L. A., Yokoyama, A., and Ohki, M. (2001). Activation of AML1-mediated transcription by MOZ and inhibition by the MOZ-CBP fusion protein. Embo J 20, 7184-7196.

Klapholz, B., Dietrich, B. H., Schaffner, C., Heredia, F., Quivy, J. P., Almouzni, G., and Dostatni, N. (2009). CAF-1 is required for efficient replication of euchromatic DNA in Drosophila larval endocycling cells. Chromosoma 118, 235-248.

Le, S., Davis, C., Konopka, J. B., and Sternglanz, R. (1997). Two new S-phase-specific genes from Saccharomyces cerevisiae. Yeast 13, 1029-1042.

Liang, J., Prouty, L., Williams, B. J., Dayton, M. A., and Blanchard, K. L. (1998). Acute mixed lineage leukemia with an inv(8)(p11q13) resulting in fusion of the genes for MOZ and TIF2. Blood 92, 2118-2122.

Marheineke, K., and Krude, T. (1998). Nucleosome assembly activity and intracellular localization of human CAF-1 changes during the cell division cycle. J Biol Chem 273, 15279-15286.

Meijsing, S. H., and Ehrenhofer-Murray, A. E. (2001). The silencing complex SAS-I links histone acetylation to the assembly of repressed chromatin by CAF-I and Asf1 in Saccharomyces cerevisiae. Genes Dev 15, 3169-3182.

Mello, J. A., Sillje, H. H., Roche, D. M., Kirschner, D. B., Nigg, E. A., and Almouzni, G. (2002). Human Asf1 and CAF-1 interact and synergize in a repair-coupled nucleosome assembly pathway. EMBO Rep 3, 329-334.

Moggs, J. G., Grandi, P., Quivy, J. P., Jonsson, Z. O., Hubscher, U., Becker, P. B., and Almouzni, G. (2000). A CAF-1-PCNA-mediated chromatin assembly pathway triggered by sensing DNA damage. Mol Cell Biol 20, 1206-1218.

Nabatiyan, A., and Krude, T. (2004). Silencing of chromatin assembly factor 1 in human cells leads to cell death and loss of chromatin assembly during DNA synthesis. Mol Cell Biol 24, 2853-2862.

Osada, S., Sutton, A., Muster, N., Brown, C. E., Yates, J. R., 3rd, Sternglanz, R., and Workman, J. L. (2001). The yeast SAS (something about silencing) protein complex contains a MYST-type putative acetyltransferase and functions with chromatin assembly factor ASF1. Genes Dev 15, 3155-3168.

Perez-Campo, F. M., Borrow, J., Kouskoff, V., and Lacaud, G. (2009). The histone acetyl transferase activity of monocytic leukemia zinc finger is critical for the proliferation of hematopoietic precursors. Blood 113, 4866-4874.

Polo, S. E., Theocharis, S. E., Klijanienko, J., Savignoni, A., Asselain, B., Vielh, P., and Almouzni, G. (2004). Chromatin assembly factor-1, a marker of clinical value to distinguish quiescent from proliferating cells. Cancer Res 64, 2371-2381.

Prado, F., Cortes-Ledesma, F., and Aguilera, A. (2004). The absence of the yeast chromatin assembly factor Asf1 increases genomic instability and sister chromatid exchange. EMBO Rep 5, 497-502.

Reese, B. E., Bachman, K. E., Baylin, S. B., and Rountree, M. R. (2003). The methyl-CpG binding protein MBD1 interacts with the p150 subunit of chromatin assembly factor 1. Mol Cell Biol 23, 3226-3236.

Sarraf, S. A., and Stancheva, I. (2004). Methyl-CpG binding protein MBD1 couples histone H3 methylation at lysine 9 by SETDB1 to DNA replication and chromatin assembly. Mol Cell 15, 595-605.

Shibahara, K., and Stillman, B. (1999). Replication-dependent marking of DNA by PCNA facilitates CAF-1-coupled inheritance of chromatin. Cell 96, 575-585.

Smith, S., and Stillman, B. (1989). Purification and characterization of CAF-I, a human cell factor required for chromatin assembly during DNA replication in vitro. Cell 58, 15-25.

Tchenio, T., Casella, J. F., and Heidmann, T. (2001). A truncated form of the human CAF-1 p150 subunit impairs the maintenance of transcriptional gene silencing in mammalian cells. Mol Cell Biol 21, 1953-1961.

Troke, P. J., Kindle, K. B., Collins, H. M., and Heery, D. M. (2006). MOZ fusion proteins in acute myeloid leukaemia. Biochem Soc Symp, 23-39.

Tyler, J. K., Adams, C. R., Chen, S. R., Kobayashi, R., Kamakaka, R. T., and Kadonaga, J. T. (1999). The RCAF complex mediates chromatin assembly during DNA replication and repair. Nature 402, 555-560.

Tyler, J. K., Collins, K. A., Prasad-Sinha, J., Amiott, E., Bulger, M., Harte, P. J., Kobayashi, R., and Kadonaga, J. T. (2001). Interaction between the Drosophila CAF-1 and ASF1 chromatin assembly factors. Mol Cell Biol 21, 6574-6584.

Voegel, J. J., Heine, M. J., Tini, M., Vivat, V., Chambon, P., and Gronemeyer, H. (1998). The coactivator TIF2 contains three nuclear receptor-binding motifs and mediates transactivation through CBP binding-dependent and -independent pathways. Embo J *17*, 507-519.

Ye, X., Franco, A. A., Santos, H., Nelson, D. M., Kaufman, P. D., and Adams, P. D. (2003). Defective S phase chromatin assembly causes DNA damage, activation of the S phase checkpoint, and S phase arrest. Mol Cell *11*, 341-351.

Yin, H., Glass, J., and Blanchard, K. L. (2007). MOZ-TIF2 repression of nuclear receptor-mediated transcription requires multiple domains in MOZ and in the CID domain of TIF2. Mol Cancer *6*, 51.

Zabaronick, S. R., and Tyler, J. K. (2005). The histone chaperone anti-silencing function 1 is a global regulator of transcription independent of passage through S phase. Mol Cell Biol *25*, 652-660.

3

The Use of Reductive Methylation of Lysine Residues to Study Protein-Protein Interactions in High Molecular Weight Complexes by Solution NMR

Youngshim Lee[1], Sherwin J. Abraham[2] and Vadim Gaponenko[1,*]
[1]*Department of Biochemistry and Molecular Genetics,*
University of Illinois at Chicago, Chicago, IL
[2]*Department of Molecular and Cellular Physiology, Stanford University,*
School of Medicine, Beckman Center, Stanford, CA
USA

1. Introduction

While solution state NMR is very well suited for analysis of protein-protein interactions occurring with a wide range of affinities, it suffers from one significant weakness, known as the molecular weight limitation. This limitation stems from the efficient nuclear relaxation processes in macromolecules larger than 30 kDa (Wider & Wüthrich, 1999). These relaxation processes cause rapid decay of NMR signals. Although the use of transverse relaxation optimized spectroscopy (TROSY) approaches has made solution state NMR of large proteins and protein-protein complexes more feasible, it is still limited by the ability to produce isotope enriched proteins (Pervushin et al., 1997). However, there is a significant number of proteins for which no convenient system for stable isotope incorporation exists. We recently utilized reductive methylation methodology to demonstrate that it is possible to introduce ^{13}C-enriched methyl groups into lysine residues in otherwise unlabeled proteins with the purpose of studying protein-ligand and protein-protein interactions by NMR (Abraham et al., 2008).

Reductive methylation is commonly used to improve crystallization of proteins (Schubot & Waugh, 2004). Studies show that success of protein crystallization improves significantly through reductive methylation of solvent exposed lysines due to a reduction in surface entropy. Reductive methylation does not alter significantly protein structures and native protein-protein interactions (Gerken et al., 1982; Kurinov et al., 2000; Rayment, 1997; Walter et al., 2006). Despite clear advantages offered by reductive methylation, this technique remains underutilized in solution NMR. Here we show that reductive methylation allows characterization of high molecular weight protein-protein complexes that is not achievable using traditional NMR approaches.

* Corresponding Author

For reductive methylation of NMR protein samples, ^{13}C-enriched carbonyl compound (e.g. ^{13}C-formaldehyde) and reducing agents are required. The primary amine of lysine in polypeptide molecules acting as a nucleophile attacks the carbonyl group of formaldehyde. This reaction results in formation of an intermediate imine through the carbonyl-condensation process. The intermediate imine subsequently reacts with a proton donor to give rise to the higher order amine (Scheme 1). The solvent exposed lysine residues are frequently dimethylated when a sufficient amount of formaldehyde is present.

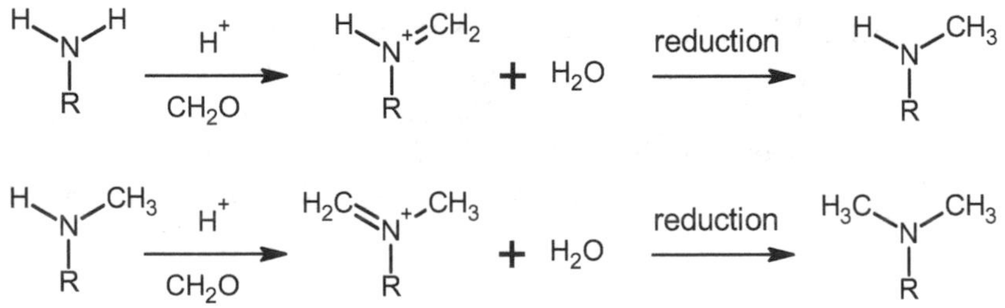

Scheme 1.

The reductive methylation technique offers several advantages. First, proteins purified from their native hosts can be directly used for enrichment with stable isotopes. In this way, the protein molecules are likely to retain their correct fold and post-translational modifications. Second, since only a small amount of ^{13}C-labeled formaldehyde is used in the reaction the reductive methylation procedure is significantly more economical than the traditional isotope enrichment protocols. Finally, the use of ^{13}C-labeled methyl groups in lysines offers an opportunity to observe NMR signals with favorable relaxation properties in large molecular weight proteins due to reduced order parameters for lysine side-chains (Abraham et al., 2009). In this report we not only demonstrate that observation of NMR signals in high molecular weight non-isotope enriched proteins is possible but also that investigation of conformational changes due to binding in protein-protein complexes is amenable to solution state NMR through reductive methylation.

2. Cardiac muscle proteins: Actin, tropomyosin, and troponin complex

Muscle contraction is caused by cyclic interaction between myosin and actin filaments. In cardiac muscle, regulation of contraction is controlled by the troponin complex and tropomyosin which bind to the actin filament (Galińska-Rakoczy et al., 2008; Kobayashi et al., 2008; Kobayashi & Solaro, 2005). The actin filaments consist of polymerized actin (F-actin) molecules which contain myosin binding sites. At rest, the myosin binding site is concealed by tropomyosin forming a coiled-coil dimer that lies in the two grooves of actin. Seven actin molecules interact with one tropomyosin dimer. Each tropomyosin dimer also binds one troponin complex composed of three subunits: troponin C, troponin I, and troponin T. The N-terminal domain of troponin C has a calcium binding pocket. The troponin complex, together with tropomyosin, regulate muscle contraction in a Ca^{2+}-dependent manner. This is accomplished by altering accessibility of actin binding sites to myosin. Being a Ca^{2+} sensor, troponin functions as an on/off switch for muscle contraction. Muscle contraction occurs when Ca^{2+} binds to the regulatory site in troponin C. Conversely, the muscle relaxes when Ca^{2+} dissociates. When Ca^{2+} concentration is high, Ca^{2+} binding to troponin C induces a structural change in the troponin complex that causes relocation of tropomyosin away from the actin groove. Due to tropomyosin relocation, the myosin binding site on actin is exposed and cross-bridge formation is initiated between actin and myosin. Troponin I is known to inhibit myosin cross-bridge formation by inducing relocation of tropomyosin. Troponin T associates with troponin C and I to form the complete troponin complex. Troponin T also binds to tropomyosin and actin to inhibit myosin binding to thin filaments.

Alpha-helical coiled-coil tropomyosin assembles into filaments in the end-to-end configuration and interacts with actin polymers. When bound to polymerized actin, tropomyosin filament spans seven consecutive actin monomers forming a 369 kDa complex. One troponin binds to each tropomyosin coiled-coil dimer such that the molecular ratio for actin, Tm, and troponin is 7:2:1. There are two kinds of interaction between the troponin complex and actin-tropomyosin. One is Ca^{2+}-independent binding through troponin T, anchoring the troponin complex to actin-tropomyosin. The other is Ca^{2+}-dependent regulatory interactions through inhibitory C-terminal half of troponin I, turning muscle contraction "on" and "off". The cytoplasmic Ca^{2+} concentration is essential for muscle contraction. However, allosteric regulation of the troponin complex is also known to be an important contributor. Solution NMR can detect conformational changes in protein molecules and thus is a good tool to study the allosteric regulation. We utilize the reductive methylation technique because the thin fiber is a large protein-protein complex containing molecules that are difficult to produce as recombinant proteins for enrichment with stable isotopes.

2.1 Conformation of reductively methylated cardiac troponin C free and as part of the cardiac troponin complex in the presence and absence of Ca^{2+}

2.1.1 Methods of preparation of reductively methylated troponin complex

Reductive methylation of troponin C for NMR experiments was performed using ^{13}C-enriched formaldehyde and borane-ammonia complex ($NH_3.BH_3$) as a reducing agent. Briefly, 20 μL of 1 M borane-ammonia complex and 40 μL of ^{13}C formaldehyde were added to 1 mL of troponin C in methylation buffer (10 mM HEPES pH 7.6, 50 mM $MgCl_2$, 50 mM $CaCl_2$, and 1 mM β-mercaptoethanol). The reaction mix was incubated at 4 °C with stirring for 2 h. The procedure was repeated one more time with a final addition of 10 μL of ^{13}C formaldehyde and was incubated at 4 °C with stirring overnight. The reaction was stopped by adding 200 mM glycine and the undesired reaction products and excess reagents were removed by extensive dialysis against 10 mM Tris/HCl pH 7.6, 50 mM $MgCl_2$, 50 mM $CaCl_2$, and 1 mM β-mercaptoethanol. To obtain the troponin complex, troponin I and troponin T were added to methylated troponin C in the 1:1:1 molar ratio. To obtain larger molecular weight complexes tropomyosin was added to the troponin complex containing methylated troponin C in the 2:1 molar ratio.

The NMR experiments were performed on samples containing 20 μM troponin C (either alone or in complex) in NMR buffer containing 40 mM Tris-HCl (pH 10.0), 50 mM KCl, 1 mM β-mercaptoethanol, and either 50 mM $CaCl_2$ or 50 mM $MgCl_2$. All ^{1}H-^{13}C heteronuclear single-quantum correlation (HSQC) spectra were acquired on the 600 MHz Bruker Avance spectrometer fitted with a cryoprobe using 128 indirect points at 25 °C. The data were processed using NMRPipe software (Delaglio et al., 1995).

2.1.2 Results

NMR ^{1}H-^{13}C HSQC experiments were performed on free reductively methylated troponin C in the presence and absence of Ca^{2+} and on the cardiac troponin complex containing reductively methylated troponin C in the presence and absence of Ca^{2+}. The results of these experiments are shown in Figure 1. All of the acquired spectra display the expected 12 signals representing methyl groups on eleven lysines in troponin C and one on the N-terminal

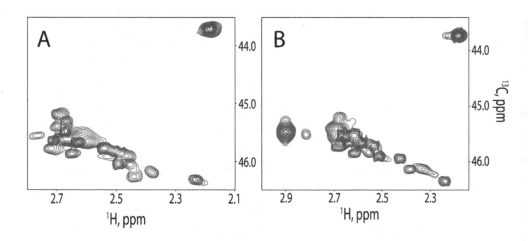

Fig. 1. An overlay wof ^{13}C-^{1}H HSQC spectra of reductively methylated 20 μM troponin C (blue) and the troponin complex consisting of full length troponin C, troponin I, and troponin T (red). The spectra in (A) were recorded in the presence of 50 mM Ca^{2+}. The spectra in (B) were recorded in the absence of Ca^{2+} and the presence of 50 mM Mg^{2+}. The spectra were acquired at 600 MHz at 25 °C with 256 indirect points. The buffer conditions are 40 mM Tris-HCl (pH 10.0), 50 mM KCl, 1 mM β-mercaptoethanol, 50 mM $CaCl_2$ (A) or 50 mM $MgCl_2$ (B).

primary amine. Comparison of spectra of free troponin C in the presence and absence of Ca^{2+} (Fig. 1A and 1B) reveals significant differences in methyl chemical shift values. These chemical shift perturbations indicate expected structural rearrangements in the N-terminal domain of troponin C caused by Ca^{2+} binding. Comparison of NMR spectra of free Ca^{2+}-bound troponin C with Ca^{2+}-bound troponin C in the troponin complex reveals significant perturbations in nine out of twelve methyl chemical shifts (Fig. 1A). This observation suggests involvement of troponin C in intermolecular interactions with components of the troponin complex. In the absence of Ca^{2+} only five out of twelve signals experience significant chemical shift perturbations (Fig. 1B). One possible explanation of this is that in the absence of Ca^{2+}, troponin C is less extensively engaged in protein-protein interactions within the troponin complex. Together, we demonstrate that using reductive methylation it is possible to characterize protein-protein interactions within the troponin complex by NMR despite the high molecular weight of the protein system.

2.2 NMR experiments with reductively methylated 369kDa actin-tropomyosin complex suggest that a global conformational rearrangement is induced in polymerized actin upon tropomyosin binding

2.2.1 Methods of preparation of reductively methylated actin-tropomyosin complex

Globular actin was dialyzed into 10mM phosphate buffered saline, pH 7.4, 0.1mM $MgCl_2$, 1mM dithiothreitol, 0.1mM ATP, and 0.01% NaN_3, to make actin filaments. Initially 20mM borane ammonia complex and 40mM ^{13}C-formaldehyde (20% w/w in H_2O) were added into 0.7mL of 60 µM F-actin and the mixture was stirred for 2 hours at 4°C. Addition of borane ammonia complex and ^{13}C-formaldehyde was repeated and mixture was incubated for another 2 hours at 4°C. After incubation, 10 mM borane ammonia complex was added to the mixture. The mixture was incubated at 4°C with stirring overnight. To quench the reaction, the 50 µL of 2M Tris-HCl was added. To study the change in actin structures upon binding of Tm, Tm is added into ^{13}C methylated F-actin to make 7.5 µM of final Tm concentration whereas the concentration of F-actin is 37µM. The molar ratio of actin and tropomyosin was 5 to 1. The samples were dialyzed against 10mM phosphate buffered saline, pH 7.4, with 1mM $MgCl_2$, 0.1mM ATP, 0.01% NaN_3 and 10% D_2O was added for further NMR experiments. All NMR experiments were carried out on Bruker Avance 600 or 900 NMR spectrometers equipped with cryogenic probes. The 2D ^1H-^{13}C edited HSQC experiments were processed with NMRPipe software (Delaglio et al., 1995).

2.2.2 Results

To assess conformational changes occurring in polymerized actin upon binding tropomyosin we performed a reductive methylation reaction on actin and carried out ^1H-^{13}C HSQC experiments on actin alone and on actin in the presence of tropomyosin (Fig. 2). In the spectrum of polymerized actin seven out of nineteen expected signals were observable. Significant chemical shift changes in four out of seven signals in actin were detected upon Tm binding. Lysines are evenly distributed in the actin structure with no accumulation in any one particular area. Therefore, the data shown here indicates that binding of tropomyosin causes a global conformational change in the structure of polymerized actin.

This observation is contrary to many computational models that propose that tropomyosin binding sites in actin are small and global changes do not occur in the actin-tropomyosin complex.

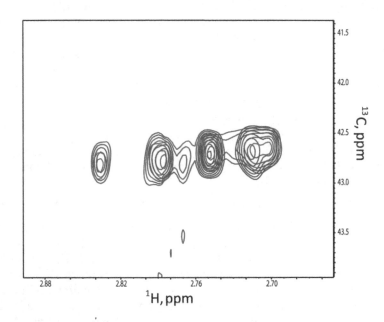

Fig. 2. An overlay of ^{1}H- ^{13}C HSQC spectra of reductively methylated polymerized actin (red) and actin-tropomyosin complex (blue). The spectra were acquired at 900 MHz at 25 °C with 256 indirect points. The buffer conditions are 10 mM phosphate buffered saline (pH 7.4), 150 mM KCl, 50 mM MgCl$_2$, and 1 mM ATP.

3. Conclusion

In conclusion, we have described an important novel application of the reductive methylation methodology to observation of conformational changes in high molecular weight protein-protein complexes by NMR. Using cardiac troponin C as a model system, for which structural information is available, we confirmed that the proposed methodology allows detection of conformational rearrangements in cardiac troponin C upon Ca^{2+} binding. This was done in the context of the full-length troponin complex. Similar experiments would have been very difficult to perform using conventional NMR approaches due to the high molecular weight limitation. We also show that reductive methylation can be used to discover novel conformational changes in a 369 kDa actin-tropomyosin complex. For the first time we show that actin undergoes a global conformational change upon tropomyosin binding. This appears to be the only way such molecular events can be observed. The available computational models were unable to predict this phenomenon. Electron microscopy images of the cardiac thin fiber are too low resolution to detect a conformational change in actin. Crystallization of polymerized actin is not feasible due to heterogeneity of actin fibers. In addition, there is no good procedure for production of recombinant actin that would allow traditional approaches for stable isotope enrichment for NMR. The functional significance of actin conformational rearrangements upon binding of tropomyosin is still under investigation. However, the discovery that these conformational changes occur in actin is a significant step forward.

4. Acknowledgment

We acknowledge Dr. Tomoyoshi Kobayashi at the University of Illinois at Chicago for providing purified actin, tropomyosin, and protein samples for the troponin complex. We also acknowledge support from the National Cancer Institute grant R01CA135341 to Vadim Gaponenko.

5. References

Abraham, S. J.; Hoheisel, S. & Gaponenko, V. (2008). Detection of protein-ligand interaction by NMR using reductive methylation of lysine residues. *J. Biomol. NMR*, Vol. 42, No. 2, pp. 143-148.

Abraham, S. J.; Kobayashi, T.; Solaro, R. J. & Gaponenko, V. (2009). Differences in lysine pKa values may be used to improve NMR signal dispersion in reductively methylated proteins. *J. Biomol. NMR*, Vol. 43, No. 4,, pp. 239-246.

Delaglio F.; Grzesiek S., Vuister G. W., Zhu G., Pfeifer J. & Bax A. (1995) NMRPipe: a multidimensional spectral processing system based on UNIX pipes. *J. Biomol. NMR*, Vol. 6, No 3, pp. 277-293.

Galińska-Rakoczy A.; Engel P., Xu C., Jung H., Craig R., Tobacman L. S, & Lehman W. (2008). Structural basis for the regulation of muscle contraction by troponin and tropomyosin. *J. Mol. Biol.*, Vol. 379, No. 5, pp. 929–935.

Gerken T. A.; Jentoft J. E., Jentoft N. & Dearborn D. G. (1982) Intramolecular interactions of amino groups in ^{13}C reductively methylated hen egg-white lysozyme. *J. Biol. Chem.*, Vol. 257, No. 6, pp. 2894–2900.

Kobayashi T.; Jin L. & de Tombe P. P. (2008) Cardiac thin filament regulation. *Pflugers Arch.* Vol. 457 No. 1, pp. 37–46.

Kobayashi T. & Solaro R. J. (2005) Calcium, thin filaments and the integrative biology of cardiac contractility. *Annu. Rev. Physiol.* Vol. 67, pp. 39-67.

Kurinov I. V.; Mao C., Irvin J. D. & Uckun F. M. (2000) X-ray crystallographic analysis of pokeweed antiviral protein-II after reductive methylation of lysine residues. *Biochem. Biophys. Res. Commun.*, Vol. 275, No. 2, pp. 549–552.

Pervushin K.; Riek R., Wider G. & Wüthrich K. (1997) Attenuated T_2 relaxation by mutual cancellation of dipole-dipole coupling and chemical shift anisotropy indicates an avenue to NMR structures of very large biological macromolecules in solution. *Proc. Natl. Acad. Sci. USA* Vol. 94, No. 23, pp. 12366-12371.

Rayment I. (1997) Reductive alkylation of lysine residues to alter crystallization properties of proteins. *Methods Enzymol.* Vol. 276, pp. 171–179.

Schubot F. D. & Waugh D. S. (2004) A pivotal role for reductive methylation in the *de novo* crystallization of a ternary complex composed of *Yersinia pestis* virulence factors YopN, SycN and YscB. *Acta Crystallogr., Sect D; Biol. Crystallogr.* Vol. 60, pp. 1981-1986.

Walter T. S.; Meier C., Assenberg R., Au K. F., Ren J., Verma A., Nettleship J. E., Owens R. J., Stuart D. I. & Grimes J. M. (2006) Lysine methylation as a routine rescue strategy for protein crystallization. *Structure* Vol. 14, No. 11, pp. 1617–1622.

Wider G. & Wüthrich K. (1999) NMR spectroscopy of large molecules and multimolecular assemblies in solution. *Curr. Opin. Struct. Biol.* Vol. 9, No. 5, pp. 594-601.

4

Regulation of Protein-Protein Interactions by the SUMO and Ubiquitin Pathways

Yifat Yanku and Amir Orian
Technion-Israel Institute of Technology
Israel

1. Introduction

Post-transcriptional modifications of proteins by ubiquitin and ubiquitin-like proteins (UBLs) such as SUMO (Small Ubiquitin-related Modifier) regulate the function of protein-networks, enable cells to respond to signaling cues during development and to cope with the changing environment during adult life. The ubiquitin and SUMO pathways have profound impacts on protein stability, localization, protein-protein interactions and function.

In this chapter we will review mechanistic and biological aspects of protein-protein interactions that are regulated by ubiquitin and SUMO. We will describe the covalent tagging of proteins by ubiquitin and SUMO, and the enzymatic machineries that regulate these modifications. Subsequently, we will discuss how ubiquitylated or SUMOylated proteins are recognized by ubiquitin and SUMO recognition motifs present on interacting proteins. We will also illuminate how these non-covalent interactions regulate diverse cellular processes such as DNA repair, transcription, signaling, and autophagy in health and disease. Finally, we will address the crosstalk between the ubiquitin and SUMO pathways by SUMO-Targeted Ubiquitin Ligases (STUbLs).

2. Covalent modification of proteins by ubiquitin and SUMO

Ubiquitylation is a post-transcriptional modification where ubiquitin, a 76 amino acids long polypeptide, is covalently attached to proteins. Originally, ubiquitylation was considered as a "death-tag" targeting proteins for degradation by the 26S proteasome. However, over the last two decades non-proteolytic roles of ubiquitylation have also been found to impact protein function, cellular localization and protein-protein interactions. Furthermore, in addition to ubiquitin, ubiquitin-like proteins were identified and collectively termed UBLs. These proteins include among others SUMO (Small Ubiquitin Like Modifier), Nedd8 (Neural precursor cell expressed developmentally down-regulated 8), ISG15 (interferon stimulated gene 15) and FAT10, and all share at least one of the ubiquitin canonical folds (Hershko, 1983; Hochstrasser, 2009).

Among UBLs, the most studied modifier is SUMO. Vertebrates possess four different SUMO isoforms, termed SUMO1-4. While SUMO1-3 are efficiently conjugated to target proteins by specific SUMO enzymes, it is less clear if SUMO4 is conjugated to proteins *in vivo*. SUMO conjugation of target proteins results in a change in their activity, affecting protein localization, and modulates composition of protein complexes. In some cases SUMOylation may act as a priming modification that promotes ubiquitylation via a specific specialized sub-type of E3 ubiquitin ligases termed SUMO-Targeted Ubiquitin Ligases (Abed et al., 2011b; Praefcke et al., 2011). Within the scope of this chapter we will focus solely on the interactions mediated by ubiquitin and SUMO.

2.1 SUMO and Ubiquitin pathways

Ubiquitylation or SUMOylation are both mediated by a tripartite enzymatic cascade comprised of specific sets of enzymes. Ubiquitylation is mediated by the E1 ubiquitin-activating enzyme, E2-ubiquitin-conjugating enzyme (Ubc), and an E3 ubiquitin ligase (Hershko 1983; Pickart 2001). SUMOylation is similarly carried out by a different set of specialized SUMO specific E1, E2 and E3 enzymes. Both Ubiquitin and SUMO are covalently conjugated via their C-terminus Gly residue to free NH_2 amine group of the target protein that may reside on the ε-amino group of a Lys residue along the protein sequence, or on the amino terminus of the protein. Post-transcriptional regulation by ubiquitin and UBLs are tightly regulated, as covalent modification by ubiquitin or SUMO requires ATP. In the human genome only two genes coding for E1, ubiquitin-activating enzymes have been found. It is estimated that about one hundred E2s (ubiquitin-activating enzymes), and hundreds or more E3 ubiquitin-protein ligases exist. Interestingly, in plants the ubiquitin and SUMO pathways are greatly expanded. Collectively, these observations probably reflect the high degree of specificity and regulatory role of these pathways (Hershko & Ciechanover, 1998; Weissman et al., 2011).

E3 ligases are the "match makers" that directly recognize the targeted protein substrate. Ubiquitin E3 ligases can be classified into two main functional classes. The first is the HECT (Homologous to the E6-AP Carboxyl Terminus) domain ligases that accept ubiquitin from an E2 enzyme in the form of a thio-ester via a Cys residue in their catalytic domain, thus forming a thio-ester bond directly with the ubiquitin molecule. The second and most abundant class is the RING (Really Interesting New Gene) finger E3 ligases which utilizes a metal binding domain harbouring Zn+2 ions to facilitate ubiquitylation (Deshaies & Joazeiro, 2009). RING ligases are a diverse subclass, encompassing several hundreds of proteins in the human genome. This large family of ligases is further divided into modular sub-classes: 1. Single subunit ligases such as c-Cbl and parkin, that directly bind to both the E2 enzymes and the targeted ubiquitylation protein substrate, and 2. RING E3 ligases that function as a multi-protein-complex that recruits substrates via separate subunits. Examples for this subclass are the well-characterized APC/C (Anaphase promoting complex; McLean et al., 2011) and, the Cullin-based RING E3 ligases (SCF). A subclass of RING-like ligases is the U-box ligases, that contain a modified RING motif lacking canonical cysteine residues for Zn2+ coordination (Hindley et al., 2001, Patterson, 2002).

While only one ubiquitin isoform exists, three functional SUMO isoforms have been characterized: SUMO1 and SUMO2/3 that share 97% sequence identity. SUMO pathway components are much less diverse, comprising of only one known E2 conjugating enzyme termed Ubc9 and a few known E3 ligases. A key difference between SUMO1 and SUMO2/3 is the lack of a SUMOylation motif in SUMO1. Therefore, SUMO2/3 can form high molecular weight SUMO polymers with greater affinity than SUMO1. Similarly to ubiquitin ligation, SUMOylation is mostly facilitated via one of two main mechanisms: by recruiting an E2-SUMO to the substrate or by enhancing the conjugation of SUMO to a substrate already bound to the E2 (Ulrich, 2009). Like E3 ubiquitin ligases, SUMO E3 ligases are diverse and have been categorized into three families. The first class is PIAS like (Protein Inhibitor of Activated STAT–signal transducer and activator of transcription) proteins which posses an SP-RING domain functioning similarly to the ubiquitin RING domain and interacting with Ubc9. The second class is the RanBP2/Nup358 (Nuclear pore proteins Ran binding protein 2 and nucleoporin 358) like proteins, which harbour tandem elements that are capable of binding both Ubc9 and SUMO. A third group includes proteins like polycomb group protein Pc2, TOPORS (Topoisomerase I-binding RING finger protein) and likely HDAC4 (histone deacetylase 4) whose molecular mechanism of SUMO ligation is not fully understood (Gareau & Lima, 2010; Hannoun et al., 2010).

2.1.1 Mono ubiquitylation and SUMOylation

Mono-ubiquitylation is formed in most cases by a covalent attachment of the carboxy terminal glycine of the ubiquitin polypeptide to the ε-amino group within the side chain of Lys residues on the target protein. In some cases the attachment site could be the free NH2-group of the protein's first amino acid. Mono ubiquitylation or mono SUMOylation are mediated by ubiquitin-conjugating enzymes (Ubcs, E2s) or in the case of SUMO by the single SUMO conjugating enzyme, Ubc9. A single protein can be modified by several mono-ubiquitin monomers, resulting in multi-mono-ubiquitylation.

Mono-ubiquitylation most commonly has a regulatory function as a cellular signal, marking transmembrane receptors for recycling in the lysosome, or alternatively marking specific histone tails, thereby impacting chromatin structure. Yet, it can also serve as a degradation signal. As for SUMO, the three SUMO isoforms differ in their conjugation ability, but all three forms can be conjugated to form mono-SUMOylated substrates. Like ubiquitin, SUMO is bound to the target protein via a covalent attachment of its C-terminal carboxyl group to the ε-amino group on the Lys residue of the modified protein. Subsequently, SUMOylated and ubiquitylated proteins are recognized by a specific interaction motif that functions as a 'receptor' for these proteins on their interaction partners. In both cases multiple Lys residues within the target substrate can undergo ubiquitin/UBL modifications, generating multi-ubiquitylated or SUMOylated proteins (Hurley, 2006; Gareau & Lima, 2010).

2.1.2 Poly ubiquitylation and SUMOylation

Successive rounds of ubiquitylation or SUMOylation of a single covalently attached mono-ubiquitin or mono-SUMO molecules will generate poly-ubiquitin or poly-SUMO chains. Ubiquitin- or SUMO-protein ligase enzymes (E3), together with distinct E2s are essential for catalyzing poly-ubiquitylation, SUMOylation, and determine substrate specificity as well as govern chain structure. Following mono-ubiquitiylation, E2 and E3 ligases conjugate subsequent ubiquitin units, forming additional iso-peptide bonds between the carboxyl group of the newly added ubiquitin molecule and the ε-amino group of the Lys residue of the already covalently attached ubiquitin molecule. Recent work suggests that these chains serve as a versatile "code" that regulates protein fate, and that the internal structure and length of the chain directly impacts its recognition by "reader" proteins (Weismann et al., 2011).

Poly-ubiquitylation may be linked through any of the seven Lys residues present in the ubiquitin molecule - Lys^6, Lys^{11}, Lys^{27}, Lys^{29}, Lys^{33}, Lys^{48} and Lys^{63} or the ubiquitin N-terminal Met (Met^1). Additional ubiquitin chains can be linked through the same Lys residue within ubiquitin, forming homotypic chains. One unique form are linear poly-ubiquitin chains where the C-terminal Gly of the ubiquitin molecule is sequentially linked to the N-terminal Met of the next ubiquitin molecule. In addition, chains linked through different Lys side chains can form mixed poly-ubiquitin chains (i.e. harbouring alternating Lys linkage types) or even branched trees of ubiquitin molecules in which more than one Lys residue is extended. Thus, while only one isoform of ubiquitin exists, diverse arrays of ubiquitin chains are formed, dictating different globular structures. Aside from the role in proteasomal degradation mediated by chains with a Lys^{48} linkage, other types of poly-ubiquitylation chains regulate protein-protein interactions in processes such as DNA repair or immune signalling. For example, linear poly-ubiquitin chains are important for the regulation of NFκB signalling (as shortly described below, and Harper & Schulman, 2006; Ikeda & Dikic, 2008; Iwai & Tokunaga, 2009; Weismann et al., 2011; Kim et al., 2011; Behrends & Harper, 2011).

K63-linked ubiquitin chains have been reported to function as scaffolds for the recruitment of other signaling proteins upon cytokine stimulation, and recently an emerging unique role for ubiquitin conjugation in TNF-R signaling was characterized. TNF receptor-1 (TNF-R) activation results in K^{63}-linked ubiquitin chains. These chains are specifically generated by two ubiquitin RING E3 ligases named Haeme-Oxidized-IRP2-ubiquitin-Ligase-1 (HOIL-1) and Haeme-Oxidized-IRP2-ubiquitin-Ligase-1-Interacting-Protein (HOIP). HOIL-1 and HOIP along with a third protein, SHARPIN, form the linear ubiquitin chain assembly complex (LUBAC) that catalyzes linear head-to-tail ubiquitylation by ligating the N-terminal Met^1 residue of the ubiquitin molecule to the C-terminal Gly residue of another ubiquitin molecule. TNF- induced LUBAC complex binds to an activator of the NFκB pathway named NFκB Essential Modifier (NEMO). LUBAC conjugates linear poly-ubiquitin chains to NEMO and enhances the interaction between NEMO and the TNF-R signaling complex. Since NEMO is required for efficient activation of the TNF-R signaling complex, LUBAC activity influences activation of the NFκB pathway. Taken together, linear ubiquitylation in this case serves as a survival machinery required for the proper activation of the TNF-R signaling complex, NFκB gene induction, and protection from TNF-induced cell death (Haas et al., 2009; Gerlech et al., 2011; Iwai, 2011; Niu et al., 2011).

As for SUMO, three SUMO1-3 genes encode for proteins that differ from one another in their ability to form SUMO chains. Only SUMO2/3 possess a Lys residue within a consensus motif ΨKXE that facilitates the formation of SUMO chains. SUMO-1 lacks this consensus site and therefore formation of poly-SUMO chains is less favorable. Yet, *in vitro* it can form polymeric chains and can serve as the chain terminator of SUMO chains. Of the eight Lys residues encoded in the SUMO molecule, SUMO chains are predominantly formed via Lys^{11}. SUMOylation of proteins was thought to enhance transcriptional repression; however, new findings suggest a more diverse function for SUMOylation. SUMOylation was found to affect sub-cellular localization and is also involved in intra-nuclear localization, transport and apoptosis, as well as in targeting proteins for ubiquitin-mediated degradation (Ulrich 2008; Matic et al., 2008; Ulrich, 2009).

2.1.3 Ubiquitin and SUMO chain editing

Covalent ubiquitylation or SUMOylation is a reversible process in which de-ubiquitylating enzymes (DUBs) or sentrins/SUMO specific proteases (SENPs), promote the cleavage of the iso-peptide bond and release ubiquitin or SUMO molecules, respectively. About 80 known DUBs, and less than 10 SENPs, are devoted to removing covalent ubiquitin or SUMO modifications. DUBs serve to perform three distinct roles in the cell; First, DUBs can cleave some of the ubiquitin molecules that are transcribed as a linear fusion chain for future conjugation processes. Second, DUBs mediate the removal of ubiquitin from tagged proteins prior to their degradation, allowing for recycling of ubiquitin molecules for future conjugation processes. Third, DUBs can trim ubiquitin chains, subsequently changing their length and structure (Komander, et al., 2009a). Editing of ubiquitin chains by specific ubiquitin peptidases may impact their recognition by the proteasome, or affect protein-protein interaction. Interaction between DUBs and ubiquitylated proteins is mediated through ubiquitin interacting motifs within the DUBs (UIMs and UBD, ubiquitin binding domains, see below and chapter 3.1.1). DUB-mediated chain editing is essential for regulation of chromatin structure, and is involved in DNA damage repair pathways as well as endosomal targeting of membrane bound receptors (Katz, 2010).

One of the well-studied DUBs is the tumor suppressor CYLD (cylindromatosis associated DUB). CYLD is a negative regulator of Wnt, NFκB and JNK signaling pathways in immunity and inflammation. Among CYLD's targets are substrates with Lys[60]-linkage ubiquitylation chains like TRAF6, BCL3, PLK1 that regulate cell cycle proliferation and apoptosis (Massoumi, 2010). CYLD also forms a protein complex with the ubiquitin ligase Itch. Together this editing complex inhibits TAK1, which is required for termination of the immune response (Wertz, 2011). Importantly, mutations in CYLD are associated with cylindromatosis, a predisposition to benign tumors of the hair follicle in the skin and other secretory glands. Another work established that CYLD functions as a tumor suppressor and its loss is associated with cancer (Bignell et al., 2000). In this regard, DUBs are emerging as excellent targets for small molecule inhibitors. For example, recent work from the Dixit group revealed that genetically or chemically targeting USP1 induces muscle stem-cell differentiation, and can serve as a molecular target for therapy of osteosarcoma that is highly resistant to conventional chemotherapy (Williams, 2011).

Like ubiquitiylation, SUMOylation is also a reversible process, and a family of seven sentrin-specific proteases (SENPs) catalyzes de-SUMOylation. Family members differ from one another based on SUMO chain specificity and cellular localization that is determined by distinct N-terminal domains. Biochemical and genetic experiments revealed that SENPs show high degrees of specificity. Towards example, the SENP6 and SENP7 have greater affinity de-conjugation of di- and poly-SUMO2/3 chains than SUMO1 (Lima & Reverter, 2008). SENPs harbor a Ulp domain at their C-terminus which facilitates the cleavage of the isopeptidic bond between the SUMO molecule and the Lys group on the modified protein. SENP-mediated de-conjugation plays an important role in the regulation of developmental and signaling processes. It is also important for tightly regulating the levels of free SUMO in the cell. The regulatory function of SENPs is biologically relevant in development. SENP3/5 are required for ribosomal biogenesis, and targeting SENP2 during cardiac development impairs the expression of key cardiac factors Gata2 and Gata6 (Yun et al, 2008; Kang et al., 2010). Specifically, in SENP2 null embryos, SUMOylated polycomb group Pc2/CBX4 complex accumulates on the promoters of PcG target genes, leading to repression of Gata4 and Gata6 transcription. SENPs also play a key role in tumorigenesis; expression of SENP1 transforms prostate cancer cells and activates Androgen Receptor (AR) signaling. In addition, SENP3 regulates angiogenesis via its impact on HIF1α-associated coactivator p300, and elevated mRNA levels of SENP6, 7 are linked to breast cancer (Cheng, 2010; Bawa-Khalfe & Yeh, 2010).

3. Recognition motifs for ubiquitin and SUMO ligases

Are there preferred (consensus) sites for ubiquitylation? While recognition motifs for recruitment of ubiquitin ligases ("Degrons") exist, it appears that site-specific ubiquitylation is more promiscuous. This variability may stem from the different structural requirements for the diverse interactions of heterogeneous substrates with a variety of E2-E3 complexes. In contrast, the acceptor Lys residues in many SUMOylated proteins reside within a consensus motif ΨKXE (Ψ- hydrophobic residue; X, any amino acids; E, glutamic or aspartic), forming a unique conformation, which interacts directly with the specific hydrophobic groove on the Ubc9 enzyme. In addition to the canonical SUMOylation consensus sequence, longer consensus sequences with specific characteristics have been identified. Among them are the inverted form of the canonical sequence, a hydrophobic cluster motif enriched with a consecutive sequence of large hydrophobic residues, a phosphorylation dependent motif – PSDM (ΨKXEXXS^P) and a negatively charged amino-acid motif - NDSM (ΨKXEXXEEEE). The existence of this highly characterized consensus motif correlates with to the existence of a single unique E2-SUMO conjugating enzyme. Yet, some proteins can be SUMOylated without the presence of the characterized consensus sequence (Ulrich, 2009).

3.1 Recognition of ubiquitylated and SUMOylated proteins

Regardless of the final fate of the modified proteins, covalent tagging by ubiquitin or SUMO is "sensed" by protein motifs that subsequently mediate protein-protein interactions involved in numerous cellular processes. In this section we will discuss the currently known domains that recognize mono and poly-ubiquitin/UBL chains. While our understanding of these interactions is in its infancy, it is the focus of intensive research. For simplicity we will focus on "sensing " ubiquitin and SUMO monomers and polymers.

Proteins modified by ubiquitylation or SUMOylation interact non-covalently with other proteins via ubiquitin binding domains (UBD), or SUMO interacting motifs (SIM), respectively. These motifs are present in many proteins and thereby mediate multiple interactions, and have the potential to induce conformational changes and to impact the avidity of existing protein complexes or to form new protein complexes (see Fig 1).

3.2 Ubiquitin binding domains

UBDs are motifs that enable the association of proteins with either mono ubiquitin or ubiquitin polymers. More than 20 domains have already been identified, and more than 150 different human proteins harbor a versatile combinations of UBDs. Most UBDs interact with ubiquitylated proteins via a hydrophobic patch that include Leu8, Ile44 and Val44 within the ubiquitin molecule on one hand, and a α-helix motif of the UBD on the other hand.

Intensive structure analysis identified that UBD includes, among others, zinc finger interacting domains (ZnF/PAZ/UBZ), plekstrin homology fold (PH fold), and ubiquitin association domain (UBA) as well as ubiquitin-conjugating-like domains (Dikic et al., 2009).

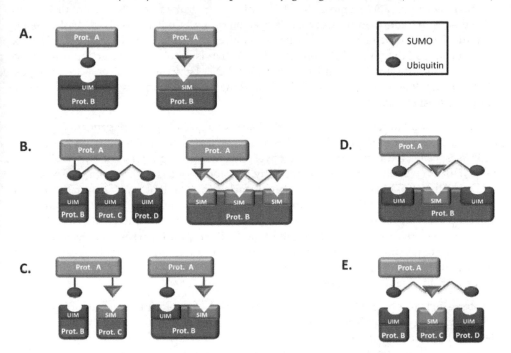

Fig. 1. **Regulation of protein-protein interactions by Ub/SUMO signalling**
(**A**). Mono-ubiquitin or mono-SUMO modified substrates are recognized by Ubiquitin Interacting Motif (UIM/UBDs), or by SUMO interacting motifs (SIMs). (**B**). Poly-ubiquitin or poly-SUMO chains can mediate interaction with several proteins in tandem, or with a single protein harboring several UIM/SIM domains. (**C**). Combinatorial modifications by mono ubiquitin or SUMO can be sensed by multiple proteins each harboring a discrete interaction motif, or via multiple domains within a single protein. (**D-E**). Similar to mono ubiquitylation and SUMOylation, modification by UB/SUMO mixed-chains can be recognized by multiple domains either within a single protein (D), or in the context of several proteins within a protein complex (E).

Different UBDs recognize and interact with different ubiquitin polymers. Some UBDs bind specifically to mono-ubiquitylated proteins. For example the mammalian Eap45 subunit of the endosomal sorting complex harbors a UBD domain termed GLUE domain that interacts with the hydrophobic patch of the ubiquitin molecule (Hirano et al., 2006).

Interestingly, the classical α-helix ubiquitin-interacting motif UIM, such as the one that is present in the proteasome S5a subunit and yeast Rpn10 as well as RAP80, can be found in some cases in an "inverted" orientation to form "inverted UIM" (IUIM/MIU). In most cases, each UBD interacts with a single ubiquitin molecule. Yet, ZnF binding domains, such as the one found in the guanine nucleotide exchange factor RABEX; interact with multiple surfaces on ubiquitin. Hence, in this case a single ubiquitin molecule can interact with three ZnF motifs simultaneously (Penengo et al., 2006). Tandem repeats of UBDs may dictate chain specific interaction. For example, RAP80, which is recruited together with BRCA1 to damage sites, harbors two adjacent UIMs that binds Lys[63] but not of Lys[48] Linked-poly-ubiquitin chains (Hicke et al., 2005; Harper & Schulman, 2006; Komander 2009b; Dikic et al. 2009). In contrast, the ubiquitin receptor RAD23A, that is required for targeting proteins to the proteasome, has a C-terminal UBA domain. RAD23 UBA domain has a 6.3 fold higher affinity to Lys[48] than to Lys[63] chains and a 70 fold higher affinity to Lys[48] chains than to free ubiquitin (Raasi et al., 2005).

The specificity of different UBDs toward chain linkage is greatly dependent on UBDs present in DUBs. For example, while isoT is dedicated for the de-conjugation of Lys[48] chains, CYLD is specific for Lys[63]. Recently a novel UBD that recognizes linear poly-ubiquitin chains (UBAN) was characterized. Importantly, the ability of a cell to respond to TNFα and to activate the IKK kinase complex is compromised in cells that have a mutated UBAN domain within the NEMO protein that is required for activation of the IKK complex (Rahighi et al., 2009; Lo et al., 2009). Likewise, recent work has shown that the ESCRT sorting complex that is involved in targeting the EGF receptor for lysosomal degradation is based on a combination of various UBDs on different ubiquitin receptors. Together these ubiquitin receptors form a large protein complex harboring a high avidity interaction surface with an ubiquitylated cargo (Raiborg & Stenmark, 2009).

An interesting case where binding to ubiquitin chains via ubiquitin receptors plays a key role is autophagy. Autophagy is used by macrophages as a defense mechanism against infection by invading intracellular bacteria. A molecular link between autophagy and ubiquitylation was established following the identification of autophagy receptors, which simultaneously bind both ubiquitin and autophagy-specific ubiquitin-like modifiers (Atg8). Several ubiquitin-related autophagy pathways have already been characterized such as the ubiquitin-NDP52-LC3 pathway, which targets group-A-Streptococcus, *Salmonella typhimurium*, or the ubiquitin-p62-LC3 pathway, which targets *Mycobacterium tuberculosis* and *Listeria monocytogenes*. *Listeria* is a gram-positive pathogen that expresses several virulence proteins including a hemolysin (listeriolysin O, LLO). LLO bacterial proteins in macrophages infected with *Listeria* were found to form small aggregates associated with poly-ubiquitin chains and to undergo selective autophagy by the p62-LC3 pathway (Ogawa et al., 2011). A recent report determined that targeting the SUMO conjugating E2, Ubc9, for degradation and subsequently inhibiting SUMOylation mediate part of the virulence of *Listeria*. Furthermore the Dikic lab recently showed that invading *Salmonella* are coated with poly-ubiquitin chains ligated by a yet to be identified ligase. Subsequently, the ubiquitin chain binding proteins p62 and NDP52 bind to the poly-ubiquitin chains and recruit the protein Optineurin (OPTN) that upon its phosphorylation by Tank Binding Kinase (TBK) connects the coated pathogen to LC3 autophagy receptors allowing the engulfment and autophagy of the pathogen. Thus, an emerging network of interactions between the SUMO pathway, ubiquitin chains, ubiquitin binding proteins and ubiquitin like proteins (Atg8) plays a key role in elimination of bacteria and innate immunity (Wild et al., 2011; Weidberg et al., 2011).

3.3 SUMO interacting motifs

In analogy to ubiquitylation, SUMOylation is also "sensed" by a specific protein motif, termed SUMO Interacting Motif (SIM). SIM motifs are characterized by a sequence motif of hydrophobic amino acids (V/I) X (V/I) (V/I). The SIM domain interacts with the hydrophobic patch on SUMO. This hydrophobic interaction is re-enforced with other weak non-covalent interactions that are formed between basic residues on the SUMO and the acidic residues flanking the SIM domain (Gareau & Lima, 2010). SIM–mediated recruitment plays a key role in transcriptional repression. SUMO/SIM-mediated binding is required for the recruitment of histone de-acetylases and histone de-methylases to co-repressor complexes, as well as impacts the activities of chromatin remodeling factors. Examples for these interactions are the SIM/SUMO dependent recruitment of HDAC2 to SUMOylated Elk2, and the SUMO2/3 and SIM dependent recruitment of the Lys demethylase (LSD) to the CoRest co-repressor complex. In this case, CoREST1 binds directly and non-covalently SUMOylated REST (NRSF) to bridge HDAC2 and LSD (Yang & Sharrocks, 2004; Gill, 2005; Ouyang & Gill, 2009). Furthermore, SUMO/SIM interactions are likely to impact nucleosome remodeling as the recruitment of the ATP-remodeling complex protein Mi-2 to the transcription factor SP3 is SIM/SUMO dependent (Stielow et al., 2006).

SIMs and UBDs are targets for posttranscriptional modification (PTMs). These PTMs impact the structural properties of SIMs, UBDs or their immediate vicinity, resulting in a change in the binding properties of the modified domains. For example a CKII-mediated phosphorylation site near the SIM domain in the co-repressor DAXX shifts specificity between SUMO prologs. The DAXX (Fas death domain associated protein) is a transcriptional co-repressor that binds to a variety of transcription factors at the promoter sites of anti-apoptotic genes. The SIM domain within DAXX facilitates a non-covalent interaction with other SUMOylated proteins. DAXX binding to SUMO-1 but not SUMO2/3 is enhanced by CKII phosphorylation of DAXX Ser[737, 739] surrounding its SIM domain, enhancing its recruitment and subsequent transcriptional repression of these anti-apoptotic genes (Chang et al., 2011; Mukhopadhyay & Matunis 2011). Thus, the observations that SIMs and likely UBDs are targeted to PTMs by signaling pathways provide evidence for another layer of regulation that establish a direct crosstalk between signaling pathways and ubiquitin/SUMO signals.

4. Cross talk between ubiquitylation, SUMOylation, and the function of SUMO-targeted ubiquitin ligases

While both ubiquitin and SUMO pathways have been well studied individually, the long-speculated nature of the crosstalk between SUMO and ubiquitin pathways has been molecularly enigmatic. Importantly the interplay between SUMOylation and ubiquitylation can be a critical determinant in signaling, transcription, and cancer (Karscher 2006). For example, the equilibrium between SUMOylation and ubiquitylation can influence the balance between p53 nuclear localization and stabilization, cytoplasmic export and degradation, as well as regulating the activity and stability of Hypoxia Induced Factor (HIF; Lee et al., 2006; Carter et al., 2007; Carbia-Nagashima et al., 2007).

The crosstalk between Ubiquitin and SUMO is mediated at several levels. First, SUMOylation or ubiquitylation on the same Lys residue can differentially regulate the activity and fate of several proteins. For example, ubiquitylation of Lys[164] of the Proliferating Small Nuclear Antigen (PCNA), which is required for replication and DNA damage response, enhances the recruitment of translesion error-prone DNA polymerases. Yet, genetic evidence suggests that SUMOylation at this site by the SUMO ligase Siz-1 at S-phase promotes PCNA association with the Srs2 helicase and restricts the helicase activity (Bergink & Jentch 2009). Second, enzymes within each pathway are targets for modification by the other pathway. For example, the ubiquitin ligase E2-25k undergoes SUMOylation at its core domain that inhibits its activity. Another example is the DUB USP25 that is regulated by both SUMOylation and ubiquitylation. In this case SUMOylation inhibits its function, and ubiquitylation at the same site enhances its enzymatic activity. An interesting case is the ubiquitin/SUMO ligase TOPORS. TOPORS is unique as it can catalyze the formation of either ubiquitin or SUMO chains. Importantly, a phospho-switch induced by the polo like kinase, PLK1 results site-specific phosphorylation of TOPORS, inhibiting its ability to SUMOylate, but enhancing its ubiquitylation activity (Yang et al., 2009).

However, until recently it was not clear how does the cell directly "sense" and integrate the ubiquitin and SUMO signals at the single protein level. A first direct and enzymatic link between the two pathways was established by the identification of SUMO-targeted ubiquitin ligases (STUbLs). STUbLs are a unique group of RING proteins: they bind non-covalently to the SUMO moiety of SUMOylated proteins via several SIM motifs, and subsequently target the SUMOylated protein for ubiquitylation via a RING domain [Sun et al., 2007; Geoffroy &Hay, 2009; Abed et al 2011b]. STUbLs impact protein stability, localization, and are required for the maintenance of genomic integrity, transcription and are involved cancer. Thus, STUbLs integrate the SUMO and ubiquitin pathway and generate a SUMO/ubiquitin dual signal that may serve as an additional level of regulation of protein-protein interactions.

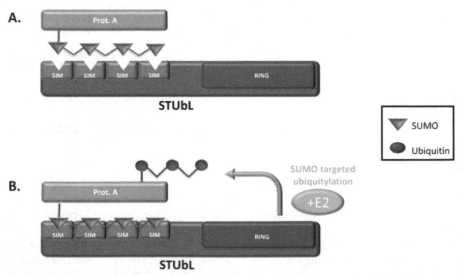

Fig. 2. Classical mode of action of STUbL: STUbL interact with SUMOylated-proteins via its SUMO Interacting Motifs (SIM). Subsequently, dimers of STUbL proteins interact with charged E2 –Ub complex, and catalyze poly-ubiquitylation via the RING domain.

4.1 Characterization of STUbL proteins

STUbLs are highly conserved in eukaryotes. Members of the STUbL family include for example: the yeast *S. pombe* Slx8-Rfp; *S. cerevisiae* Slx5–Slx8; *H. sapiens* RNF4; *D. discoideum* MIP1; and the *D. melanogaster* Degringolade (Dgrn). Yet, no clear STUbL orthologs exist in the worm *C. elegans*. These members are structurally and functionally conserved, as RNF4 protein can substitute for the yeast and fly genes in functional assays (Abed et al., 2011a Abed et al, 2011b; Barry et al., 2011; Praefcke et al., 2011). Recent structural work from the Hay lab uncovered that RNF4, and likely other STUbLs, function as dimers and that dimer formation is actively required to facilitate SUMO-dependent ubiquitin conjugation (Plechanovova' et al. 2011).

Several observations link STUBLs to protein degradation: 1. STUbLs bind and ubiquitylate SUMO chains, 2. STUbLs enhance the degradation of SUMOylated proteins 3. Genetic ablation of STUbL genes in yeast, flies, and cancer cells results in accumulation of poly-SUMOylated proteins. Among the most studied substrates of RNF4 are the promyelocytic leukaemia protein, PML and its derived oncogene PML-RAR. An elegant set of experiments by several groups established that arsenic-induced phosphorylation enhances poly SUMOylation of PML and recruitment of RNF4. Subsequently, RNF4-dependent ubiquitylation targets SUMOylated and ubiquitylated PML for degradation via the 26S proteasome (Lallemand-Breitenbach et al., 2008; Tathem et al., 2008). This is highly relevant to the treatment of acute promyelocytic leukemia, where a combination of Retinoic Acid (RA) with arsenic treatment can result in 90% cure (Lallemand, 2011). Other bona-fide human substrates of RNF4 are the kinetochore proteins CENP-I and VHL. In addition, proteomic analysis using RNF4 as a bait identified a wide spectrum of proteins that are bound by RNF4 (Makhopadhyay et al., 2010; Tatham et al., 2011). While the exact nature of these interactions requires further characterization, GO analysis already points out that SUMO-dependent ubiquitylation by RNF4 is relevant to a large verity of protein complexes and involves diverse cellular process.

An interesting issue is the recognition of proteins by STUbLs. While by their definition STUbLs recognize SUMOylated substrates via their SIM motifs, recent reports suggest a more complex picture. For example, the recognition of the Mat2α repressor by Slx5: Slx8 does not require substrate SUMOylation, but does require intact SIM motifs (Xia et al., 2010) Furthermore, the binding of the *Drosophila* STUbL Dgrn to the Notch-related HES repressor proteins is independent of SUMOylation, and is mediated by Dgrn's RING domain and not the SIM motif. Yet, the SIM motifs are required for ubiquitylation *in vitro* and for Dgrn's impact *in vivo* (Abed et al., 2011a, Barry et al., 2011). Thus, it is highly likely that specific interactions are determined *in vivo* by a dual recognition machinery, where the SIM motif interacts with the poly-SUMO chain and the RING domain interacts with other non-SUMO determinants in the target protein or adjacent proteins. In addition, recent work suggests that RNF4 can interact with proteins such as Nip45 that harbour two SUMO Like Domains (SLDs), but are not SUMOylated, thus expanding the spectrum of RNF4 targets (Sekiyama et al., 2010).

4.2 Cellular processes regulated by STUbLs

STUbLs are required for normal development, for the cell's ability to cope with genotoxic stress, and to maintain genome stability (Prudden et al., 2007; Nagi et al. 2008; Nagi et al.,

2011and Barry et al., 2011). During mouse development, RNF4 is highly expressed in the stem cell compartment of the developing gonads and brain (43). RNF4 was also identified as a gene that is specifically expressed in hematopoietic, embryonic, and neural progenitor cells, likely representing its role in "stemness" (Galili et al., 2000; Ramalho-Santos et al., 2002). During early *Drosophila* development Dgrn localizes to centrosomes, and *dgrn* null embryos accumulate SUMOylated proteins. *dgrn* null embryos show genomic instability phenotypes; they fail to incorporate DNA into centrosomes, assemble aberrant mitotic spindles and exhibit chromosomal bridges at anaphase. Cells in *dgrn* null embryos show high SUMO content and fall from the embryo surface into the center of the syncytium (Barry et al., 2011). These findings fit well with those reported for the yeast STUbLs, as yeast lacking *Slx5:Slx8* display genomic instability, and are hypersensitive to replication stress. For example, yeast deficient in *Slx5: Slx8* fail to replicate DNA upon hydroxy-urea treatment (Prudden 2007, Rouse, 2009). While the protein substrates of STUbLs in this context are still unknown, the observations that many of the proteins involved in the DNA damage response are ubiquitylated, SUMOylated, or contain SLD motifs suggest that STUbLs are targeting specific regulatory "nodes" in the DNA response network. Since genomic instability is a hallmark of cancer cells, proteins such as STUbLs may be the "Achilles heel " in specific cancerous settings. Therefore we predict that STUbLs inhibition will results in collapse of the tumorigenic network and cancer elimination.

Interestingly, and prior to the identification of RNF4 as a dedicated STUbL protein, RNF4 was identified by the Palvimo lab as a potent transcriptional regulator that functions both as a co-activator or a co-repressor depending on the cellular context. For example RNF4 was shown to be essential for androgen and steroid receptor-mediated target gene activation (Moilanen et al., 1998; Poukka et al., 2000;). In addition, we found that in transcription the consequences of Dgrn activity are not strictly limited to targeting SUMOylated proteins for degradation (Abed et al., 2011a; Barry et al., 2011). Importantly, Dgrn/RNF4-mediated ubiquitylation impacts the affinity between proteins, inhibiting the interaction of a given protein with one protein but not affecting its ability to bind other protein. Specifically, we found that during fly development Dgrn serves as a molecular selector that determines co-repressor recruitment as described below. Dgrn-mediated ubiquitylation of the HES-related repressor Hairy inhibits its ability to interact with its co-repressor Groucho but not with other Hairy co-repressors such as dCtBP. In addition, we find that Dgrn specifically targets SUMOylated Gro for sequestration. Yet, the exact cellular and molecular details surrounding sequestration require further exploration. Accordingly, Dgrn antagonize Hairy/Groucho-mediated repression in transcription and function in cells and *in vivo*. Genome wide association studies using DamID profiling unveiled that the activity of Dgrn is relevant genome wide. Thus, Dgrn serves as a "molecular selector" that determines protein-protein interactions. We found that this "selector" activity of Dgrn/RNF4 is likely relevant also to HES independent processes and in other types of co-factor switches, and may directly impact chromatin structure (Abed et al., 2011a; Hu et al., 2010; Orian unpublished). We speculate that this activity of STUbLs will be highly relevant not only in transcription, but in the regulation of protein-protein interactions in other cellular process such as the selective recruitment of proteins to DNA repair foci.

5. Conclusion and future challenges

We focused on ubiquitin and SUMO signalling as a mode to regulate protein-protein interactions. We predict that the lessons learned during the last decades regarding ubiquitin and SUMO will be highly relevant for other UBLs and non UBL modifications. An important concept that emerges from these studies is that combinatorial posttranscriptional modifications by ubiquitin/UBLs serves as a molecular tool to regulate and establish diverse and selective, signal-induced protein-protein interactions. Furthermore, proteins that have ubiquitin like or SUMO like domains, and the ability of specific enzymes to catalyse different and distinct chains of ubiquitin/UBL proteins, further add to this diversity. This complexity is also reflected at the level of "reader" proteins that contain several UBD/UBL recognition motifs and that bind only a discrete combinatorial ubiquitin/UBL signal. Thus, we can envision how a relatively small number of signalling pathways and a limited pool of ubiquitin/UBLs can generate discrete protein-protein interactions. We predict that a key objective for future studies will be to understand the enzymatic machinery that dictates selective recruitment In distinct cellular processes. The identification of these enzymes has direct implications beyond basic research. It will pave the way to design highly selective inhibitors that will impact specific pathways with minimal side effects, features that are desired in the clinic such as in the case of cancer treatments.

6. Acknowledgment

We thank Tom Schultheiss and members of the Orian lab for discussions and comments on the manuscript. This work was supported by; ISF grants (F.I.R.ST 1215/07 and 418/09), ICRF grant (2011-3075-PG), a special ICA concert in the name of Menashe Mani, and the Rappaport Research Fund to AO.

7. References

Abed, M., Barry, K. C., Kenyagin, D., Koltun, B., Phippen, T. M., Delrow, J. J., Parkhurst, S. M. & Orian, A. (2011a). Degringolade, a SUMO-targeted ubiquitin ligase, inhibits Hairy/Groucho-mediated repression. *EMBO J*, Vol. 30, No. 7, (April 2011), pp. 1289-1130, ISSN 1460-2075

Abed, M., Bitman-Lotan, E. & Orian, A. (2011b). A fly view of a SUMO-targeted ubiquitin ligase. *Fly (Austin)*, Vol. 5, No. 4, (October 2011), PMID 21857164, ISSN 1933-6942

Barry, K. C., Abed, M., Kenyagin, D., Werwie, T. R., Boico, O., Orian, A. & Parkhurst, S. M. (2011). The Drosophila STUbL protein Degringolade limits HES functions during embryogenesis. *Development*, Vol. 138, No. 9, (April 2011), pp. 1759-1769, ISSN 1477-9129

Bawa-Khalfe, T. & Yeh, E. T. (2010). SUMO Losing Balance: SUMO Proteases Disrupt SUMO Homeostasis to Facilitate Cancer Development and Progression. *Genes Cancer*, Vol. 1, No. 7, (June 2010), pp. 748-752, ISSN 1947-6027

Behrends, C. & Harper, J. W. (2011). Constructing and decoding unconventional ubiquitin chains. *Nat Struct Mol Biol*, Vol. 18, No. 5, (May 2011), pp. 520-528, ISSN 1545-9985

Bergink, S. & Jentsch, S. (2009). Principles of ubiquitin and SUMO modifications in DNA repair. *Nature*, Vol. 458, No. 7237, (March 2009), pp. 461-467, ISSN 1476-4687

Bignell, G. R., Warren, W., Seal, S., Takahashi, M., Rapley, E., Barfoot, R., Green, H., Brown, C., Biggs, P. J., & Lakhani, S. R. (2000). Identification of the familial cylindromatosis tumour-suppressor gene. *Nat Genet*, Vol. 25, No. 2. (June 2000), pp. 160-165, ISSN 1061-4036

Carbia-Nagashima, A., Gerez, J., Perez-Castro, C., Paez-Pereda, M., Silberstein, S., Stalla, G. K., Holsboer, F. & Arzt, E. (2007). RSUME, a small RWD-containing protein, enhances SUMO conjugation and stabilizes HIF-1alpha during hypoxia. *Cell*, Vol. 131, No. 2, (October 2007), pp. 309-323, ISSN 0092-8674

Carter, S., Bischof, O., Dejean, A. & Vousden, K. H. (2007). C-terminal modifications regulate MDM2 dissociation and nuclear export of p53. *Nat Cell Biol*, Vol. 9, No. 4, (April 2007), pp. 428-435, ISSN 1465-7392

Chang, C. C., Naik, M. T., Huang, Y. S., Jeng, J. C., Liao, P. H., Kuo, H. Y., Ho, C. C., Hsieh, Y. L., Lin, C. H., & Huang, N. J. (2011). Structural and functional roles of Daxx SIM phosphorylation in SUMO paralog-selective binding and apoptosis modulation. *Mol Cell*, Vol. 42, No. 1, (April 2011), pp. 62-74, ISSN 1097-4167

Cheng, J., Kang, X., Zhang, S. & Yeh, E. T. (2007). SUMO-specific protease 1 is essential for stabilization of HIF1alpha during hypoxia. *Cell*, Vol. 131, No. 3, (November 2007), pp. 584-595, ISSN 0092-8674

Deshaies, R. J. & Joazeiro, C. A. (2009). RING domain E3 ubiquitin ligases. *Annu Rev Biochem*, Vol. 78, pp. 399-434, ISSN 1545-4509

Dikic, I., Wakatsuki, S. & Walters, K. J. (2009). Ubiquitin-binding domains - from structures to functions. *Nat Rev Mol Cell Biol*, Vol. 10, No. 10, (October 2009), pp. 659-671, ISSN 1471-0080

Galili, N., Nayak, S., Epstein, J. A. & Buck, C. A. (2000). Rnf4, a RING protein expressed in the developing nervous and reproductive systems, interacts with Gscl, a gene within the DiGeorge critical region. *Dev Dyn*, Vol. 218, No. 1, (May 2000), pp. 102-111, ISSN 1058-8388

Gareau, J. R. & Lima, C. D. (2010). The SUMO pathway: emerging mechanisms that shape specificity, conjugation and recognition. *Nat Rev Mol Cell Biol*, Vol. 11, No. 12, (December 2010). pp. 861-871, ISSN 1471-0080

Geoffroy, M. C. & Hay, R. T. (2009). An additional role for SUMO in ubiquitin-mediated proteolysis. *Nat Rev Mol Cell Biol*, Vol. 10, No. 8, (August 2009), pp. 564-568, ISSN 1471-0080

Gerlach, B., Cordier, S. M., Schmukle, A. C., Emmerich, C. H., Rieser, E., Haas, T. L., Webb, A. I., Rickard, J. A., Anderton, H., & Wong, W. W. (2011). Linear ubiquitination preventsinflammation and regulates immune signalling. *Nature*, Vol. 471, No. 7340, (March 2011), pp. 591-596, ISSN 1476-4687

Gill, G. (2005). Something about SUMO inhibits transcription. *Curr Opin Genet Dev*, Vol. 15, No.5, (October 2005), pp. 536-541, ISSN 0959-437X

Haas, T. L., Emmerich, C. H., Gerlach, B., Schmukle, A. C., Cordier, S. M., Rieser, E., Feltham, R., Vince, J., Warnken, U., & Wenger, T. (2009). Recruitment of the linear ubiquitin chain assembly complex stabilizes the TNF-R1 signaling complex and is required for TNF-mediated gene induction. *Mol Cell*, Vol. 36, No. 5, (December 2009), pp. 831-844, ISSN 1097-4164

Hannoun, Z., Greenhough, S., Jaffray, E., Hay, R. T. & Hay, D. C. (2010). Post-translational modification by SUMO. *Toxicology*, Vol. 278, No. 3, (December 2010), pp. 288-293, ISSN 1879-3185

Harper, J. W. & Schulman, B. A. (2006). Structural complexity in ubiquitin recognition. *Cell* Vol. 124, No. 6, (March 2006), pp. 1133-1136, ISSN 0092-8674

Hershko, A. (1983). Ubiquitin: roles in protein modification and breakdown. *Cell*, Vol. 34, No. 1, (August 1983), pp. 11-12, ISSN 0092-8674

Hershko, A. & Ciechanover, A. (1998). The ubiquitin system. *Annu Rev Biochem*, Vol. 67, pp. 425-479, ISSN 0066-4154

Hicke, L., Schubert, H. L. & Hill, C. P. (2005). Ubiquitin-binding domains. *Nat Rev Mol Cell Biol*, Vol. 6, No. 8, (August 2005), pp. 610-621, ISSN 1471-0072

Hindley, C. J., McDowell, G. S., Wise, H. & Philpott, A. (2011). Regulation of cell fate determination by Skp1-Cullin1-F-box (SCF) E3 ubiquitin ligases. *Int J Dev Biol*, Vol. 55, No. 3, pp. 249-260, ISSN 1696-3547

Hirano, S., Suzuki, N., Slagsvold, T., Kawasaki, M., Trambaiolo, D., Kato, R., Stenmark, H. & Wakatsuki, S. (2006). Structural basis of ubiquitin recognition by mammalian Eap45 GLUE domain. *Nat Struct Mol Biol*, Vol. 13, No. 11, (November 2006), pp. 1031-1032, ISSN 1545-9993

Hochstrasser, M. (2009). Origin and function of ubiquitin-like proteins. *Nature*, Vol. 458, No. 7237, (March 2009), pp. 422-429, ISSN 1476-4687

Hu, X. V., Rodrigues, T. M., Tao, H., Baker, R. K., Miraglia, L., Orth, A. P., Lyons, G. E., Schultz, P. G. & Wu, X. (2010). Identification of RING finger protein 4 (RNF4) as a modulator of DNA demethylation through a functional genomics screen. *Proc Natl Acad Sci USA*, Vol. 107, No. 34, (August 2010), pp. 15087-15092, ISSN 1091-6490

Hurley, J. H., Lee, S. & Prag, G. (2006). Ubiquitin-binding domains. *Biochem J*, Vol. 399, No. 3, (November 2006), pp. 361-372, ISSN 1470-8728

Ikeda, F. & Dikic, I. (2008). Atypical ubiquitin chains: new molecular signals. 'Protein Modifications: Beyond the Usual Suspects' review series. *EMBO Rep*, Vol. 9, No. 6, (June 2008), pp. 536-542, ISSN 1469-3178

Iwai, K. (2011). Linear polyubiquitin chains: a new modifier involved in NFkappaB activation and chronic inflammation, including dermatitis. *Cell Cycle*, Vol. 10, No. 18, (September 2011), pp. 3095-3104, ISSN 1551-4005

Iwai, K. & Tokunaga, F. (2009). Linear polyubiquitination: a new regulator of NF-kappaB activation. *EMBO Rep*, Vol. 10, No. 7, (July 2009), pp. 706-713, ISSN 1469-3178

Kang, X., Qi, Y., Zuo, Y., Wang, Q., Zou, Y., Schwartz, R. J., Cheng, J. & Yeh, E. T. (2010). SUMO-specific protease 2 is essential for suppression of polycomb group protein-mediated gene silencing during embryonic development. *Mol Cell*, Vol. 38, No. 2, (April 2010), pp. 191-201, ISSN 1097-4164

Katz, E. J., Isasa, M. & Crosas, B. (2010). A new map to understand deubiquitination. *Biochem Soc Trans*, Vol. 38, No. Pt1, (February 2010), pp. 8-21, ISSN 1470-8752

Kerscher, O., Felberbaum, R. & Hochstrasser, M. (2006). Modification of proteins by ubiquitin and ubiquitin-like proteins. *Annu Rev Cell Dev Biol*, Vol. 22, pp. 159-180, ISSN 1081-0706

Kim, W., Bennett, E. J., Huttlin, E. L., Guo, A., Li, J., Possemato, A., Sowa, M. E., Rad, R., Rush, J., & Comb, M. J. (2011). Systematic and quantitative assessment of the ubiquitin-modified proteome. *Mol Cell*, Vol. 44, No. 2, (October 2011), pp. 325-340, ISSN 1097-4164

Komander, D., Clague, M. J. & Urbe, S. (2009a). Breaking the chains: structure and function of the deubiquitinases. *Nat Rev Mol Cell Biol*, Vol. 10, No. 8, (August 2009), pp. 550-563, ISSN 1471-0080

Komander, D., Reyes-Turcu, F., Licchesi, J. D., Odenwaelder, P., Wilkinson, K. D. & Barford, D. (2009b). Molecular discrimination of structurally equivalent Lys 63-linked and linear polyubiquitin chains. *EMBO Rep*, Vol. 10. No. 5, (April 2009), pp. 466-473, ISSN 1469-3178

Lallemand-Breitenbach, V., Jeanne, M., Benhenda, S., Nasr, R., Lei, M., Peres, L., Zhou, J., Zhu, J., Raught, B. & de The, H. (2008). Arsenic degrades PML or PML-RARalpha through a SUMO-triggered RNF4/ubiquitin-mediated pathway. *Nat Cell Biol*, Vol. 10, No. 5, (May 2008), pp. 547-555, ISSN 1476-4679

Lallemand-Breitenbach, V., Zhu, J., Chen, Z. & de The, H. (2011). Curing APL through PML/RARA degradation by As(2)O(3). *Trends Mol Med*. PMID 22056243, ISSN 1471-499X

Lee, M. H., Lee, S. W., Lee, E. J., Choi, S. J., Chung, S. S., Lee, J. I., Cho, J. M., Seol, J. H., Baek, S. H., & Kim, K. I. (2006). SUMO-specific protease SUSP4 positively regulates p53 by promoting Mdm2 self-ubiquitination. *Nat Cell Biol*, Vol. 8, No. 12, (December 2006), pp. 1424-31, ISSN 1465-7392

Lima, C. D. & Reverter, D. (2008). Structure of the human SENP7 catalytic domain and poly-SUMO deconjugation activities for SENP6 and SENP7. *J Biol Chem*, Vol. 283, No. 46, (November 2008), pp. 32045-32055, ISSN 0021-9258

Lo, Y. C., Lin, S. C., Rospigliosi, C. C., Conze, D. B., Wu, C. J., Ashwell, J. D., Eliezer, D. & Wu, H. (2009). Structural basis for recognition of diubiquitins by NEMO. *Mol Cell*, Vol. 33, No. 5, (February 2009), pp. 602-615, ISSN 1097-4164

Massoumi, R. (2010). Ubiquitin chain cleavage: CYLD at work. *Trends Biochem Sci*, Vol. 35, No. 7, (July 2010), pp. 392-399, ISSN 0968-0004

Matic, I., van Hagen, M., Schimmel, J., Macek, B., Ogg, S. C., Tatham, M. H., Hay, R. T., Lamond, A. I., Mann, M. & Vertegaal, A. C. (2008). In vivo identification of human small ubiquitin-like modifier polymerization sites by high accuracy mass spectrometry and an in vitro to in vivo strategy. *Mol Cell Proteomics*, Vol. 7, No, 1 (January 2007), pp. 132-144, ISSN 1535-9476

McLean, J. R., Chaix, D., Ohi, M. D. & Gould, K. L. (2011). State of the APC/C: organization, function, and structure. *Crit Rev Biochem Mol Biol*, Vol. 46, No. 2, (April 2011), pp. 118-136, ISSN 1549-7798

Moilanen, A. M., Poukka, H., Karvonen, U., Hakli, M., Janne, O. A. & Palvimo, J. J. (1998). Identification of a novel RING finger protein as a coregulator in steroid receptor-mediated gene transcription. *Mol Cell Biol*, Vol. 18, No. 9, (September 1998), pp. 5128-5139, ISSN 0270-7306

Mukhopadhyay, D., Arnaoutov, A. & Dasso, M. (2010). The SUMO protease SENP6 is essential for inner kinetochore assembly. *J Cell Biol*, Vol. 188, No. 5, (March 2010), pp. 681-692, ISSN 1540-8140

Mukhopadhyay, D. & Matunis, M. J. (2011). SUMmOning Daxx-mediated repression. *Mol Cell*, Vol. 42, No. 1, (April 2009), pp. 4-5, ISSN 1097-4164

Nagai, S., Davoodi, N. & Gasser, S. M. (2011). Nuclear organization in genome stability: SUMO connections. *Cell Res*, Vol. 21, No. 3, (March 2011), pp. 474-485, ISSN 1748-7838

Nagai, S., Dubrana, K., Tsai-Pflugfelder, M., Davidson, M. B., Roberts, T. M., Brown, G. W., Varela, E., Hediger, F., Gasser, S. M. & Krogan, N. J. (2008). Functional targeting of DNA damage to a nuclear pore-associated SUMO-dependent ubiquitin ligase. *Science*, Vol. 322, No. 5901, (October 2008), pp. 597-602, ISSN 1095-9203

Niu, J., Shi, Y., Iwai, K. & Wu, Z. H. (2011). LUBAC regulates NF-kappaB activation upon genotoxic stress by promoting linear ubiquitination of NEMO. *EMBO J*, Vol. 30, No. 18), pp. 3741-3753, ISSN 1460-2075

Ogawa, M., Yoshikawa, Y., Mimuro, H., Hain, T., Chakraborty, T. & Sasakawa, C. (2011). Autophagy targeting of Listeria monocytogenes and the bacterial countermeasure. *Autophagy*, Vol. 7, No. 3, (March 2011), pp. 310-314, ISSN 1554-8635

Ouyang, J. & Gill, G. (2009). SUMO engages multiple corepressors to regulate chromatin structure and transcription. *Epigenetics*, Vol. 4, No. 7, (October 2010), pp. 440-444, ISSN 1559-2308

Patterson, C. (2002). A new gun in town: the U box is a ubiquitin ligase domain. *Sci STKE*, Vol. 2002, No. 116, (January 2002), pp. pe4, ISSN 1525-8882

Penengo, L., Mapelli, M., Murachelli, A. G., Confalonieri, S., Magri, L., Musacchio, A., Di Fiore, P. P., Polo, S. & Schneider, T. R. (2006). Crystal structure of the ubiquitin binding domains of rabex-5 reveals two modes of interaction with ubiquitin. *Cell*, Vol. 124, No. 6. (March 2006), pp. 1183-1195, ISSN 0092-8674

Pickart, C. M. (2001). Mechanisms underlying ubiquitination. *Annu Rev Biochem*, Vol. 70, pp. 503-533, ISSN 0066-4154

Plechanovova, A., Jaffray, E. G., McMahon, S. A., Johnson, K. A., Navratilova, I., Naismith, J. H. & Hay, R. T. (2011). Mechanism of ubiquitylation by dimeric RING ligase RNF4. *Nat Struct Mol Biol*, Vol. 18, No. 9, (August 2011), pp. 1052-1059, ISSN 1545-9985

Poukka, H., Aarnisalo, P., Santti, H., Janne, O. A. & Palvimo, J. J. (2000). Coregulator small nuclear RING finger protein (SNURF) enhances Sp1- and steroid receptor-mediated transcription by different mechanisms. *J Biol Chem*, Vol. 275, No. 1, (January 2000), pp. 571-579, ISSN 0021-9258

Praefcke, G. J., Hofmann, K. & Dohmen, R. J. (2011). SUMO playing tag with ubiquitin. *Trends Biochem Sci.*, (October 2011), ISSN 0968-0004

Prudden, J., Pebernard, S., Raffa, G., Slavin, D. A., Perry, J. J., Tainer, J. A., McGowan, C. H. & Boddy, M. N. (2007). SUMO-targeted ubiquitin ligases in genome stability. *EMBO J*, Vol. 26, No. 18, (September 2007), pp. 4089-4101, ISSN 0261-4189

Raasi, S., Varadan, R., Fushman, D. & Pickart, C. M. (2005). Diverse polyubiquitin interaction properties of ubiquitin-associated domains. *Nat Struct Mol Biol*, Vol. 12, No. 8, pp. 708-714, ISSN 1545-9993

Rahighi, S., Ikeda, F., Kawasaki, M., Akutsu, M., Suzuki, N., Kato, R., Kensche, T., Uejima, T., Bloor, S., & Komander, D. (2009). Specific recognition of linear ubiquitin chains by NEMO is important for NF-kappaB activation. *Cell*, Vol. 136, No. 6, (March 2009), pp. 1098-1109, ISSN 1097-4172

Raiborg, C. & Stenmark, H. (2009). The ESCRT machinery in endosomal sorting of ubiquitylated membrane proteins. *Nature*, Vol. 458, No. 7237, (March 2009), pp. 445-452, ISSN 1476-4687

Ramalho-Santos, M., Yoon, S., Matsuzaki, Y., Mulligan, R. C. & Melton, D. A. (2002). "Stemness": transcriptional profiling of embryonic and adult stem cells. *Science*, Vol. 298, No. 5593, (October 18), pp. 597-600, ISSN 1095-9203

Rouse, J. (2009). Control of genome stability by SLX protein complexes. *Biochem Soc Trans*, Vol. 37, No. Pt3, pp. 495-510, ISSN 1470-8752

Sekiyama, N., Arita, K., Ikeda, Y., Hashiguchi, K., Ariyoshi, M., Tochio, H., Saitoh, H. & Shirakawa, M. (2010). Structural basis for regulation of poly-SUMO chain by a SUMO-like domain of Nip45. *Proteins*, Vol. 78, No. 6, pp. 1491-1502, ISSN 1097-0134

Stielow, B., Sapetschnig, A., Kruger, I., Kunert, N., Brehm, A., Boutros, M. & Suske, G. (2008). Identification of SUMO-dependent chromatin-associated transcriptional repression components by a genome-wide RNAi screen. *Mol Cell*, Vol. 29, No. 6, (April 2008), pp. 742-754, ISSN 1097-4164

Sun, H., Leverson, J. D. & Hunter, T. (2007). Conserved function of RNF4 family proteins in eukaryotes: targeting a ubiquitin ligase to SUMOylated proteins. *EMBO J*, Vol. 26, No. 18, (September 2007), pp. 4102-4112, ISSN 0261-4189

Tatham, M. H., Geoffroy, M. C., Shen, L., Plechanovova, A., Hattersley, N., Jaffray, E. G., Palvimo, J. J. & Hay, R. T. (2008). RNF4 is a poly-SUMO-specific E3 ubiquitin ligase required for arsenic-induced PML degradation. *Nat Cell Biol*, Vol. 10, No. 5, (May 2008), pp. 538-546, ISSN 1476-4679

Tatham, M. H., Matic, I., Mann, M. & Hay, R. T. (2011). Comparative proteomic analysis identifies a role for SUMO in protein quality control. *Sci Signal*, Vol. 4, No. 178, pp. rs4, ISSN 1937-9145

Ulrich, H. D. (2008). The fast-growing business of SUMO chains. *Mol Cell*, Vol. 32, No. 3, (November 2008), pp. 301-305, ISSN 1097-4164

Ulrich, H. D. (2009). The SUMO system: an overview. *Methods Mol Biol*, Vol. 497, pp. 3-16, ISSN 1064-3745

Weidberg, H. & Elazar, Z. (2011). TBK1 mediates crosstalk between the innate immune response and autophagy. *Sci Signal*, Vol. 4, No. 187, (August 2008), pp. pe39, ISSN 1937-9145

Weissman, A. M., Shabek, N. & Ciechanover, A. (2011). The predator becomes the prey: regulating the ubiquitin system by ubiquitylation and degradation. *Nat Rev Mol Cell Biol*, Vol. 12, No. 9, (August 2011), pp. 605-620, ISSN 1471-0080

Wertz, I. E. (2011). It takes two to tango: a new couple in the family of ubiquitin-editing complexes. *Nat Immunol*, Vol. 12, No. 12, (November 2011), pp. 1133-1135, ISSN 1529-2916

Wild, P., Farhan, H., McEwan, D. G., Wagner, S., Rogov, V. V., Brady, N. R., Richter, B., Korac, J., Waidmann, O., & Choudhary, C. (2011). Phosphorylation of the autophagy receptor optineurin restricts Salmonella growth. *Science*, Vol. 333, No. 6039, (May 2011), pp. 228-233, ISSN 1095-9203

Williams, S. A., Maecker, H. L., French, D. M., Liu, J., Gregg, A., Silverstein, L. B., Cao, T. C., Carano, R. A. & Dixit, V. M. (2011). USP1 deubiquitinates ID proteins to preserve a mesenchymal stem cell program in osteosarcoma. *Cell*, Vol. 146, No. 6, (September 2011), pp. 918-930, ISSN 1097-4172

Xie, Y., Rubenstein, E. M., Matt, T. & Hochstrasser, M. (2010). SUMO-independent in vivo activity of a SUMO-targeted ubiquitin ligase toward a short-lived transcription factor. *Genes Dev*, Vol. 24, No. 9, (May 2010), pp. 893-903, ISSN 1549-5477

Yang, S. H. & Sharrocks, A. D. (2004). SUMO promotes HDAC-mediated transcriptional repression. *Mol Cell*, Vol. 13, No. 4, (February 2004), pp. 611-617, ISSN 1097-2765

Yang, X., Li, H., Zhou, Z., Wang, W. H., Deng, A., Andrisani, O. & Liu, X. (2009). Plk1-mediated phosphorylation of Topors regulates p53 stability. *J Biol Chem*, Vol. 284, No. 28, (July 2009), pp. 18588-18592, ISSN 0021-9258

Yun, C., Wang, Y., Mukhopadhyay, D., Backlund, P., Kolli, N., Yergey, A., Wilkinson, K. D. & Dasso, M. (2008). Nucleolar protein B23/nucleophosmin regulates the vertebrate SUMO pathway through SENP3 and SENP5 proteases. *J Cell Biol*, Vol. 183, No. 4, (November 2008), pp. 589-595, ISSN 1540-8140

Zhang, X. D., Goeres, J., Zhang, H., Yen, T. J., Porter, A. C. & Matunis, M. J. (2008). SUMO-2/3 modification and binding regulate the association of CENP-E with kinetochores and progression through mitosis. *Mol Cell*, Vol. 29, No. 6, (March 2008), pp. 729-741, ISSN 1097-4164

5

The TPR Motif as a Protein Interaction Module – A Discussion of Structure and Function

Natalie Zeytuni and Raz Zarivach
Department of Life Sciences, Ben-Gurion University of the Negev and the National Institute for Biotechnology in the Negev (NIBN)
Beer sheva,
Israel

1. Introduction

Many biological functions involve the formation of protein-protein complexes. Indeed, protein–protein interactions are considered to be the center of all functional living cells. Proteins can interact with each other by various structural, chemical and physical means. Some of these interactions can be highly specific, accompanied by high affinity, while some proteins are more flexible and bind diverse proteins as ligands. A common group of proteins that participate in protein-protein interactions and serve as multi-protein complex mediators are the tetra-trico-peptide repeat (TPR) proteins. TPR-containing proteins were found to be involved in many diverse processes in eukaryotic cells, including synaptic vesicle fusion (Young et al. 2003), peroxisomal targeting and import (Brocard et al. 2006; Fransen et al. 2008) and mitochondrial and chloroplast import (Baker et al. 2007; Mirus et al. 2009). In addition, TRPs are required for many bacterial pathways, such as outer membrane assembly (Gatsos et al. 2008), bio-mineralization of iron oxides in magnetotactic bacteria (Zeytuni et al. 2011) and pathogenesis (Edqvist et al. 2006; Tiwari et al. 2009). In addition, mutations in TPR-containing proteins have been linked to a variety of human diseases, such as chronic granulomatous disease and Leber's congenital amaurosis (D'Andrea & Regan 2003).

To date, more than 5,000 TPR-containing proteins were identified in different organisms by bioinformatics tools. Of these, more than 100 structures have been determined and deposited in the Protein Data Bank. These structures demonstrate the tendency of TPR motifs to exist as an independent fold or as a segment of a fold within a protein. The available structures allow the study of a protein-protein interaction platform at atomic resolution. In general, TPR-containing proteins can serve as a study case for protein interactions as they display great binding variety.

In this chapter, we describe the basic TPR sequence and structure, as well as several examples of diverse TPR binding properties. These include:

1. Various amino acid sequences which give rise to the binding of diverse ligands that participate in different biological processes.
2. Variety of structural features and conformations that serve as TPR ligands
3. The role of TPR protein curvature angle in ligand specificity.

4.   Protein surface electrostatic potential distribution and the contribution of such distribution to the diverse binding properties of TPRs.
5.   Multiple binding pockets which allows TPR-containing proteins to serve as mediators in multi-protein complexes.
6.   The broad distribution of homo-oligomerization states in TPR proteins.
7.   Ligand computational docking.
8.   TPR proteins and biotechnology.

2. TPR sequence and basic structure

The TPR represents a structural motif consisting of 34 amino acids, sharing a degenerate consensus sequence defined by a pattern of small and large hydrophobic amino acids. In this consensus sequence, there are no completely invariant residual positions. The consensus pattern of conserved residues involves positions 4, 7, 8, 11, 20, 24, 27 and 32, in reference to the single motif N-terminal (Fig. 1A). Residues type is highly conserved only at positions 8 (Gly or Ala), 20 (Ala) and 27 (Ala). The rest of the consensus positions displays a preference for small, large or aromatic amino acids rather than for a specific residue. In addition, important structural characteristic conservation can be found at the turns located between two helical segments which contain helix-breaking residues (D'Andrea & Regan 2003). Nowadays, TPR consensus sequences can be identified by most general sequence analysis programs, such as the Simple Modular Architecture Research Tool (SMART) or the PROSITE dictionary of protein sites and motif patterns. At the same time TPRpred is a specially designated tool that uses the profile representation of the known repeats to detect TPR motifs and other patterns of protein repeats.

The canonical unit of the TPR motif adopts a basic helix-turn-helix fold (Fig. 1B). Adjacent TPR units packed in parallel create a series of repeating anti-parallel α-helices that give rise to an overall super-helix structure. This super-helical twist is affected by the type of residue found between adjacent TPR motifs. The unique super-helix fold forms a pair of concave and convex curved surfaces (Fig. 1C). Concave and convex surfaces display some extent of flexibility, as well as amino acid variety, which permits the binding of diverse ligands, usually via the concave surface.

In today's era of complete genome sequencing, one can predict TPR-containing protein dispersion throughout different organisms. Indeed, TPR proteins were found to be common in all forms of life, namely bacteria, archaea and eukaryotes. In nature, TPR motifs can be found in tandem arrays of 3-16 sequential motifs within a given protein (Fig. 1D). Moreover, Kajander et al. (2007) designed and determined the structure of a non-natural recombinant TPR-containing protein incorporating 20 sequential motifs. By using an increasing number of repeat-containing proteins, these authors found a positive correlation between the number of repeats and protein thermo-stability.

3. Ligand binding diversity

TPR-containing proteins bind diverse ligands in different binding pockets. These ligands usually do not share sequence or secondary structure similarities. Despite the lack of a set of defined rules, binding is usually highly specific, with TPR-containing proteins able to identify their ligand within the dense cellular environment. To obtain such diverse binding,

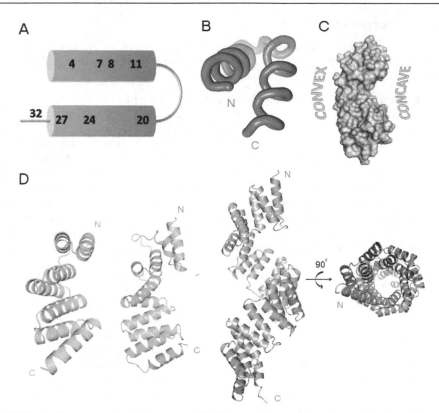

Fig. 1. TPR motif sequence and structure. (A) Schematic representation of the secondary structure arraignment of the 34 amino acids in a TPR motif. Numbers represent the conserved positions of amino acids within the motif. (B) The basic helix-turn-helix fold of a TPR motif canonical unit. (C) Surface representation of a TPR-containing protein displaying concave and convex surfaces. (D) Representative TPR-containing protein structures. From left, the TPR domain of Hop, containing three sequential motifs, MamA, containing five sequential TPR motifs, and the super-helix forming O-linked N-acetylglucosamine transferase TPR domain, containing 11 sequential TPR motifs, are shown in side and top view, in blue, green and orange respectively. Images B, C and D were generated by Pymol software (www.pymol.org).

several TPR-distinct folds serve as an interaction platform. This platform can display different surface residues in each binding surface, which later interacts in a specific manner with the ligand. Additionally, residue type influences the electrostatic nature of the binding surface, with, for example, arginine and lysine contributing positive charges, whereas glutamic acid and aspartic acid contribute negative charges. In addition, residues of different hydrophobicity and size can support hydrophobic interactions between the TPR protein and its ligand and therefore enhance protein-ligand specificity. Overall, the available TPR protein structures showing interaction with their ligands show that binding specificity cannot be attributed to a single force and that it is usually achieved by a combination of factors, such as residue type, hydrophobic pockets, charge and electrostatics.

3.1 Ligand sequence diversity

In this section, we describe two relatively simple TPR ligand interactions to demonstrate the multiple interaction types involved in binding and the correlation between TPR binding surfaces to the amino acid sequence of their ligands.

The first released structures of TPR domains bound to their ligand peptides were the two domains of the Hop protein (Scheufler et al. 2000). Hop is an adaptor protein which mediates the association of the molecular chaperones, Hsp70 and Hsp90. The TPR1 domain of Hop specifically recognizes the C-terminal heptapeptide of Hsp70, whereas the TPR2A domain binds the C-terminal pentapeptide of Hsp90 (Figs. 2A&B). Both C-terminal peptides share EEVD sequence ends and were found to bind in an extended conformation. An extended conformation allows the display of a maximal surface to the TPR domain and facilitates specific recognition of short amino acid stretches with sufficient affinity. Examination of solved crystal structures revealed that all electrostatic contacts between TPR domains and both peptides involve the EEVD sequence motif. These interactions include three classes of hydrogen bonding interaction, namely sequence-independent interactions with the peptide backbone, specific interactions with peptide side chains and the carboxylate of the C -terminal residue interaction with the TPR (Figs. 2C&D). The three strong hydrogen bonds formed between the peptide C-terminal carboxylate and the TPR1 and TPR2A domains conserved residues, Lys 8 (Lys 229), Asn 12 (Asn 233) and Asn 43 (Asn 264) respectively, allows the formation of a two carboxylate clamp. This clamp ensures the proper docking of the peptide ends to the TPR domains. Additional peptide residues located at the N-terminal, relative to EEVD motif, are exclusively engaged in hydrophobic and van der Waal's interactions. These contacts were found to be critical for peptide binding with a physiologically relevant high affinity. Other TPR domains which are known to bind Hsp70/Hsp90 proteins also contains identical residues which form the carboxylate clamp, suggesting that these TPR domains bind to the C-terminal carboxylate via a similar network of electrostatic interactions. The Hop and Hsp70/Hsp90 case studies presented here also demonstrate the importance of sequence conservation between TPR domains and their ligands involved in similar cellular functions.

Other example for the importance of TPR-ligand sequence conservation can be found in the peroxin 5 (PEX5) protein from the protozoan parasite, *Trypanasomas brucei*, the agent of human African trypanosomiasis (sleeping sickness). PEX5 is a cytosolic receptor which promotes cargo translocation across the glycosomic membrane, with the glycosome being a peroxisome-like organelle which hosts the metabolic reactions of the parasite. Two domains comprise PEX5, with the C-terminal domain consisting of 7 TPR motifs. The PEX5 C-terminal domain binds either the type 1 (PTS1) or type 2 (PTS2) peroxisomal targeting signal. The more common PTS1 sequence is a C-terminal tripeptide with the SKL amino acid sequence, or variants thereof, such as AKL or SHL. Additional residues upstream to the SKL sequence have also been implicated in binding to PEX5. The crystal structure of the C-terminal domain of PEX5 with five different PST1 fragments revealed that the protein does not fold as a sequential TPR protein with classic super-helical fold, but rather as two distinct sub-domains (Sampathkumar et al. 2008). The N-terminal sub-domain comprises TPR motifs 1-3 while the C-terminal sub-domain comprises TPR motifs 5-7. The fourth TPR motif that serves to interconnect the two sub-domains is only partial ordered and cannot be seen in the electron density resulted from the X-ray determination (Fig. 3A). The disordered nature of the fourth TPR motif implies its flexible tendencies and involvement in ligand-induced conformational changes that can promote cargo translocation. The PTS1 peptide is bound within the cavity found between the two sub-domains and interacts with residues from both, although the major binding contribution is attributed to the C-terminal sub-domain. The five PEX5 structures bound to five different peptides containing PTS1 sequences indicate that the ligand recognition mechanism involves three critical factors. The first is recognition of the C-terminal

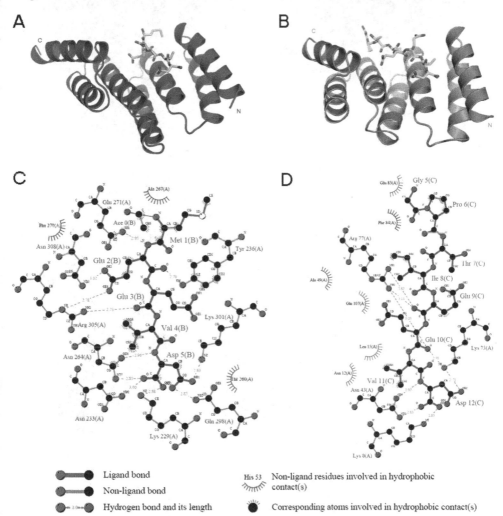

Fig. 2. TPR domains of the Hop protein and their interacting Hsp70/Hsp90 partner peptides. (A) The TPR2A domain of the Hop protein in complex with a Hsp90-derived peptide. (B) The TPR1 domain of the Hop protein in complex with a Hsp70-derived peptide. (C) Schematic representation of the TPR2A-hsp90 peptide interaction. (D) Schematic representation of the TPR1-Hsp70 peptide interaction. Images A and B were generated using Pymol software (www.pymol.org). Images C and D were generated using LigPlot software (Wallace et al. 1995).

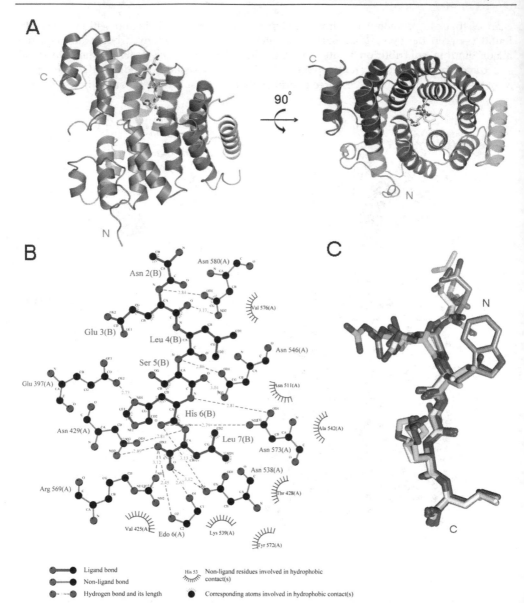

Fig. 3. TPR domains of PEX5 protein and their interacting PST1 partner peptides.
(A) TPR domains of PEX5 in complex with a PST1 peptide (NFNELSHLC). PEX5 is
represented as a cartoon, whereas the N-terminal domain, the fourth TPR motif and
C- terminal domain are colored in blue, green and orange, respectively. The PST1 peptide is
represented as sticks in yellow. (B) Schematic representation of the PEX5-PST1 peptide
interactions. (C) Superposition of five PST1 peptides bound to PEX5, in green, blue, pink,
yellow and light pink. Images A and C were generated using Pymol software
(www.pymol.org). Image B was generated using LigPlot software (Wallace et al. 1995).

PST1 carboxylate in a similar manner to the Hop and Hsp70/Hsp90 complex by two PEX5 Asn residues and a single Arg residue, the second is hydrophobic embedding of the PTS1 C-terminal residue side chain and the third is multiple PTS1 backbone interactions with PEX5 Asn side chains (Fig. 3B). Overall, PTS1 peptides are bound in a similar manner despite substantial differences in amino acid composition. In addition, the spatial positions of the five Asn residues and a single Arg residue of PEX5 involved in backbone binding of PTS1 peptides are similar, emphasizing their significant function in diverse ligand binding (Fig. 3C).

3.2 Ligand secondary structure and length diversity

Secondary structure of a TPR-bound ligand varies between the coiled extended conformation to an α-helix or both. As mentioned previously, an extended conformation maximizes the ligand surface to the TPR domain and facilitates specific recognition of short amino acid stretches with sufficient affinity. Two examples of extended conformation peptide binding are Hop bound to Hsp70/Hsp90 peptides (Fig. 4A) and PEX5 bound to PTS1 peptides. Three TPR-containing proteins bound to their ligands display binding of long peptides that adopt both helical and extended conformations. The first structure of such binding is PscG-PscE in complex with the PscF peptide (Quinaud et al. 2007). PscG-PscE–PscF proteins are members of the bacterial Type III secretion system and play a specific role in formation of the needle that transports virulence effectors into target cell cytoplasm. PscG displays a three TPR motif fold with an additional C-terminal helix, even though its TPR fold could not have been predicted from its sequence. PscG also interacts with PscE through a hydrophobic platform formed by the N-terminal TPR motif of PscG. PscF is composed of two sub-domains, an extended coil (13 amino acids-long) and a C-terminal helix (17 amino acids-long). PscG and PscE fold into a cupped-hand form, whereas the amphipathic C-terminal helix of PscF is bound to the concave surface of PscG. The N-terminal of PscF is bound to the PscG convex surface (Fig. 4B).

The second structure displaying binding of a long peptide which adopts both helical and extended conformations is the TPR-containing protein, APC6, in complex with CDC26 (Wang et al. 2009). APC6 and CDC26 are both members of the multi-subunit anaphase-promoting complex (APC), an essential cell-cycle regulator. The crystallized APC6 TPR domain contains eight full TPR motifs and an additional C-terminal helix. APC6 adopts a solenoid-like structure, wrapping around the entire length of N-terminal region of CDC26 (26 amino acids). The bound CDC26 forms a rod-like structure with the first 12 amino acids adopting an extended conformation and the last 14 amino acids forming a helix (Fig. 4C). Interestingly, as the CDC26 peptide interacts with APC6, an additional non-TPR C-terminal helix with geometry that mimics two helices in a TPR motif appears. This inter-molecular TPR mimic continues the sinuous form of the overall structure and packs against the eight TPR motifs of APC6 to form a four-helix bundle.

A third structure displaying binding of two helices connected by a loop to a TPR-containing protein is Fis1 in complex with Caf4 (Zhang et al. 2007). Both Fis1 and Caf4 participate in the formation of yeast mitochondrial fission complex that controls the shape and physiology of the mitochondria. The Fis1 protein core is composed of two TPR motifs with two additional helices at each motif end and another N-terminal helix arm packed against the hydrophobic groove formed by the protein core. The Caf4 peptide adopts a U-fold with two helices formed at each end connected by a loop. The unique U-fold of Caf4 allows large scale interactions between the peptide and both the Fis1 core at the concave and convex surfaces (Fig. 4D).

Although these are only a few examples of TPR binding modes reflecting secondary structure conformations and peptides lengths, they describe well the diverse nature of ligand types that theTPR platform can bind.

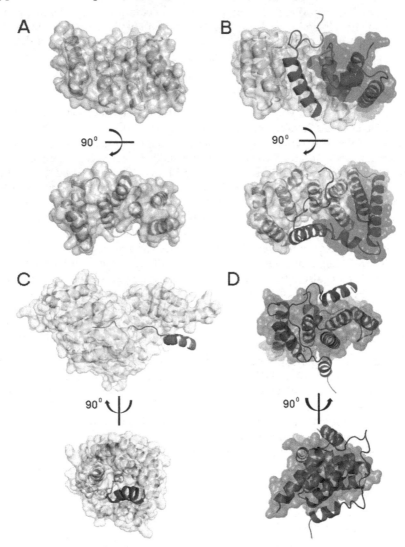

Fig. 4. TPR ligand secondary structure and length diversity. (A) TPR2A domain of Hop is shown in surface representation with secondary structure indicated in light pink, bound to Hsp90, in green. (B) A surface representation of a PscG-PscE dimer shown with secondary structure indicated in light blue and purple, respectively. PscF, in pink, is bound to the PscG-PscE dimer. (C) APC6 is shown in surface representation with secondary structure indicated in light green, bound to CDC26, in red. (D) Caf4 is shown in surface representation with secondary structure indicated in pink, bound to Fis1, in dark blue. Images were generated using Pymol software (www.pymol.org).

3.3 Curvature angle role and diversity

Major diversity can be seen in the overall shape and curvature angle displayed by TPR-containing proteins. There are many factors that determine the overall shape and curvatures angle of a protein. Among these are the number of repeating motifs, their arrangement as sequential motifs or separation into sub-domains, the presence of helix-breaking residues within turns between motifs, as well as residues type, protein function and others. Despite the apparent significance, no extensive study addressing the overall shape and curvature angle displayed by TPR-containing proteins has been reported. Indeed, articles presenting new TPR protein structures hardly refer to this property. The majority of TPR protein structures were determined by X-ray crystallography, a method that favors ordered proteins and can only tolerate a low extant of protein flexibly within the crystal. Since proteins and protein interactions are usually flexible and require the partners to be rigid, the use of X-ray crystallography to study protein interactions can be quite challenging. To overcome this challenge, structural biologists often use a relatively short peptide from the whole binding partner that participates directly in recognition and binding in their experiments. By using a small portion of the binding partner, the overall flexibility is reduced throughout the crystal, a trait that later improves crystallization and crystal quality. Due to method limitations, we usually obtain a close look at the detailed interaction and the forces participating in such interactions but cannot see the complete interaction or the spatial positions of the partners, nor detect remote stabilizing interactions. This incomplete observation of interactions between partners creates substantial difficulty in determining the exact role and importance of the curvature angle. Perhaps in the future, when new structure determination techniques emerge, a better understanding of the role of the curvature angle in TPR protein recognition and binding will be achieved.

4. TPR homo-oligomerization

TPR-containing proteins serve as mediators in the formation of multi-protein complexes. Within these complexes, TPR proteins or domains can be found as homo-oligomers, with oligomerization serving as a crucial factor for realizing proper protein function in the cell. TPR proteins displays a broad range of oligomerization states, such as monomers (Scheufler et al. 2000; Sampathkumar et al. 2008) and dimers (Lunelli et al. 2009; Z. Zhang et al. 2010). It is of note that a complex containing 24-26 TPR protein monomers has been described (Zeytuni et al. 2011). The protein surfaces involved in oligomer formation are diverse, as are the inter-molecule interaction types seen between monomers. In the following section, three interesting protein oligomerization forms will be presented.

The first TPR-containing protein displaying a dimer formation to be considered is invasion plasmid gene C (IpgC), a chaperone that binds two essential virulence factors of the pathogenic bacteria, *Shigella* (Lunelli et al. 2009). IpgC binds the invasion plasmid antigens, IpaB and IpaC which are responsible for epithelial cell invasion, membrane lysis of the phagocytic vacuole, contact hemolysis and macrophage cell death. The IpgC chaperone contains three TPR motifs, with two additional helices at the protein N-terminal and another helix at the protein C-terminal. The protein functional unit that allows efficient substrate binding is a dimer, with dimerization occuring in an asymmetric manner. Such asymmetric dimerization presents an interesting binding mode in which the first helix of one of the monomers binds on the convex surface presented by the other monomer (Fig. 5A). While the

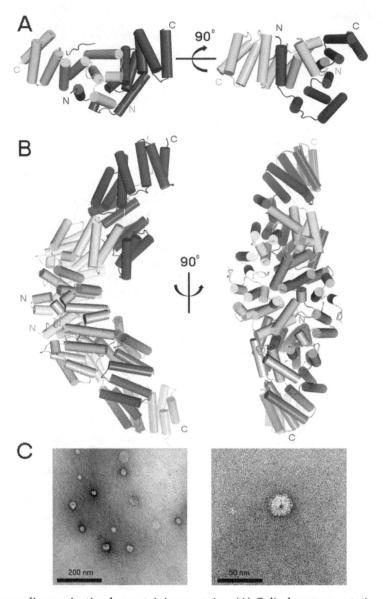

Fig. 5. Homo-oligomerization by -ontaining proteins. (A) Cylinder representation of a IpgC dimer, in green and purple, where one monomer is bound to an IpgB peptide, in pink. (B) The N-terminal domain of Cut9 promotes dimerization. Cylinder representation of the N-terminal domain of each monomer, shown in cyan and light pink. C-terminal domains are in blue and light purple. Hcn1 peptides are colored in green and yellow. (C) Transmission electron microscopy images of negatively stained MamA protein complexes. Images A, B were generated using Pymol software (www.pymol.org). The images shown 4C were taken by Zeytuni N. as described in Zeytuni et al. (2011).

first helix of one of the monomers displays inter-molecule binding, the first helix of the bound monomer appears to be found in a more packed, unbound conformation. The formation of a dimer interface at the TPR concave surface allows the binding of the IpaB peptide at the concave surface in an extended conformation. Moreover, the mode of IpgC dimerization demonstrates the significance of a convex surface as an additional protein interaction platform.

A second TPR-containing protein displaying dimer formation is *Schizosaccharomyces pombe* Cut9, the yeast homolog of human APC6. The Cut9 structure was determined while bound to Hcn1, the yeast homolog of human CDC26 in a similar binding conformation as APC6 in complex with CDC26 (Z. Zhang et al. 2010). Cut9 is composed of 14 TPR motifs forming a contiguous super-helix divided into two functionally and structurally distinct domains, namely an N-terminal domain comprising the first seven TPR motifs and a C-terminal domain comprising the last seven TPR motifs. The Cut9 subunits homo-dimerize through their N-terminal domains to generate a shallow 'V'-shaped molecule (Fig. 5B). The more globular N-terminal dimerization module forms the apex of the 'V'-shape, with narrower C-terminal TPR domains projecting away from the dimer interface. The majority of Cut9 interactions with Hcn1 involve the C-terminal TPR-containing domain. Dimerization of the N-terminal domains forms a tight interface in which the concave surface of each N-terminal domain encircles its dimer counterpart in an inter-lock clasp-like arrangement. In these interactions, the first two TPR motifs of a single monomer interact with residues lining the inner groove formed by the seven TPR motifs of the opposite Cut9 N-terminal domain.

An additional example involving dimerization is the TPR-containing protein, MamA, from *Magnetospirillium* magnetotactic bacteria species. MamA forms large homo-oligomeric complex of 24-26 monomers. This complex is presumed to serve as a wide platform for protein interaction during the iron-oxide bio-mineralization by the magnetotactic bacteria (Zeytuni et al. 2011). MamA contains five TPR motifs with an additional N-terminal putative TPR motif that was found to be responsible for oligomerization and complex formation. Through binding of the first helix of a single monomer to a binding surface displayed on a different monomer, a round-shaped complex of 14-20 nm in diameter with a central pore cavity is formed (Fig. 5C). The structural details of the monomer-monomer interaction remain unclear, since crystallization trials of MamA in complex with peptides of the first and/or second helices proved unsuccessful.

Overall, these three examples of homo-oligomrization by TPR-containing proteins describe additional TPR diversity and further establish the TPR motif as a broad platform for protein interactions.

5. Predicting TPR-ligand interactions

Today, available bioinformatics tools and the well-defined TPR profile allow us to identify TPR-containing proteins with great accuracy through analysis of amino acid sequence. However, these tools still cannot predict TPR-interacting partners and/or the region of interaction. As discussed in section 3.3, the majority of TPR-containing protein structures were determined by X-ray crystallography, a methodology in which the crystallization probability of protein complexes is significantly lower than the probability of crystallization of a single protein. Therefore, the majority of determined TPR proteins were crystallized in a non-complex form, not bound to their interacting partner, even if that partner was previously identified. In order to predict ligand binding, one should discriminate between two cases, the first involving an unidentified partner protein and the second involving an unknown binding region. In this section, we discuss these two cases and provide several examples of each.

5.1 Interacting protein prediction

A genomic approach for TPR-interacting protein identification was first presented by D'Andrea and Regan (2003). In their study, the authors generated a list of all TPR-containing proteins predicted from the *Saccharomyces cerevisiae* genome. Later, they used the generated list of 22 predicted TPR-containing proteins in protein-protein interaction databases to identify potential binding partners. Their search revealed about 80 potential interacting proteins, some of which are known to participate in multi-protein complex formation. Nevertheless, these authors could not exclude the possibility that these interacting proteins might not all interact directly with TPR domains within the multi-protein complexes.

Another approach uses information derived from the structures of unbound TPR-containing proteins. Certain properties of the binding partner can be deduced from a simple examination of a TPR protein structure, especially from its concave binding pocket, specifically its dimensions, residue composition and electrostatic potential. Although the concave surface serves as the common binding area, the convex surface can also participate in binding and should not be excluded from predictions.

Our first example is derived from the structure of the super-helical TPR domain of O-linked *N*-acetylglucosamine transferase (OGT). The TPR domain of OGT contains 11 motifs with an additional C-terminal helix and forms a homo-dimer through interactions at the convex surface. The inner surface of the elongated super-helix is highly conserved and contains an asparagine ladder. This asparagine ladder bears marked similarity to the array of conserved asparagines in the ARM-repeat importin α- and β-catenin proteins. In both ARM-repeat proteins, the asparagine side chains contribute to binding of the target peptide by forming hydrogen bonds with the peptide backbone. This structural similarity suggests that a similar binding mechanism for the OGT protein. In addition, the extensive surface generated by OGT is likely to represent several overlapping binding pockets which can accommodate multiple substrates (Fig. 6A). Furthermore, partner binding can rely on a mechanism similar to the mode of binding described earlier, in the case of CDC26 bound to APC6 (see Fig. 4C) (Wang et al. 2009).

Another example of TPR partner prediction is derived from the bacterial YrrB protein structure. YrrB is a *Bacillus subtilis* protein containing five TPR motifs with an additional C-terminal helix. The YrrB structure reveals a unique, highly negatively charged deep concave surface containing an aspartic acid array that can accommodate the binding of positively charged residues (Fig 6B). In order to discover new details on the role of the YrrB protein, functional gene cluster localization analysis was performed. Such analysis suggested that YrrB plays a role in mediating complex formation among RNA sulfuration components, RNA processing components and aminoacyl-tRNA synthetases. An opposite charge distribution to the YrrB protein was found in the TPR-containing protein, MamA, a protein that displays a highly positive concave binding surface, implying the electrostatic charge nature of ligand bindng (Fig. 6C). Although sharing a similar fold, MamA and YrrB demonstrate yet another variation among TPR-containing proteins, namely charge distribution. In general, binding partner identification is not straightforward and can be challenging, as the information obtained from the genetic approach can point to indirect interactions. Indeed, the structural approach can only provide clues for identifying the specific region of the partner involved in binding.

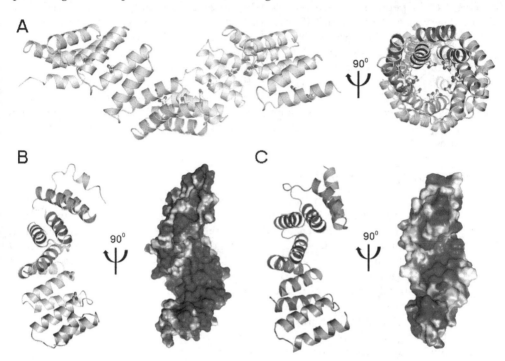

Fig. 6. Structural analysis of concave binding surface. (A) The TPR domain of O-linked *N*-acetylglucosamine transferase is presented as a grey cartoon, while the asparagine ladder within the inner surface of the super-helix is presented as yellow sticks. (B) *left* - YrrB in light orange cartoon. The aspartic acid ladder is presented as green sticks. *right* –Electrostatic surface potential representation, where blue and red represent positive and negative electrostatic potentials, respectively. The YrrB concave surface displays a highly negative potential distribution. (C) *Left* - MamA in light pink cartoon representation. *Right* – Electrostatic surface potential representation, where blue and red represent positive and negative potentials, respectively. The MamA concave surface displays a highly positive potential distribution. Images were generated using Pymol software (www.pymol.org). Surface electrostatic potential calculations were performed using the APBS plug-in (www.poissonboltzmann.org/apbs).

5.2 Interaction region prediction

Interaction region prediction requires at least two components, the TPR protein structure and the identified binding sequence of the partner. Over the past few years, several attempts to dock ligand peptides onto TPR-containing proteins have been made. These attempts mainly included manual docking of the ligand peptide onto the available TPR protein structure. The resulting models did not, however, consider side chain flexibility and conformational changes due to peptide binding and used limited energy minimization tools (Gatto et al. 2000; Kim et al. 2006). However, with the remarkable development of computational docking servers, more accurate models can be generated for TPR protein–peptide interactions. To date, however, no study has employed these advanced servers to demonstrate TPR-peptide docking, although successful docking have been recorded with other proteins involved in protein-protein interactions, such as PDZ domains and others (London et al. 2010; Raveh et al. 2010; Raveh et al. 2011). Overall, the prediction of interaction region is considered to be more accurate than binding partner prediction, since the former uses algorithms that consider chemical restrains, as well as energy minimization of the final model. In the near future, these tools might aid in overcoming challenges associated with crystallizing proteins with their binding partners and could provide important insight for molecular understanding of binding and recognition.

6. TPR design and biotechnology

The basic TPR fold resulting from its consensus sequence can be considered as a protein scaffold. Redesigning this stable basic scaffold by grafting functional residues involved in binding recognition and specificity enables the introduction of novel binding activities. TPR design includes three major steps. The first stage includes stable consensus scaffold generation by an alignment of natural TPR motifs (D'Andrea et al. 2003). Later, the minimal number of repeating motifs required for thermodynamic stability are determined (Kajander et al. 2007) and finally, functional residues are grafted onto the generated scaffold. An example for a successful designing process was the designed TPR module that binds the C-terminal peptide of Hsp90. This module has been designed by grafting Hsp90-binding residues from natural TPR proteins onto a consensus TPR scaffold to bind Hsp90 with greater affinity and specificity than natural co-chaperones. Introduction of this designed protein into breast cancer cells inhibited Hsp90 activity, presumably by out-competing the interaction of Hsp90 with its natural co-chaperones. Hsp90 inhibition resulted in misfolding and degradation of Hsp90-dependent proteins, such as HER2, and led to cancer cell death (Cortajarena et al. 2008).

TPR binding properties can also be used for specific identification of tagged proteins and can act as a functional substitute for antibodies in a wide range of applications (Jackrel et al. 2009). Conjugation of a peptide ligand sequence to the N or C termini of a desirable protein allows its identification by a TPR-containing protein that can bind the ligand peptide with sufficient affinity. This TPR-containing protein can be conjugated to a reporter protein, such as horseradish peroxidase, biotin or green fluorescent protein and can be later used in one or two-step western blot detection systems, replacing any requirement for antibodies. In addition, conjugation of a TPR-containing protein or domain directly to resin or beads can be used for affinity purification. Immobilizing a TPR-containing protein onto a resin permits the specific binding and enrichment of desirable proteins conjugated to a peptide ligand. Interaction dissociation at the elution step does not require extremely harsh pH conditions as may be needed with the use of antibodies.

7. Concluding remarks

The protein-protein interaction platform generated by TPR motifs can support the binding of diverse ligands. The elegant super-helical fold of TPR-containing proteins presents several binding surfaces that can promote the formation of multi-protein complexes. These binding surfaces can bind chemically distinct peptides in a variety of conformations with sufficient affinity. Therefore, it is not surprising that TPR-containing proteins are widespread across all kingdoms of life, where they participate in diverse cellular processes. From a molecular point of view, the diverse nature of interactions presented by TPR protein structures demonstrates multiple chemical forces involved in ligand binding and can promote the design of novel protein-protein interactions.

Overall, TPR-containing proteins hold great promise for protein engineering, therapeutics and biotechnology, as the basic TPR scaffold can be redesigned to modulate binding specificity and/or affinity towards desirable peptide ligands. As such, TPR-containing proteins can be inhibited by a designed ligand with higher affinity, serve as scaffolds to present proteins in nano-technological applications and more.

8. References

Baker, M. J., Frazier, A. E., Gulbis, J. M., and Ryan, M. T. (2007). Mitochondrial protein-import machinery: correlating structure with function. *Trends in Cell Biology* 17, 456-64.

Brocard, C., and Hartig, A. (2006). Peroxisome targeting signal 1: is it really a simple tripeptide? *Biochimica et Biophysica Acta* 1763, 1565-73.

Cortajarena, A. L., Yi, F., and Regan, Lynne (2008). Designed TPR modules as novel anticancer agents. *ACS Chemical Biology* 3, 161-6.

D'Andrea, L., and Regan, L (2003). TPR proteins: the versatile helix. *Trends in Biochemical Sciences* 28, 655-662.

Edqvist, P. J., Bröms, J. E., Betts, H. J., Forsberg, A., Pallen, M. J., and Francis, M. S. (2006). Tetratricopeptide repeats in the type III secretion chaperone, LcrH: their role in substrate binding and secretion. *Molecular Microbiology* 59, 31-44.

Fransen, M., Amery, L., Hartig, A., Brees, C., Rabijns, A., Mannaerts, G. P., and Veldhoven, P. P. Van (2008). Comparison of the PTS1- and Rab8b-binding properties of Pex5p and Pex5Rp/TRIP8b. *Biochimica et Biophysica Acta* 1783, 864-73.

Gatsos, X., Perry, A. J., Anwari, K., Dolezal, P., Wolynec, P. P., Likić, V. a, Purcell, A. W., Buchanan, S. K., and Lithgow, T. (2008). Protein secretion and outer membrane assembly in Alphaproteobacteria. *FEMS Microbiology Reviews* 32, 995-1009.

Gatto, G. J., Geisbrecht, B. V., Gould, S. J., and Berg, J. M. (2000). A proposed model for the PEX5-peroxisomal targeting signal-1 recognition complex. *PROTEINS: Structure, Function, and Genetics* 38, 241-6.

Jackrel, M. E., Valverde, R., and Regan, Lynne (2009). Redesign of a protein-peptide interaction: characterization and applications. *Protein Science* 18, 762-74.

Kajander, T., Cortajarena, A. L., Mochrie, S., and Regan, Lynne (2007). Structure and stability of designed TPR protein superhelices: unusual crystal packing and implications for natural TPR proteins. *Acta Crystallographica. Section D*, Biological crystallography 63, 800-11.

Kim, K., Oh, J., Han, D., Kim, E. E., Lee, B., and Kim, Y. (2006). Crystal structure of PilF: functional implication in the type 4 pilus biogenesis in Pseudomonas aeruginosa. *Biochemical and Biophysical Research Communications* 340, 1028-38.

London, N., Movshovitz-Attias, D., and Schueler-Furman, O. (2010). The structural basis of peptide-protein binding strategies. *Structure* (London, England: 1993) 18, 188-99..

Lunelli, M., Lokareddy, R. K., Zychlinsky, A., and Kolbe, M. (2009). IpaB-IpgC interaction defines binding motif for type III secretion translocator. *Proceedings of the National Academy of Sciences of the United States of America* 106, 9661-6.

Mirus, O., Bionda, T., Haeseler, A. von, and Schleiff, E. (2009). Evolutionarily evolved discriminators in the 3-TPR domain of the Toc64 family involved in protein translocation at the outer membrane of chloroplasts and mitochondria. *Journal of Molecular Modeling* 15, 971-82.

Quinaud, M., Plé, S., Job, V., Contreras-Martel, C., Simorre, J.-P., Attree, I., and Dessen, A. (2007). Structure of the heterotrimeric complex that regulates type III secretion needle formation. *Proceedings of the National Academy of Sciences of the United States of America* 104, 7803-8.

Raveh, B., London, N., Zimmerman, L., and Schueler-Furman, O. (2011). Rosetta FlexPepDock ab-initio: simultaneous folding, docking and refinement of peptides onto their receptors. *PloS One* 6, e18934.

Raveh, B., London, N., and Schueler-Furman, O. (2010). Sub-angstrom modeling of complexes between flexible peptides and globular proteins. *Proteins* 78, 2029-40.

Sampathkumar, P., Roach, C., Michels, P. a M., and Hol, W. G. J. (2008). Structural insights into the recognition of peroxisomal targeting signal 1 by Trypanosoma brucei peroxin 5. *Journal of Molecular Biology* 381, 867-80.

Scheufler, C., Brinker, a, Bourenkov, G., Pegoraro, S., Moroder, L., Bartunik, H., Hartl, F. U., and Moarefi, I. (2000). Structure of TPR domain-peptide complexes: critical elements in the assembly of the Hsp70-Hsp90 multichaperone machine. *Cell* 101, 199-210.

Tiwari, D., Singh, R. K., Goswami, K., Verma, S. K., Prakash, B., and Nandicoori, V. K. (2009). Key residues in Mycobacterium tuberculosis protein kinase G play a role in regulating kinase activity and survival in the host. *The Journal of Biological Chemistry* 284, 27467-79.

Wallace, A. C., Laskowski, R. A., and Thornton, J. M. (1995). LIGPLOT: a program to generate schematic diagrams o protein-ligand interactions. *Protein Engineering* 8, 127-134.

Wang, J., Dye, B. T., Rajashankar, K. R., Kurinov, I., and Schulman, B. a (2009). Insights into anaphase promoting complex TPR subdomain assembly from a CDC26-APC6 structure. *Nature Structural & Molecular Biology* 16, 987-9.

Young, J. C., Barral, J. M., and Ulrich Hartl, F. (2003). More than folding: localized functions of cytosolic chaperones. *Trends in Biochemical Sciences* 28, 541-547.

Zeytuni, N., Ozyamak, E., Ben-Harush, B., Davidov, G., Levin, M., Gat, Y., Moyal, T., Brik, A., Komeili, A., Zarivach, R., (2011). Self-recognition mechanism of MamA, a magnetosome-associated TPR-containing protein, promotes complex assembly. *Proceedings of the National Academy of Sciences of the United States of America* 108, 480-7.

Zhang, Y., and Chan, D. C. (2007). Structural basis for recruitment of mitochondrial fission complexes by Fis1. *Proceedings of the National Academy of Sciences of the United States of America* 104, 18526-30.

Zhang, Z., Kulkarni, K., Hanrahan, S. J., Thompson, A. J., and Barford, D. (2010). The APC/C subunit Cdc16/Cut9 is a contiguous tetratricopeptide repeat superhelix with a homo-dimer interface similar to Cdc27. *EMBO Journal* 29, 3733-44.

6

Functional Protein Interactions in Steroid Receptor-Chaperone Complexes

Thomas Ratajczak[1,2,*], Rudi K. Allan[1,2],
Carmel Cluning[1,2] and Bryan K. Ward[1,2]
*[1]Laboratory for Molecular Endocrinology, Western Australian Institute
for Medical Research and the UWA Centre for Medical Research,
The University of Western Australia, Nedlands WA,
[2]Department of Endocrinology & Diabetes, Sir Charles Gairdner Hospital,
Hospital Avenue, Nedlands WA,
Australia*

1. Introduction

Heat shock protein 90 (Hsp90) is unique in that it chaperones a select group of client proteins and assists their folding in preparation for key regulatory roles in cellular signalling. Steroid receptors are among the most extensively studied Hsp90 chaperone substrates and belong to the large nuclear receptor superfamily of hormone-activated transcription factors that respond to hormonal cues through conformational changes induced by hormone binding within the ligand-binding domain (LBD). In an ATP-dependent assembly process, high affinity hormone binding is achieved through the direct interaction of the steroid receptor LBD with Hsp90 and specific Hsp90-associated chaperones. After synthesis, steroid receptors enter the Hsp90 chaperoning pathway by initial assembly with Hsp40, followed by incorporation of Hsp70 and Hip. The binding of Hop and Hsp90 then generates an intermediate receptor complex which is further modified by the release of Hsp70 and Hop, allowing a transition of the receptor to hormone-binding competency. Recruitment of p23 leads to formation of mature receptor complexes capable of binding hormone with high affinity and characterized by the additional presence of one of the immunophilin cochaperones, FKBP51, FKBP52, CyP40 and PP5. This dynamic assembly of receptors to a hormone-activatable state, together with a selective functionality of receptors associated with specific Hsp90-immunophilin complexes provides mechanisms through which Hsp90 and the immunophilin cochaperones may regulate hormone-induced signalling events. This may occur directly by enhancing hormone binding as has been observed for AR, GR and PR associated with Hsp90-FKBP52 complexes or indirectly by facilitating nuclear import of receptor as seen

---

* Corresponding Author

subsequent to the hormone-induced exchange of FKBP51 by FKBP52 in GR-Hsp90 complexes. For more in depth summaries related to the mechanism and functional consequences of steroid receptor assembly with the Hsp90 chaperone machine, readers are referred to recent reviews (Echeverria & Picard, 2010; Picard, 2006; Pratt & Toft, 2003; Ratajczak *et al.*, 2003; Riggs *et al.*, 2004; Smith & Toft, 2008).

It is understood that ligand binding induces conformational changes within the steroid receptor LBD, facilitating release of Hsp90 and its cochaperones and exposing elements required for homodimerization, nuclear translocation and DNA binding. The mechanisms through which Hsp90 chaperone machinery regulates the physiological response to steroid hormones mediated by steroid receptors remain unclear. In early work, multiple approaches that included deletion analyses, peptide competition studies and use of the *in vitro* receptor-Hsp90 heterocomplex assembly system present in rabbit reticulocyte lysate were aimed at defining the regions within steroid receptors and Hsp90 responsible for interaction (Pratt & Toft, 1997). These revealed that the GR LBD was essential for formation of apo-GR-Hsp90 heterocomplexes and defined a ~100-amino acid minimal segment (human GR residues 550-653) required for high-affinity Hsp90-binding. The region contains the so-called signature sequence (human GR residues 577-596) that is conserved among steroid receptors, and may contribute to the stability of receptor-Hsp90 interaction. Despite the identification of this core Hsp90 interaction domain, other results suggested a role for nearly all of the LBD in GR association with Hsp90. Similar studies with PR and ERα also concluded that several regions throughout the LBD participate in the assembly of receptor-Hsp90 complexes, although for ERα the much less stable association of the LBD with Hsp90 requires a short upstream sequence (human ERα residues 251-71), located at the C-terminal end of the DNA-binding domain, to confer increased stability. Since Hsp90 has not been shown to bind directly to this upstream sequence, it has been proposed that the region may alternatively serve as a contact site for Hsp90 cochaperones (e.g. FKBP52) (Pratt & Toft, 1997).

Studies by the Toft laboratory, with mutants of chicken Hsp90α translated *in vitro* in reticulocyte lysate, have shown that the PR-Hsp90 interaction can tolerate deletion of the first 380-residues within the 728-amino acid chicken Hsp90α sequence to produce a hormone-activatable receptor. By contrast, selected regions (amino acids 381-441 and 601-677) in the C-terminal half of chicken Hsp90α were shown to be particularly important for PR-Hsp90 binding, with their deletion also interfering with receptor hormone responsiveness (Sullivan & Toft, 1993). An alternate approach by Baulieu and coworkers, in which human GR was coexpressed in baculovirus-infected insect cells with wild type or mutant chicken Hsp90α containing selective internal deletions (ΔA: 221-290; ΔB: 530-581; ΔZ: 392-419), revealed a loss of GR-Hsp90 interaction upon deletion of region A within the N-terminal domain, whereas deletions of regions B and Z afforded aggregated receptor-Hsp90 complexes in which receptor was unable to bind hormone (Cadepond *et al.*, 1993). An extension of these studies by the same laboratory, to chicken ERα and human MR, also concluded that deletion of the A domain in chicken Hsp90α negates interaction with both receptors (Binart *et al.*, 1995). None of the deletions affected ERα hormone-binding capacity, but MR failed to bind aldosterone with removal of region B. Although these investigations led to conflicting conclusions in relation to the role of the Hsp90 N-terminal domain in receptor interaction, it is appreciated that the introduced modifications may have caused structural perturbations leading to a disruption of Hsp90 functions elsewhere in the protein, possibly hampering valid interpretation of the results (Pratt & Toft, 1997).

Recent developments have led to the crystallographic analysis of steroid receptors, as well as Hsp90 and several of its cochaperones. At the same time, the use of the yeast two-hybrid system has revealed novel interactions between specific steroid receptors and selected cochaperones involved in the Hsp90 chaperoning pathway. Additionally, further insight is now available into the mechanism(s) that underlie the potentiation of AR, GR and PR by FKBP52. This review provides a summary of this recent progress with a focus on steroid receptor, Hsp90 and cochaperone contact domains that mediate interactions important for steroid receptor function.

2. Hsp90-steroid receptor interactions

2.1 GR LBD sub-regions required for assembly of apo-GR-Hsp90 complexes; GR structure

Further endeavours to identify sequences within the GR LBD critical for Hsp90 recognition were undertaken jointly by the Simons and Pratt laboratories. In initial studies using COS-7 cell-expressed receptor chimeras comprising glutathione S-transferase (GST) fused to the N-terminal end of an intact rat GR LBD and testing for recovered Hsp90, they found that a 7-residue amino-terminal truncation of the LBD eliminated both Hsp90 and steroid binding (Xu et al., 1998). This allowed them to determine the 7-amino acid sequence, TPTLVSL, (equivalent to amino acids 547-553 in rat GR and residues 529-535 in human GR, see Fig. 4), to be essential for the GR-Hsp90 interaction. Alignment of this sequence with the corresponding region in other steroid receptors revealed a conserved hydrophobic domain contained within helix 1 of the receptor LBD structure. It was proposed that the sequence defined a structure important for the unfolding of the hormone binding pocket, permitting steroid access and resulting in the exposure of a hydrophobic contact domain for stable Hsp90 interaction (Xu et al., 1998). Extending the 7-amino acid sequence to include Leu554 in rat GR (Leu536 in human GR), gave the sequence TPTLVSLL and led to the recognition of the LXXLL protein-protein interaction motif within helix 1 (Giannoukos et al., 1999). Such motifs have previously been shown to mediate interactions between transcriptional coactivators and members of the steroid/nuclear receptor super family (Ratajczak, 2001). Mutation of the first two leucine residues within the motif (L550S/L553S in rat GR) caused an increased rate of steroid dissociation, resulting in a dramatic loss of transcriptional activity. From a predicted GR structure, the GR LBD was seen as a "hinged pocket" with helices 1-6 comprising one side of the steroid-binding domain. In this model, the LXXLL motif within helix 1 was proposed to function as a hydrophobic clasp, helping to close one end of the steroid binding pocket by forming intramolecular contacts with residues in helices 8 and 9 on the opposite arm of the pocket, as well as residues in helix 3 and the intervening loop between helices 3 and 4 (Giannoukos et al., 1999). The LXXLL motif was proposed then to play a key role in stabilizing GR LBD tertiary structure and would, as a consequence, make important contributions to steroid binding activity.

A mutational study of specific rat GR LBD residues within the previously defined minimal high affinity binding segment for Hsp90 revealed that alanine substitution of the conserved

Pro643 (analogous to human GR Pro625) profoundly reduced both the stability of the GR-Hsp90 heterocomplex, as well as transcriptional activity, despite retaining almost normal hormone-binding affinity (Caamano *et al.*, 1998). The negative effect on transcriptional function was related to a defect in nuclear translocation for the mutated receptor. Together the results strengthened the case for the requirement of Hsp90 as a critical component of steroid receptor signalling and identified an essential role for proline residue 643, located within an exposed hydrophobic loop between helices 5 and 6 in the receptor, in maintaining the apo-GR-Hsp90 interaction.

The x-ray structure of the human GR LBD, liganded to dexamethasone, resembles those for AR and PR, bound to their respective agonists and confirmed a helical sandwich arrangement for the steroid binding pocket (Bledsoe *et al.*, 2002). Pro625 was shown to be a key residue of a novel receptor dimerization interface involving reciprocal hydrophobic interactions between the helix 5-6 loop residues, Pro625 and Ile628 from each LBD and a hydrophobic bond network between the LBDs involving residues within the helix 1-3 loops (see Fig. 4). Since Pro625 is also central to the stability of GR-Hsp90 heterocomplexes, the finding suggested an overlap between the interface for receptor dimerization and an important contact domain for Hsp90. Indeed, this may form part of the mechanism that allows the Hsp90 chaperone complex to restrict transactivation of receptor in the absence of hormone (Picard, 2006). In comparison to GR, studies have revealed that ERα is less reliant on Hsp90 regulatory control over its hormone-dependent function (Picard *et al.*, 1990), allowing the ERα LBD to mediate dimerization in the absence of hormone *in vivo* (Aumais *et al.*, 1997). ERα homodimer formation in the LBD is mediated through helix 10, thus differing in configuration to that of GR (Bledsoe *et al.*, 2002). It is of interest that for ERα, substitution of a valine residue for Gly400, also within the helix 5-6 loop of the ERα LBD, induces a conformational change that destabilizes the receptor LBD, promoting a stronger, more stable association with Hsp90, similar to that for GR and rendering receptor transactivation more hormone-dependent (Aumais *et al.*, 1997).

2.2 Hsp90 structure; Amphipathic helices 1 and 2 in the Hsp90 C-terminal domain with potential for GR-binding

The x-ray crystal structure of the C-terminal dimerization domain of htpG, the *Escherichia coli* Hsp90, was recently solved by Agard and coworkers, revealing a dimerization motif defined by a four-helix bundle interface derived from the interaction of helices 4 and 5 of one monomer with equivalent helices from a second monomer (Harris *et al.*, 2004). The structure also identified helix 2, a flexible, solvent exposed amphipathic helix, as a potential chaperone substrate-binding site. Hydrophobic residues within helix 2 are strongly conserved in Hsp90 homologues across species, suggesting an important underlying function. This was supported by other studies in which deletion of a region encompassing the corresponding helix 2 sequence in yeast Hsp82 impaired viability (Louvion *et al.*, 1996), while the point mutation, A587T, which defines the start of the helix, compromised the ability of Hsp82 to promote GR activity and caused a general reduction in Hsp90 function (Nathan &Lindquist, 1995). Core hydrophobic residues within the helix 2 sequence were observed to share sequence similarity with helix 12 of steroid receptors, leading to a proposal that Hsp90 helix 2 acts as a receptor helix 12 mimic in apo-receptor-Hsp90 complexes, occupying the normal activation function 2 (AF2) position of helix 12 following hormone binding (Jackson et al., 2004). Structural elucidation of full-length yeast Hsp90 (Ali et al., 2006) allowed the recognition of helix 1, also consisting of a solvent-exposed, hydrophobic surface within the Hsp90 C-terminal domain, as a possible contact site for protein-protein interactions (Fang et al., 2006). The highly conserved hydrophobic sequence of this helix closely matches the LXXLL recognition motif of the Steroid Receptor Coactivator/p160 family of coactivators that modulate receptor transcriptional activity by interacting with the AF2 agonist-induced hydrophobic groove of nuclear receptors (Ratajczak, 2001).

2.3 Flexible positioning of receptor LBD helix 12; Hsp90 helix 2 induces apo-GR helix 12 to adopt the GR-RU486 antagonist conformation

Recent studies by Darimont and coworkers have confirmed that Hsp90 helix 2 stabilizes unliganded GR by engaging apo-GR at the position normally occupied by receptor helix 12 in response to hormonal activation and forcing the flexible helix 12 to bind to the hydrophobic groove, at the same time preventing receptor interaction with coactivators (Fang et al., 2006). The resulting structure corresponds to the native conformation of unliganded GR, with an orientation of helix 12 similar to that in antagonist (RU486)-bound GR (Fang et al., 2006; Kauppi et al., 2003). On agonist binding, hormone-induced conformational changes within the LBD of holo-GR promote the replacement of Hsp90 helix 2 by receptor helix 12, causing loss of Hsp90 chaperone machinery and establishing the AF2 contact domain for coactivator interaction. Alternatively, the new structure might facilitate Hsp90 helix 1 binding to the hydrophobic groove. Since Hsp90 helices 1 and 2 are proximally located at the Hsp90 C-terminus, this exchange of receptor-Hsp90 interactions, which is partly determined by the dynamics of receptor helix 12, may likely be achieved within the one receptor-Hsp90 complex (Fig. 1).

The hormone-induced progression from apo- to holo-GR-Hsp90 complexes, through changes in the mode of receptor-Hsp90 interaction resulting from altered receptor LBD conformation, provides a suitable model for visualising the transition between inactive and active receptor that may also involve the participation of Hsp90 cochaperones such as FKBP51 and FKBP52. Although FKBP51 is the preferred cochaperone in mature GR-Hsp90 complexes (Barent et al., 1998; Nair et al., 1997), FKBP52 has been shown to promote increased GR hormone binding affinity and to potentiate the transcriptional activity of the receptor (Riggs et al., 2003). It is possible that the observed hormone-induced interchange of FKBP51 by FKBP52 in GR-Hsp90 complexes, resulting in the favoured nuclear translocation of receptor complexes (Davies et al., 2002), might be initiated by a change in GR LBD conformation elicited by the transfer of receptor interaction from Hsp90 helix 2 to helix 1, both helices being close to the common TPR binding site for immunophilin cochaperones in the C-terminal region of Hsp90. Unique steroid receptor LBD conformations then might be an important determinant of receptor preferences for specific immunophilin cochaperones within receptor-Hsp90 complexes (e.g. FKBP51 in GR, PR and MR complexes (Barent et al., 1998; Nair et al., 1997); PP5, the major cochaperone in GR complexes (Silverstein et al., 1997) and CyP40, the prevalent immunophilin in ER complexes (Ratajczak et al., 1990)), allowing these cochaperones to potentially modulate receptor function (Ratajczak et al., 2003; Smith & Toft, 2008).

apo-GR:Hsp90   holo-GR: Hsp90

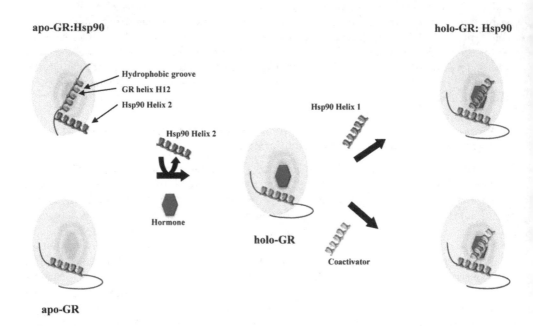

apo-GR

Hsp90 helix 2 binds apo-GR at the position normally occupied by GR helix H12, forcing H12 to dock within the hydrophobic groove, thus stabilizing the unliganded hormone-binding pocket. With hormone binding, GR H12 replaces Hsp90 helix 2 providing contacts for AF2-interacting coactivators or for Hsp90 helix 1.

Fig. 1. Hsp90 interactions with apo-GR and holo-GR.

3. Hsp90/Hsp70-cochaperone interactions

3.1 TPR cochaperones

Folding of newly synthesized peptides to functionally mature proteins, such as steroid receptors, is actively regulated by Hsp70 and Hsp90 with their cochaperones in what is known as the Hsp70/Hsp90-based chaperone machinery (Pratt & Toft, 2003). Cochaperones can regulate the nucleotide status, and thus function, of Hsp70 and Hsp90, and deliver non-native proteins to their respective polypeptide-binding domains for folding. Those cochaperones that regulate Hsp70 include Hsp40, Hsc70-interacting protein (Hip), Hsp-organizing protein (Hop) and small glutamine-rich TPR protein (SGT), while Hsp90 is regulated by cochaperones that include Hop, p23, PP5, CyP40, FKBP51 and FKBP52. C-terminal of Hsp70-interacting protein (CHIP) is another cochaperone that regulates both Hsp70 and Hsp90. Fig. 2 shows the domain architecture of the immunophilin and other TPR cochaperones with an established role in Hsp70 and/or Hsp90 chaperone function.

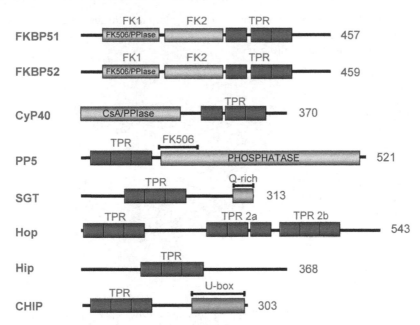

TPR domains are depicted in red whilst other specialized functional domains are highlighted in other various colours and labelled accordingly. Abbreviations: FKBP, FK506-binding protein; PPIase, peptidylprolyl isomerase; TPR, tetratricopeptide repeat; CyP40, cyclophilin 40; CsA, cyclosporin A; PP5, protein phosphatase 5; SGT, small glutamine-rich TPR protein; Hop, Hsp-organizing protein; Hip, Hsc70-interacting protein; CHIP, C-terminal of Hsp70-interacting protein.

Fig. 2. Schematic presentation of the domain structures of TPR-containing proteins associated with the Hsp70/Hsp90 chaperone machinery.

Since the crystallization of the PP5 TPR domain, the structures of several other steroid receptor-associated TPR-containing proteins have been solved. There are now full-length structures available for bovine CyP40, human FKBP52, PP5 and Hop, human and squirrel monkey FKBP51, and mouse CHIP, as well as the structure of the human SGT TPR domain. It is known that TPR domains in these proteins can mediate interactions with Hsp70 and/or Hsp90 (Angeletti *et al.*, 2002; Smith, 2004), but in addition to their Hsp-recognition domains, each also possesses other localized functional domains important for their own conformation and/or the regulation of associated proteins.

3.1.1 CyP40, FKBP51 and FKBP52

CyP40 and the two FKBPs have a similar structural arrangement, each possessing an N-terminal binding site for the immunosuppressants cyclosporin A or FK506, respectively, and a C-terminal TPR domain (Sinars *et al.*, 2003; Taylor *et al.*, 2001; Wu *et al.*, 2004). The cyclophilin domain of CyP40 is similar to other single-domain cyclophilins (Kallen *et al.*, 1998). In FKBP51 and FKBP52, FK506 binds to the first of two FKBP domains, termed FK1, while the second domain, called FK2, lacks drug-binding activity. Bound immunosuppressants inhibit the peptidylprolyl isomerase (PPIase) activity of the cyclophilin and FK1 domains, which may be important for target protein regulation by direct or indirect association. Fig. 3 provides a structural comparison between CyP40, FKBP51 and FKBP52 immunophilin cochaperones.

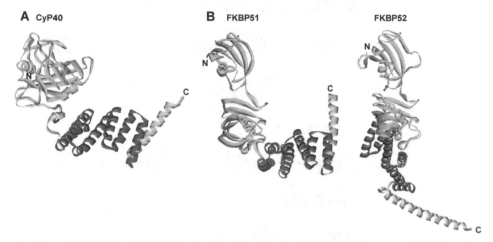

A, CyP40 and **B,** FKBP51, FKBP52. The CsA-binding domain (CyP40) and FK regions (FKBP51 and FKBP52) are shown in green. Core TPR domains for CyP40, FKBP51 and FKBP52 are depicted in red, with the final extended helices, at the C-terminal ends of each protein, shown in yellow.

Fig. 3. Ribbon representations of molecular structures of TPR-containing proteins.

3.1.2 PP5

PP5 is a phosphatase that dephosphorylates serine and threonine residues on target proteins (Barford, 1996; Cohen, 1997). Crystallisation of the full-length phosphatase in the absence of ligands or binding partners revealed the structural organization of the autoinhibited form of PP5 (Yang *et al.*, 2005). The TPR domain in PP5 is oriented to the N-terminus and is linked to a C-terminal phosphatase catalytic domain followed by a short C-terminal subdomain. In this inactive conformation, the TPR domain engages with the catalytic domain in such a way as to restrict target protein access to the enzymatic site, and this structure is stabilized by the C-terminal subdomain. Suppression of catalytic activity can be abolished by an allosteric conformational change that disrupts the TPR-catalytic domain interface, and this can be induced upon binding of polyunsaturated fatty acids or Hsp90 to the TPR domain (Chen & Cohen, 1997; Ramsey & Chinkers, 2002; Skinner *et al.*, 1997).

3.1.3 Hop

Hop plays a dual role in mature steroid receptor complex assembly by recruiting Hsp90 to preformed Hsp70-receptor complexes and inhibiting the ATPase of Hsp90 for client loading onto the chaperone for subsequent folding (Chen *et al.*, 1996b; Chen & Smith, 1998; Dittmar *et al.*, 1996; Kosano *et al.*, 1998; Prodromou *et al.*, 1999; Siligardi *et al.*, 2004). Hop has an N-terminal TPR domain (TPR1) followed by an aspartic acid/proline (DP)-rich region, and two more adjacent TPR domains (TPR2a and TPR2b) followed by a second DP-rich region.

3.1.4 Hip

Hip functions as a transient component of native steroid receptor complexes and enters the assembly cycle once Hsp70 ATPase activity has been stimulated by Hsp40 (Frydman & Höhfeld, 1997; Höhfeld *et al.*, 1995). Hip acts to stabilize the ADP-bound state of Hsp70 that is necessary for high affinity interaction with unfolded substrates (Frydman &Höhfeld, 1997; Höhfeld *et al.*, 1995). Structurally, Hip consists of an N-terminal oligomerization domain that is important for the functional maturation of GR in yeast (Nelson *et al.*, 2004), a central TPR domain and an adjacent highly charged region which are both required for Hsp70 binding (Prapapanich *et al.*, 1996b) and a C-terminal DP-rich domain that helps direct the intermediate stage recruitment of Hop-Hsp90 during assembly of steroid receptor complexes (Prapapanich *et al.*, 1998).

3.1.5 CHIP

The cochaperones described above are involved in maintaining an activatable conformation of Hsp70/Hsp90-dependent "clients", but TPR proteins also function to mediate the degradation of misfolded proteins, indicating a role in quality control (Cyr *et al.*, 2002). Selection of proteins for degradation is mediated by E3 ubiquitin ligases, and CHIP is a member of this enzymatic class (Jiang *et al.*, 2001; Murata *et al.*, 2001). CHIP has an N-terminal TPR domain and a C-terminal U-box domain that mediates its ligase activity, which promotes ubiquitylation of target substrates prior to their degradation by the proteasome.

3.1.6 SGT

Human SGT binds to viral protein U (Vpu) and Group specific Antigen, 2 proteins associated with human immunodeficiency virus-1, and the rat homologue was identified as an interactor of the non-structural protein NS-1 of the parvovirus H-1. The central TPR domain in SGT is flanked by an N-terminal dimerization domain and a C-terminal glutamine-rich domain involved in association with type 1 glucose transporter (Callahan *et al.*, 1998; Cziepluch *et al.*, 1998; Liou & Wang, 2005).

3.2 Regulation of Hsp70 and Hsp90 ATPases by TPR cochaperones

Both Hsp70 and Hsp90 require ATP for their functional association with substrates (Pratt & Toft, 2003). In the case of a steroid receptor, Hip binding to the N-terminal ATPase domain of Hsp70, possibly through a unique TPR binding site located within this region (see below), stabilizes the Hsp70-receptor complex (Frydman & Höhfeld, 1997; Höhfeld *et al.*, 1995) in a step that may be important for recognition by Hop and loading of the receptor onto Hsp90 for further processing. Hop contains three distinct TPR domains (TPR1, TPR2a, TPR2b) (Fig. 2), with TPR1 and TPR2a providing anchor points for the C-terminal EEVD peptides of Hsp70 and Hsp90, respectively. These specific interactions, coupled with domain-domain interactions, also involving its TPR domains, allow Hop to play a key role in coordinating the actions of Hsp70 and Hsp90 (Carrigan *et al.*, 2006; Chen *et al.*, 1996b; Chen & Smith, 1998; Odunuga *et al.*, 2003; Prodromou *et al.*, 1999; Ramsey *et al.*, 2009; Scheufler *et al.*, 2000). While the TPR acceptor site for Hop in the C-terminal region of Hsp90 serves to anchor the cochaperone, studies have shown that Sti1, the yeast homologue of Hop, markedly inhibits the ATPase activity of yeast Hsp90 through secondary interactions that block the ATP-binding pocket in the Hsp90 N-terminal domain (Prodromou et al., 1999). By directly competing with Sti1 for binding to Hsp90, the CyP40 yeast homologue Cpr6 can negate the Sti1-mediated blockade of Hsp90 ATPase activity following TPR protein exchange (Prodromou et al., 1999). In contrast, in vitro studies with human Hop determined that the cochaperone had no influence on the weak basal ATPase activity of human Hsp90, but significantly inhibited the increased rate of ATP hydrolysis by Hsp90 in response to interaction with the ligand binding domain of GR, an established Hsp90 client protein (McLaughlin et al., 2002). On the other hand, FKBP52, which like CyP40 binds competitively with Hop to the C-terminal TPR interaction site of Hsp90, was shown to enhance Hsp90 ATPase activity stimulated by GR (McLaughlin et al., 2002). This control over ATP utilization is important for the functional activity of newly synthesized substrates, but ATPase regulation is also required for the degradation of improperly folded substrates. CHIP can bind Hsp70 and inhibit Hsp40-stimulated Hsp70 ATPase activity, and has been reported to deplete cellular GR levels (Ballinger et al., 1999; Connell et al., 2001). Therefore, CHIP can be regarded as a degradatory cochaperone of Hsp70 and Hsp90. SGT negatively regulates Hsp70 such that the chaperone has a reduced ability to refold denatured luciferase (Angeletti et al., 2002).

3.3 Determinants of Hsp70 and Hsp90 interaction with TPR cochaperones

Deletion studies were the first to demonstrate that TPR domains mediated binding to Hsp90 (Barent et al., 1998; Chen et al., 1996a; Radanyi et al., 1994; Ratajczak & Carrello, 1996). Determination of the TPR domain structure of PP5 revealed that the packing of adjacent TPR units generated an exposed groove capable of accepting a target protein peptide (Das et al., 1998). Although TPR motifs are highly degenerate, they display a consistent pattern of key residues important for structural integrity. The two α-helical sub-domains in each TPR motif are arranged such that the groove is mainly composed of residues from the A helix of each repeat, while B helix residues are buried to form the structural backbone of the superhelix, and this groove forms a critical Hsp recognition surface.

In a PP5 mutagenesis study, Russell and coworkers carefully selected A helix residues with side-chains extended into the groove and identified four basic residues important for PP5-Hsp90 interaction (Russell et al., 1999). These amino acids are highly conserved in other Hsp90-binding TPR proteins, and mutation of aligned residues in CyP40 confirmed their importance in Hsp90 recognition (Ward et al., 2002). The key recognition sequence for the TPR domain in these proteins is the EEVD peptide located at the extreme C-terminus of Hsp90 (Carrello et al., 1999; Chen et al., 1998; Young et al., 1998), which is conserved in Hsp70. Crystallization of individual Hop TPR domains with Hsp70 and Hsp90 N-terminally extended EEVD peptides has defined the mechanism of TPR domain-peptide interaction (Scheufler et al., 2000). The TPR1 domain of Hop binds to Hsp70, while the TPR2a domain mediates Hsp90 recognition (Chen et al., 1996b; Lassle et al., 1997). The groove in each TPR domain accommodates their respective peptide in an extended conformation where the ultimate aspartate residue is tightly held by electrostatic interactions with TPR residue side-chains in a two-carboxylate clamp. Additional EEVD contacts involve hydrogen-bonding, while amino acids upstream of the EEVD enhance the affinity of the peptides for TPR domains and mediate specificity of Hsp70 and Hsp90 to TPR1 and TPR2a, respectively. Notably, Hop TPR2a provides an example of where an additional sequence within the TPR domain doesn't disrupt the overall structure. TPR2a contains an insertion between units 2 and 3 that extends the helices by a single turn but does not impact Hsp90 peptide recognition (Scheufler et al., 2000).

The Hsp90 dimerization domain, located in the C-terminal region upstream of the MEEVD peptide, contributes to TPR cochaperone recognition (Chen et al., 1998) and contains the putative binding site for novobiocin, a coumarin-based Hsp90 inhibitor (Marcu et al., 2000). In vitro studies demonstrated that novobiocin had a differential effect on Hsp90-immunophilin cochaperone interaction, suggesting that the TPR cochaperones modulate Hsp90 function through distinct contacts within the Hsp90 C-terminal domain (Allan et al., 2006).

Although EEVD interactions with the TPR domain groove are critical for Hsp binding, regions outside of the TPR domains are also important in mediating recognition. TPR domains are typically followed by a seventh α-helix that packs against and extends beyond the TPR domain and has been shown to be involved in binding Hsp90 in addition to the TPR domain. FKBP51 and FKBP52 have different affinities for Hsp90 and are assembled differentially with specific receptor complexes, and these differences map in part to sequences C-terminal of their respective TPR domains (Barent et al., 1998; Cheung-Flynn et al., 2003; Pirkl & Buchner, 2001). The charge Y motif was identified and found to be essential for FKBP-Hsp90 interaction, which was also confirmed for CyP40, but sequences further downstream in FKBP51 and FKBP52 differentially regulated Hsp90 binding (Allan et al., 2006; Cheung-Flynn et al., 2003; Ratajczak & Carrello, 1996). The acidic linker flanking the N-terminus of the CyP40 TPR domain was also shown to be important for efficient interaction (Mok et al., 2006; Ratajczak & Carrello, 1996). Although an interaction partner for Hop TPR2b has yet to be identified, mutations in TPR2b reduced Hop interaction with both Hsp70 and Hsp90, while mutations in the C-terminal DP-rich region inhibited Hop binding to Hsp70 (Chen & Smith, 1998; Nelson et al., 2003).

3.4 Alternative modes of Hsp70 and Hsp90 recognition by TPR cochaperones

Like Hop, CHIP binds to both Hsp70 and Hsp90 (Ballinger et al., 1999; Connell et al., 2001), but CHIP interacts with either of these major chaperones through a single TPR domain. Recent elucidation of the binding of Hsp90 C-terminal peptide (NH2-DDTSRMEEVD) with the CHIP TPR domain has revealed that the peptide sequence is not accommodated in an extended conformation as for Hop, but turns at the methionine residue and becomes buried within a hydrophobic pocket (Zhang et al., 2005). This pocket can accommodate either the methionine or isoleucine that lies immediately upstream of the EEVD sequence in Hsp90 and Hsp70, respectively, and the peptide is twisted, negating the role of upstream residues in conferring the same specificity seen in binding Hop TPR domains. SGT also recognizes Hsp70 and Hsp90 via its single TPR domain, but possibly through a different mechanism to that described for CHIP as SGT lacks the residues that form the hydrophobic pocket which allows the respective C-terminal peptides in the chaperones to twist (Dutta & Tan, 2008).

Hydrophobic pockets themselves may also be important structural features within TPR domains that confer Hsp specificity, as the crystal structure of Hop TPR2a with the non-cognate Hsp70 peptide shows the hydrophobic pocket to be less accommodating for the Ile

(-5) residue in the extended Hsp70 peptide than Met (-5) in the extended Hsp90 peptide, with the notable feature of a lack of bending by the Hsp70 peptide, such as with CHIP, to perhaps enhance affinity for TPR2A (Kajander et al., 2009).

General cell UNC-45 (GCUNC-45), a member of the UNC-45/Cro1/She4p (UCS) protein family, is a TPR protein that regulates PR chaperoning by Hsp90 by preventing activation of Hsp90 ATPase activity (Chadli et al., 2006). Hsp90-binding experiments in the presence of Hop revealed a novel GCUNC-45 TPR recognition site in the N-terminal domain of Hsp90, which also bound FKBP52 (Chadli et al., 2008a). Further analysis defined a non-contiguous EEVD-like motif, centered in and around the Hsp90 N-terminal ATP-binding pocket, arranged in a structural conformation that can recognize TPR domains. Nucleotide binding negatively regulates the interaction. These authors also alluded to CyP40 binding to the N-terminal interaction motif, although Onuoha and coworkers have recently confirmed CyP40 interaction only with the C-terminal domain of Hsp90 (Onuoha et al., 2008). GCUNC-45 is the first cochaperone to display a preferential association with Hsp90β over the Hsp90α isoform, resulting in functional Hsp90β-GCUNC-45 interactions that more efficiently block progression of PR chaperoning than seen with Hsp90α-GCUNC-45 complexes (Chadli et al., 2008b). An EEVD-like motif interaction with a TPR domain has also been described for androgen receptor recognition by SGT, where binding is mediated by the first 2 TPR motifs of the SGT TPR domain and the hinge region located between the DNA-binding and ligand-binding domains in the receptor (Buchanan et al., 2007).

Hip has similarly been reported to bind the Hsp70 N-terminal ATPase domain via its TPR domain (Höhfeld et al., 1995). Through this interaction, Hip, originally identified in progesterone receptor complex assembly (Prapapanich et al., 1996a; Smith, 1993), can stabilize substrate-Hsp70 binding and competitively counteract the destabilizing effects of the non-TPR cochaperone BAG1 (Bimston et al., 1998; Gebauer et al., 1997; Höhfeld & Jentsch, 1997; Takayama et al., 1997). The Hip-Hsp70 interaction also allows for the simultaneous association of Hip with Hsp70-Hop complexes (Gebauer et al., 1997; Prapapanich et al., 1996a). By analogy with the mode of GCUNC-45 interaction with Hsp90, there is the possibility that Hip targets a similar TPR recognition site in the N-terminal region of Hsp70. However, Hip is unique among the steroid receptor-associated TPR proteins in terms of Hsp recognition in that it binds Hsp70 independently of EEVD interactions (Höhfeld et al., 1995), and that efficient binding may be due to a greater requirement for additional Hsp-interaction determinants, such as the adjacent highly charged region and a C-terminal DP-repeat domain (Prapapanich et al., 1998). It is possible the mechanism of Hsp70 recognition by Hip is not unique, but may be utilized by some of the steroid-receptor TPR cochaperones to interact with binding partners in distinct cellular pathways. Dutta and Tan (2008) reported the SGT TPR domain is sufficient to bind Vpu and identified the sequence ^{31}KILRQ^{35} in Vpu as being important for this interaction.

4. p23 and Cdc37 interaction with Hsp90

p23 is an essential component involved in stabilizing mature steroid receptor-Hsp90 complexes and binds to the ATP-bound conformation of a Hsp90 dimer characterised by high affinity for client proteins (Ali et al., 2006; Felts & Toft, 2003; McLaughlin et al., 2006;

Richter *et al.*, 2004). Conformational changes that accompany ATP binding promote dimeric interaction between the N-terminal domains of the Hsp90 C-terminal dimer to form distinct binding surfaces for separate p23 molecules, thus further underpinning the ATP-bound conformation (Ali *et al.*, 2006; Karagöz *et al.*, 2010). In a recent model proposed for the Hsp90 cochaperone cycle, entry of an immunophilin cochaperone into an existing client protein-Hsp90-Sti1/Hop-Hsp70 complex forms an intermediate complex important for cycle progression. Conversion of Hsp90 to the closed conformation on ATP and subsequent p23 binding then favours the release of Sti1/Hop (Li *et al.*, 2011).

Cdc37 serves as an adaptor predominantly facilitating protein kinase interaction with Hsp90, although additional client proteins, including steroid receptors have been identified (MacLean & Picard, 2003). Similar to Hop, Cdc37 arrests the Hsp90 ATPase cycle and functions as an "early" cochaperone for the recruitment of protein kinase clients to the Hsp90 machinery. Hsp90 binding maps to the Cdc37 C-terminal region, while kinase interaction occurs via the N-terminal domain (Roe *et al.*, 2004). Hsp90 ATPase activity is coupled to an opening and closing of a molecular clamp generated by the constitutive C-terminal Hsp90 dimer at one end in combination with the ATP-dependent association of the N-terminal domains at the other (Prodromou *et al.*, 2000). A structural view of the Hsp90-Cdc37 complex shows Cdc37 located as a dimer between the N-terminal domains of the clamp, thus preventing their interaction (Roe *et al.*, 2004). With cycle progression, loss of one Cdc37 monomer leads to the formation of a stable (Hsp90)$_2$-Cdc37-kinase complex (Vaughan *et al.*, 2006; Vaughan *et al.*, 2008).

5. Receptor-cochaperone interactions

5.1 Cortisol resistance in New World primates; The key role of FKBP51; Structures of FKBP51 and FKBP52

Analysis of glucocorticoid resistance in New World primates, such as squirrel monkey, has demonstrated that the high circulating cortisol levels result from elevated expression and greatly increased incorporation of FKBP51 into GR-Hsp90 complexes, causing a significant decrease in GR hormone binding affinity (Denny *et al.*, 2000; Reynolds *et al.*, 1999; Scammell *et al.*, 2001). FKBP51 then appears to have a major role in stabilizing an inactive receptor conformation. The FK506 drug-binding pocket of FKBP51 is inaccessible to FK506 in low affinity hormone-binding GR heterocomplexes. However, incubation of receptor cytosols from squirrel monkey lymphocytes with FK506 prevented assembly of FKBP51 with GR-Hsp90 complexes, correlating with a sharp increase in receptor hormone binding and affinity. On the other hand, recognition of FK506 by FKBP52 appeared unaffected by whether the immunophilin exists as a component of mature, high affinity hormone-binding GR complexes or not (Denny *et al.*, 2000; Tai *et al.*, 1992). Furthermore, the immunosuppressant blocks FKBP52-mediated potentiation of GR activity (Riggs *et al.*, 2003). The inhibitory influence of FKBP51 on GR activity requires both FK domains, as well as Hsp90 binding, but is not reliant on FKBP51 PPIase activity (Denny *et al.*, 2005). FK506 may likely serve to sterically hinder receptor LBD interactions with the FK1 domain of FKBP51 and FKBP52 essential for inhibitory and activation effects on receptor, respectively. This differential action of FK506 may arise from distinct domain orientations that have been defined from recent structures of the two immunophilins (Sinars *et al.*, 2003; Wu *et al.*, 2004). Unique interactions between receptor and the FKBP51 and FKBP52 cochaperones have been further highlighted by results showing that deletion of the Asp195, His196, Asp197 insertion within the FK2 domain of FKBP51 compromised assembly of the immunophilin into PR complexes, whereas removal of the corresponding FK2 insertion loop from FKBP52 had no affect on receptor association (Sinars *et al.*, 2003). This raises the possibility that direct interaction of FK2 in FKBP51 with PR might favour the preferred association of FKBP51 over FKBP52 with this receptor.

5.2 Cortisol resistance in the guinea-pig; Do guinea pig GR LBD changes favour FKBP51 binding over FKBP52?

In contrast to the New World primates, the cause of glucocorticoid resistance in the guinea pig, a New World hystricomorph, has been delineated to an unstructured loop between helix 1 and helix 3 of the guinea pig GR LBD. Five amino acid substitutions in this region differentiate guinea pig GR from the human receptor, with at least four contributing to the low binding affinity phenotype (Fuller *et al.*, 2004). It has been predicted that these crucial residues (Ile538, His539, Ser540, Thr545 and Ser546) lying on the surface of the guinea pig GR LBD, disrupt a contact domain for FKBP52, favouring increased association with FKBP51 and conformational changes that compromise high affinity cortisol binding. Using a yeast-based assay (Riggs *et al.*, 2003) with rat GR substituted in the helix 1 to helix 3 loop with the guinea pig GR-specific residues, we have recently confirmed that FKBP52 can efficiently potentiate the transcriptional activity of the mutated GR, thus discounting a central role of this region in receptor-FKBP52 interaction [Cluning C and Ratajczak T, unpublished observations].

5.3 FKBP52 potentiation of AR, GR and PR

Direct interaction studies between bacterially expressed FKBP52 and GST-tagged, wild type human GR and C-terminal truncation mutants of the receptor purified from Sf9 cell extracts, identified a 35-amino acid region (hGR 465-500), between the DNA-binding domain and the LBD, to be sufficient for FKBP52 binding, with optimal interaction requiring involvement of the LBD (Silverstein *et al.*, 1999). However, recent demonstration of FKBP52 potentiation of GR activity in association with increased receptor hormone binding affinity has definitively localized the FKBP52 effect to the GR LBD (hGR 521-777) and at the same time pointed to a requirement of FKBP52 PPIase activity residing in the FK1 domain (Riggs *et al.*, 2003). Studies with FKBP52 knockout mouse strains have extended the critical physiological role of FKBP52 to cellular responses controlled by both AR (Cheung-Flynn *et al.*, 2005) and PR (Tranguch *et al.*, 2005; Yang *et al.*, 2006), while similar influences of this immunophilin cochaperone on ERα (Riggs *et al.*, 2003) and MR (Gallo *et al.*, 2007) activity have not been observed, despite the assembly of FKBP52 with Hsp90 complexes containing these receptors.

5.4 Molecular basis of FKBP52 action; Potential interaction of FKBP52 with the BF3 regulatory site

An initial understanding that FKBP52 potentiation of AR, GR and PR activity was dependent on the FK1-mediated PPIase function of the immunophilin, prompted speculation that FKBP52 might target a key proline likely to be conserved among these receptors and that this critical residue would be located on the surface of the LBD, accessible to the cochaperone and in a position where it might influence the shape of the ligand binding pocket (Cheung-Flynn et al., 2005). Although several such candidate prolines exist in the intervening loops between receptor LBD helices, a more extensive mutational analysis of the FK1 catalytic site has excluded a role for the FKBP52 PPIase activity in receptor potentiation (Riggs et al., 2007). Rather, recent evidence has identified a loop overhanging the FK1 catalytic pocket in FKBP52 that is responsible for the functional difference between FKBP52 and FKBP51 relating to AR (and GR/PR) potentiation (Riggs et al., 2007). It is proposed that a critical proline within this loop (human FKBP52 Pro119) allows specific contact with a region of the AR LBD (a structural feature that is also common to GR and PR), thus helping to stabilize an LBD conformation favourable for high affinity hormone binding and leading to efficient transcriptional activation (Riggs et al., 2007). It is speculated that a leucine substitution within the corresponding FK1 sequence of FKBP51 alters the loop conformation sufficiently to disrupt this functionally important contact. The possibility exists that in the hormone-induced transition from inactive to active states of AR-Hsp90 complexes associated with FKBP51 and FKBP52, respectively, Hsp90 orients FKBP52 to achieve unique interactions with the receptor LBD, allowing Hsp90 to facilitate optimal hormone binding and to further fine-tune the hormonal response.

Prior to investigations establishing a noncatalytic involvement of the FKBP52 PPIase domain in the modulation of receptor function, an early attempt to identify the putative proline substrate for FKBP52 isomerase activity within the AR LBD utilized AR-P723S, a proline mutant associated with androgen insensitivity syndrome (Cheung-Flynn et al., 2005). Although predicted to display basal activity, coupled with a lack of response to hormone in the presence of FKBP52, this mutant was characterized by subnormal activity in the absence of FKBP52, showing full restoration to wild type receptor activity levels with the cochaperone on exposure to hormone (Cheung-Flynn et al., 2005). Such a favoured response reflects a greater dependence of the AR-Pro723S mutant on FKBP52 for normal activity. Pro723 lies within the signature sequence conserved among all steroid receptors (Brelivet et al., 2004), close to a region directly involved in ligand binding and is situated in a solvent exposed loop between helices 3 and 4, which combine together with the mobile helix 12 to form the AF2 coactivator binding pocket (He et al., 2004; Matias et al., 2000b). For AR, AF2 initially has a preferred interaction with the AR N-terminal domain, resulting in an intramolecular fold that precedes receptor dimerization and appears critical for AR function (He et al., 2001; He et al., 2004; Schaufele et al., 2005). Pro723 also forms part of the recently identified BF-3 surface that has the ability to allosterically alter the AF2 binding pocket of AR (Estébanez-Perpiñá et al., 2007) (Fig. 4). BF-3 residues altered through natural mutations linked to androgen insensitivity and those associated with prostate cancer, either diminish or enhance AR AF2 activity, respectively, underlining the importance of the BF-3 surface for AR function (Estébanez-Perpiñá et al., 2007). FKBP52 rescue of AR-Pro723S activity might signify FKBP52 influence over some part of the BF-3 allosteric regulatory site leading to conformational changes that allow full recovery of AR activity. Indeed, Cox and coworkers have recently identified small-molecule inhibitors of FKBP52-enhanced AR function in prostate cancer cells that target a region of the AR LBD overlapping the BF3 surface (De Leon et al., 2011) (Fig. 4). Multiple residues that contribute to the FKBP52 sensitivity of AR, some of which form part of the binding site for MJC13, the lead compound, have been identified (De Leon *et al.*, 2011) (Fig. 4). Since MJC13 helps to maintain an intact AR-Hsp90-FKBP52 complex at low hormone concentrations, it is possible that the inhibitor interferes with a critical next step - a hormone-induced, FKBP52-dependent transitory change in AR conformation necessary for nuclear translocation. Sequence comparisons have revealed some conservation of BF-3 residues within the LBDs for AR, GR, MR and PR, suggesting the presence of BF-3-like regulatory domains in each receptor (Estébanez-Perpiñá *et al.*, 2007) (Fig. 4). A very limited conservation of these residues is apparent in ERα, suggesting the formation of a BF-3 type surface that is unique to this receptor (Estébanez-Perpiñá *et al.*, 2007) (Fig. 4). Both ERα and MR behave differently to AR, GR and PR, through their inability to respond to FKBP52. Certain structural differences within their LBDs distinguish these two receptors from the other members of this subfamily (De Leon *et al.*, 2011) (Fig. 4). Since FKBP52 also regulates GR and PR activity, most likely through specific BF3 surfaces, there is the potential for the development of FKBP52-specific inhibitors targeting GR and PR function to treat a range of steroid hormone-based diseases (Moore *et al.*, 2010). The BF-3 pocket is a potential target for second-site modulators that can allosterically block agonist-activated AR function to inhibit prostate cancer cell growth (Joseph *et al.*, 2009).

5.4.1 FKBP51 is an androgen-regulated gene that promotes assembly of mature AR-Hsp90 complexes

FKBP51 is recognised as a highly sensitive AR-regulated gene that functions as an important component of a feed-forward mechanism linked to the partial reactivation of AR-signalling pathways in the absence of androgens, leading to the outgrowth of androgen-independent tumours (Amler *et al.*, 2000; Febbo *et al.*, 2005; Magee *et al.*, 2006; Mousses *et al.*, 2001; Tomlins *et al.*, 2007). Sanchez and coworkers have confirmed a significantly increased expression of FKBP51, but not that of FKBP52, in most prostate cancer tissues and in androgen-dependent and androgen-independent cell lines (Periyasamy *et al.*, 2010), suggesting that FKBP51 might have a critical role in prostate cancer growth and progression. FKBP51 overexpression was found to increase the AR transcriptional response by facilitating hormone-binding competence through the assembly of the AR LBD with mature FKBP51-Hsp90-p23 complexes (Ni *et al.*, 2010), resulting in higher levels of androgen-liganded receptor and providing a pathway for AR-dependent signalling and growth in a low-androgen environment. The ability of FKBP51 to enhance AR transcription and chaperone complex assembly appears to be dependent on FKBP51 PPIase activity mediated by the FK1 domain and requires Hsp90 binding through its TPR domain (Ni *et al.*, 2010).

6. Receptor LBD contacts with other Hsp90 cochaperones

6.1 PP5; GCUNC-45; SGT

The domain structure of the Hsp90 cochaperone, PP5, a serine/threonine protein phosphatase (Chen *et al.*, 1994; Chinkers, 1994), is characterised by a C-terminal phosphatase catalytic domain and an N-terminal TPR domain that competes with FKBP51, FKBP52 and CyP40 for the TPR binding site at the Hsp90 C-terminus during assembly into mature steroid receptor-Hsp90 complexes (Banerjee *et al.*, 2008; Chen *et al.*, 1996a; Hinds Jr & Sanchez, 2008). Through its TPR domain, PP5 has also been shown to bind directly to ERα and ERβ, an interaction that targets the LBDs of these receptors, but does not require the C-terminal region incorporating

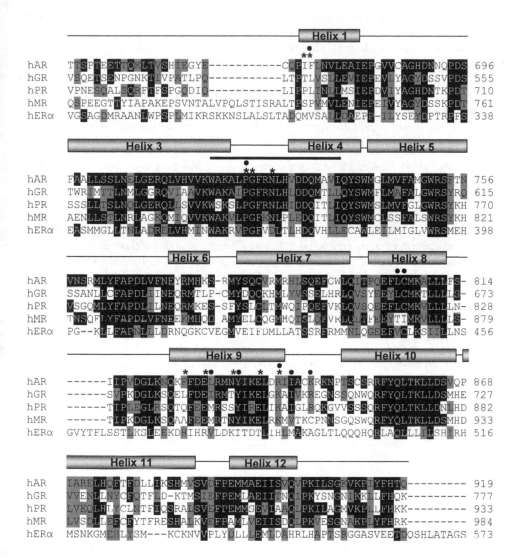

NCBI accession numbers for receptor sequences are: AR – NP000035, ERα – NP000116,
GR – NP001018087, MR – NP000892, PR – NP000917. The ERα sequence has 595 amino acids and is
shown terminated at residue 573. LBD helices are based on the structure of AR liganded to R1881
(Matias *et al.*, 2000a) (PDB ID 1E3G). The nuclear receptor signature sequence is indicated (thick black
line). Residues that map to the BF-3 allosteric regulatory site defined for AR are highlighted with an
asterisk (*). Multiple residues that contribute to the FKBP52 sensitivity of AR and form the putative
binding site for MJC13 (De Leon *et al.*, 2011) are highlighted with a black circle (•). Identical residues are
shown white against black; conserved residues (black on grey) are based on the following scheme:
(P, G), (M, C), (Y, W, F, H), (L, V, I, A), (K, R), (E, Q, N, D) and (S, T).

Fig. 4. Multiple sequence alignment of human steroid receptor LBDs.

the helix 11-12 loop and helix 12 central to AF2 function (Ikeda *et al.*, 2004). PP5 was found to function as a negative regulator of ERα transcription *in vivo* by inhibiting epidermal growth factor (EGF)-dependent phosphorylation of Ser118 in the receptor N-terminal domain. Although demonstration of a direct PP5-ERα interaction was consistent with a non-involvement of Hsp90, a role for this major molecular chaperone in the *in vivo* effects of PP5 on ERα function cannot be discounted. Similar observations have been reported for GR with evidence suggesting that PP5-dependent modulation of receptor N-terminal phosphorylation within the GR-Hsp90 apo-receptor complex is mediated through contacts between the phosphatase and receptor LBD (Wang *et al.*, 2007).

A yeast two-hybrid screen, using bait encompassing both the hinge region and LBD of human PR, liganded with the mixed antagonist RU486, identified GCUNC-45 as a PR-binding protein (Chadli *et al.*, 2006). Presence of two LXXLL motifs (similar to NR boxes of known transcriptional coregulatory proteins) within the interacting clone, corresponding to the C-terminal end of GCUNC-45, suggested a mode of interaction similar to that for receptor recognition of transcription coactivators (Ratajczak, 2001), although this remains to be confirmed. Both FKBP52 and CyP40 compete with GCUNC-45 for the N-terminal TPR site, with nucleotides causing a reduction in Hsp90 binding affinity for these cochaperones in this region and favouring their interaction with the Hsp90 C-terminus during progression of receptor to a hormone-binding state (Chadli *et al.*, 2008b). GCUNC-45 therefore, appears to have a role upstream of FKBP52 and CyP40, at an intermediate stage of the receptor activation pathway.

The Hsp70/Hsp90 cochaperone, SGT, has been shown to interact through its TPR domain with the hinge region of human AR, which contains a peptide sequence structurally resembling the EEVD binding site for TPR proteins at the extreme C-terminus of Hsp70 and Hsp90 (Buchanan *et al.*, 2007). It has been proposed that, as a component of AR-Hsp90 complexes, SGT regulates the ligand sensitivity of AR signalling by limiting receptor trafficking to the nucleus at low hormone concentrations and maintaining the receptor within the cytoplasm of the cell.

6.2 p23; Cdc37

Disruption of the p23 gene in mice has revealed that although p23 is not essential for overall perinatal development its absolute requirement for perinatal survival is linked to impaired GR function arising most likely from instability of GR-Hsp90 complexes in the absence of p23 (Grad *et al.*, 2006; Picard, 2006). These findings suggest that GR might be a key molecular target for p23. Overexpression experiments with p23 in tissue culture cells have revealed both positive and negative influences on GR function (Freeman *et al.*, 2000; Wochnik *et al.*, 2004), as well as differential effects on other steroid receptors - increasing PR activity, while decreasing the activities of AR, ERα and MR (Freeman *et al.*, 2000). In yeast, p23 has been shown to be a positive regulator of ERα transcriptional activation, being most effective at low ERα levels and hormone concentrations, consistent with the proposed role for p23 as a component of mature ERα-Hsp90 complexes (Knoblauch & Garabedian, 1999). Ectopic expression of p23 in MCF-7 breast cancer cells increased both hormone-dependent and hormone-independent ERα transcriptional activity (Knoblauch & Garabedian, 1999).

Thus, while the major impact of p23 on ERα is likely to be through an Hsp90-dependent effect on estradiol binding, p23 overexpression may also influence receptor activity independent of ligand binding and may participate in the disassembly of receptors at cognate response elements (Freeman et al., 2000; Freeman & Yamamoto, 2001; Freeman & Yamamoto, 2002). It is of interest that although p23 increases AR transcriptional activity in a variety of mammalian cell lines, partly by increasing ligand binding competence of the receptor, Hsp90 inhibitors could not abolish the AR coactivation potential of p23, consistent with an Hsp90-independent role of p23 in AR function (Querol Cano L and Bevan CL, unpublished observations).

Genetic studies in yeast have revealed that Cdc37 plays a role in AR hormone-dependent transactivation through functional interactions with the AR LBD, although the hormone-binding properties of the receptor appear to be unaffected (Fliss et al., 1997). The association with Cdc37 is specific to AR since it does not occur with closely related nuclear receptors such as GR (Rao et al., 2001). Depletion of Cdc37 using RNA interference caused growth arrest in both AR-positive and AR-negative prostate cancer cells, and in the former led to a loss of AR transcriptional activity with a concomitant decrease in androgen-dependent gene expression (Gray et al., 2007). The targeting of Cdc37 in prostate cancer causes growth inhibition that correlates with decreased signalling through multiple pathways - the extracellular signal-regulated kinase (ERK) and Akt kinase cascades, as well as reduced AR-dependent signalling (Gray et al., 2008).

7. Conclusions

We have arrived at a better understanding of the molecular mechanisms that allow the Hsp90 chaperone to modulate steroid receptor function through direct contact with receptor LBDs. Critical to this regulation is the ability of Hsp90 to coordinate and bring to receptor-Hsp90 complexes a selection of cochaperones whose specialized influences target receptor LBDs and combine, at various stages of the receptor activation pathway, to alter receptor hormone-binding status, cellular location and transcriptional activity. A number of these cochaperones may impact on steroid receptor function independently of Hsp90. Substantial gaps still remain, however in our knowledge of how the interplay between Hsp90 and its cochaperones affects receptor function. For example, while it is known the CyP40 yeast homologue, Cpr6, regulates Hsp90 ATPase activity during receptor assembly (Prodromou et al., 1999) and studies of a second yeast homologue, Cpr7, have provided some insight into the role of this immunophilin in Hsp90-dependent signalling by steroid receptors (Duina et al., 1996; Duina et al., 1998), a coherent mechanism at the molecular level has yet to be defined. From the structural similarity between CyP40 and FKBP52, both being characterized by N-terminal PPIase and C-terminal TPR domains, it is tempting to draw parallels for their mechanism of action. Within steroid receptor-Hsp90 complexes it is possible that, as for FKBP52, the CyP40 PPIase domain forms productive interactions with the receptor LBD, serving to modulate receptor conformation and function. This may be of relevance for the function of ERα, purification of which led to the isolation of CyP40 in ERα-Hsp90 complexes (Ratajczak et al., 1993) and for the regulation of AR in prostate cancer where CyP40 appears to be overexpressed (Periyasamy et al., 2010).

Hsp90 is required for the proper function of several key regulatory proteins including multiple tyrosine and serine/threonine kinases and steroid receptors, many of which are involved in promoting malignancy (Calderwood *et al.*, 2006; Pearl, 2005; Whitesell & Lindquist, 2005). The aim of targeting and pharmacological manipulation of the Hsp90 chaperoning system has led to the ongoing development and clinical evaluation of novel Hsp90 and chaperone inhibitors for potential application in therapies against selected malignancies (Donnelly *et al.*, 2010; Kim *et al.*, 2009), syndromes arising from dysfunctional protein folding and neurodegenerative diseases (Jinwal *et al.*, 2010). With growing understanding of the novel mechanisms through which Hsp90 cochaperones modulate the function of specific clients, strategies are now evolving for the targeting of chaperone-client interactions in a wide range of human diseases (De Leon *et al.*, 2011; Gray *et al.*, 2008).

8. Abbreviations

Hsp, heat shock protein; TPR, tetratricopeptide repeat; PPIase, prolylpeptidyl isomerase; FKBP, FK506-binding protein; CyP40, cyclophilin 40; PP5, serine/threonine protein phosphatase type 5; GCUNC-45, general cell UNC-45; αSGT, small glutamine-rich tetratricopeptide repeat containing protein α; AR, androgen receptor; ERα, estrogen receptor α; ERβ, estrogen receptor β; GR, glucocorticoid receptor; MR, mineralocorticoid receptor; PR, progesterone receptor; LBD, ligand-binding domain; AF2, activation function 2; GST, glutathione S-transferase.

9. Acknowledgments

The authors wish to acknowledge support from the National Health & Medical Research Council of Australia, the National Breast Cancer Foundation and the Sir Charles Gairdner Hospital Research Fund. The authors also thank colleagues for permitting citation of their data prior to publication.

10. References

Ali M. M. U., Roe S. M., Vaughan C. K., Meyer P., Panaretou B., Piper P. W., Prodromou C. & Pearl L. H. (2006). Crystal structure of an Hsp90-nucleotide-p23/Sba1 closed chaperone complex. *Nature*, Vol.440, No.7087, pp. 1013-1017.

Allan R. K., Mok D., Ward B. K. & Ratajczak T. (2006). Modulation of chaperone function and cochaperone interaction by novobiocin in the C-terminal domain of Hsp90: evidence that coumarin antibiotics disrupt Hsp90 dimerization. *J Biol Chem*, Vol.281, No.11, pp. 7161-7171.

Amler L. C., Agus D. B., LeDuc C., Sapinoso M. L., Fox W. D., Kern S., Lee D., Wang V., Leysens M., Higgins B., Martin J., Gerald W., Dracopoli N., Cordon-Cardo C., Scher H. I. & Hampton G. M. (2000). Dysregulated expression of androgen-responsive and nonresponsive genes in the androgen-independent prostate cancer xenograft model CWR22-R. *Cancer Res*, Vol.60, No.21, pp. 6134-6141.

Angeletti P. C., Walker D. & Panganiban A. T. (2002). Small glutamine-rich protein/viral protein U-binding protein is a novel cochaperone that affects heat shock protein 70 activity. *Cell Stress Chaperones*, Vol.7, No.3, pp. 258-268

Aumais J. P., Lee H. S., Lin R. & White J. H. (1997). Selective interaction of Hsp90 with an estrogen receptor ligand-binding domain containing a point mutation. *J Biol Chem*, Vol.272, No.18, pp. 12229-12235.

Ballinger C. A., Connell P., Wu Y., Hu Z., Thompson L. J., Yin L.-Y. & Patterson C. (1999). Identification of CHIP, a novel tetratricopeptide repeat-containing protein that interacts with heat shock proteins and negatively regulates chaperone functions. *Mol Cell Biol*, Vol.19, No.6, pp. 4535-4545.

Banerjee A., Periyasamy S., Wolf I. M., Hinds T. D., Yong W., Shou W. & Sanchez E. R. (2008). Control of glucocorticoid and progesterone receptor subcellular localization by the ligand-binding domain is mediated by distinct interactions with tetratricopeptide repeat proteins. *Biochemistry*, Vol.47, No.39, pp. 10471-10480.

Barent R. L., Nair S. C., Carr D. C., Ruan Y., Rimerman R. A., Fulton J., Zhang Y. & Smith D. F. (1998). Analysis of FKBP51/FKBP52 chimeras and mutants for Hsp90 binding and association with progesterone receptor complexes. *Mol Endocrinol*, Vol.12, No.3, pp. 342-354.

Barford D. (1996). Molecular mechanisms of the protein serine/threonine phosphatases. *Trends Biochem Sci*, Vol.21, No.11, pp. 407-412.

Bimston D., Song J., Winchester D., Takayama S., Reed J. C. & Morimoto R. I. (1998). BAG-1, a negative regulator of Hsp70 chaperone activity, uncouples nucleotide hydrolysis from substrate release. *EMBO J*, Vol.17, No 23, pp. 6871 6878.

Binart N., Lombès M. & Baulieu E. E. (1995). Distinct functions of the 90 kDa heat-shock protein (hsp90) in oestrogen and mineralocorticosteroid receptor activity: effects of hsp90 deletion mutants. *Biochem J*, Vol.311, 797-804.

Bledsoe R. K., Montana V. G., Stanley T. B., Delves C. J., Apolito C. J., McKee D. D., Consler T. G., Parks D. J., Stewart E. L., Willson T. M., Lambert M. H., Moore J. T., Pearce K. H. & Xu H. E. (2002). Crystal structure of the glucocorticoid receptor ligand binding domain reveals a novel mode of receptor dimerization and coactivator recognition. *Cell*, Vol.110, 93-105.

Brelivet Y., Kammerer S., Rochel N., Poch O. & Moras D. (2004). Signature of the oligomeric behaviour of nuclear receptors at the sequence and structural level *EMBO Rep*, Vol.5, No.4, pp. 423-429.

Buchanan G., Ricciardelli C., Harris J. M., Prescott J., Yu Z. C.-L., Jia L., Butler L. M., Marshall V. R., Scher H. I., Gerald W. L., Coetzee G. A. & Tilley W. D. (2007). Control of androgen receptor signaling in prostate cancer by the cochaperone small glutamine rich tetratricopeptide repeat containing protein α. *Cancer Res*, Vol.67, No.20, pp. 10087-10096.

Caamano C. A., Morano M. I., Dalman F. C., Pratt W. B. & Akil H. (1998). A conserved proline in the Hsp90 binding region of the glucocorticoid receptor is required for Hsp90 heterocomplex stabilization and receptor signaling. *J Biol Chem*, Vol.273, No.32, pp. 20473-20480.

Cadepond F., Binart N., Chambraud B., Jibard N., Schweizer-Groyer G., Segard-Maurel I. & Baulieu E. E. (1993). Interaction of glucocorticosteroid receptor and wild-type or mutated 90-kDa heat shock protein coexpressed in baculovirus-infected Sf9 cells. *PNAS*, Vol.90, No.22, pp. 10434-10438.

Calderwood S. K., Khaleque M. A., Sawyer D. B. & Ciocca D. R. (2006). Heat shock proteins in cancer: chaperones of tumorigenesis. *Trends Biochem Sci*, Vol.31, No.3, pp. 164-172.

Callahan M. A., Handley M. A., Lee Y.-H., Talbot K. J., Harper J. W. & Panganiban A. T. (1998). Functional interaction of human immunodeficiency virus type 1 Vpu and Gag with a novel member of the tetratricopeptide repeat protein family. *J Virol*, Vol.72, No.6, pp. 5189-5197.

Carrello A., Ingley E., Minchin R. F., Tsai S. & Ratajczak T. (1999). The common tetratricopeptide repeat acceptor site for steroid receptor-associated immunophilins and Hop is located in the dimerization domain of Hsp90. *J Biol Chem*, Vol.274, No.5, pp. 2682-2689.

Carrigan P. E., Sikkink L. A., Smith D. F. & Ramirez-Alvarado M. (2006). Domain:domain interactions within Hop, the Hsp70/Hsp90 organizing protein, are required for protein stability and structure. *Protein Sci*, Vol.15, No.3, pp. 522-532.

Chadli A., Bruinsma E. S., Stensgard B. & Toft D. (2008a). Analysis of Hsp90 cochaperone interactions reveals a novel mechanism for TPR protein recognition. *Biochemistry*, Vol.47, No.9, pp. 2850-2857.

Chadli A., Felts S. J. & Toft D. O. (2008b). GCUNC-45 is the first Hsp90 co-chaperone to show α/β isoform specificity *J Biol Chem*, Vol.283, No.15, pp. 9509-9512.

Chadli A., Graham J. D., Abel M. G., Jackson T. A., Gordon D. F., Wood W. M., Felts S. J., Horwitz K. B. & Toft D. (2006). GCUNC-45 is a novel regulator for the progesterone receptor/Hsp90 chaperoning pathway. *Mol Cell Biol*, Vol.26, No.5, pp. 1722-1730.

Chen M.-S., Silverstein A. M., Pratt W. B. & Chinkers M. (1996a). The tetratricopeptide repeat domain of protein phosphatase 5 mediates binding to glucocorticoid receptor heterocomplexes and acts as a dominant negative mutant. *J Biol Chem*, Vol.271, No.50, pp. 32315-32320.

Chen M. X. & Cohen P. T. W. (1997). Activation of protein phosphatase 5 by limited proteolysis or the binding of polyunsaturated fatty acids to the TPR domain. *FEBS Lett*, Vol.400, No.1, pp. 136-140.

Chen M. X., McPartlin A. E., Brown L., Chen Y. H., Barker H. M. & Cohen P. T. W. (1994). A novel human protein serine/ threonine phosphatase, which possesses four tetretricopeptide repeat motifs and localizes to the nucleus. *EMBO J*, Vol.13, No.18, pp. 4278-4290.

Chen S., Prapapanich V., Rimerman R. A., Honore B. & Smith D. F. (1996b). Interactions of p60, a mediator of progesterone receptor assembly, with heat shock proteins Hsp90 and Hsp70. *Mol Endocrinol*, Vol.10, No.6, pp. 682-693.

Chen S. & Smith D. F. (1998). Hop as an adaptor in the heat shock protein 70 (Hsp70) and Hsp90 chaperone machinery. *J Biol Chem*, Vol.273, No.52, pp. 35194-35200.

Chen S., Sullivan W. P., Toft D. O. & Smith D. F. (1998). Differential interactions of p23 and the TPR-containing proteins Hop, Cyp40, FKBP52 and FKBP51 with Hsp90 mutants. *Cell Stress Chaperones*, Vol.3, No.2, pp. 118-129.

Cheung-Flynn J., Prapapanich V., Cox M. B., Riggs D. L., Suarez-Quian C. & Smith D. F. (2005). Physiological role for the cochaperone FKBP52 in androgen receptor signaling. *Mol Endocrinol*, Vol.19, No.6, pp. 1654-1666.

Cheung-Flynn J., Roberts P. J., Riggs D. L. & Smith D. F. (2003). C-terminal sequences outside the tetratricopeptide repeat domain of FKBP51 and FKBP52 cause differential binding to Hsp90. *J Biol Chem*, Vol.278, No.19, pp. 17388-17394.

Chinkers M. (1994). Targeting of a distinctive protein-serine phosphatase to the protein kinase-like domain of the atrial natriuretic peptide receptor. *PNAS*, Vol.91, No.23, pp. 11075-11079.

Cohen P. T. W. (1997). Novel protein serine/threonine phosphatases: variety is the spice of life. *Trends Biochem Sci*, Vol.22, No.7, pp. 245-251.

Connell P., Ballinger C. A., Jiang J., Wu Y., Thompson L. J., Hohfeld J. & Patterson C. (2001). The co-chaperone CHIP regulates protein triage decisions mediated by heat-shock proteins. *Nat Cell Biol*, Vol.3, No.1, pp. 93-96.

Cyr D. M., Höhfeld J. & Patterson C. (2002). Protein quality control: U-box-containing E3 ubiquitin ligases join the fold. *Trends Biochem Sci*, Vol.27, No.7, pp. 368-375.

Cziepluch C., Kordes E., Poirey R., Grewenig A., Rommelaere J. & Jauniaux J.-C. (1998). Identification of a novel cellular TPR-containing protein, SGT, that interacts with the nonstructural protein NS1 of parvovirus H-1. *J Virol*, Vol.72, No.5, pp. 4149-4156.

Das A. K., Cohen P. T. W. & Barford D. (1998). The structure of the tetratricopeptide repeats of protein phosphatase 5: implications for TPR-mediated protein-protein interactions. *EMBO J*, Vol.17, No.5, pp. 1192-1199.

Davies T. H., Ning Y.-M. & Sanchez E. R. (2002). A new first step in activation of steroid receptors: hormone-induced switching of FKBP51 and FKBP52 immunophilins. *J Biol Chem*, Vol.277, No.7, pp. 4597-4600.

De Leon J. T., Iwai A., Feau C., Garcia Y., Balsiger H. A., Storer C. L., Suro R. M., Garza K. M., Lee S., Sang Kim Y., Chen Y., Ning Y.-M., Riggs D. L., Fletterick R. J., Guy R. K., Trepel J. B., Neckers L. M. & Cox M. B. (2011). Targeting the regulation of androgen receptor signaling by the heat shock protein 90 cochaperone FKBP52 in prostate cancer cells. *PNAS*, Vol.108, No.29, pp. 11878-11883.

Denny W. B., Prapapanich V., Smith D. F. & Scammell J. G. (2005). Structure-function analysis of Squirrel Monkey FK506-binding protein 51, a potent inhibitor of glucocorticoid receptor activity. *Endocrinology*, Vol.146, No.7, pp. 3194-3201.

Denny W. B., Valentine D. L., Reynolds P. D., Smith D. F. & Scammell J. G. (2000). Squirrel monkey immunophilin FKBP51 is a potent inhibitor of glucocorticoid receptor binding. *Endocrinology*, Vol.141, No.11, pp. 4107-4113.

Dittmar K. D., Hutchison K. A., Owens-Grillo J. K. & Pratt W. B. (1996). Reconstitution of the steroid receptor Hsp90 heterocomplex assembly system of rabbit reticulocyte lysate. *J Biol Chem*, Vol.271, No.22, pp. 12833-12839.

Donnelly A. C., Zhao H., Reddy Kusuma B. & Blagg B. S. J. (2010). Cytotoxic sugar analogues of an optimized novobiocin scaffold. *Med Chem Commun*, Vol.1, No.2, pp. 165-170.

Duina A. A., Chang H.-C. J., Marsh J. A., Lindquist S. & Gaber R. F. (1996). A cyclophilin function on Hsp90-dependent signal transduction. *Science*, Vol.274, 1713-1715.

Duina A. A., Marsh J. A., Kurtz R. B., Chang H.-C. J., Lindquist S. & Gaber R. F. (1998). The peptidyl-prolyl isomerase domain of the CyP-40 cyclophilin homolog Cpr7 is not required to support growth or glucocorticoid receptor activity in *Saccharomyces cerevisiae*. *J Biol Chem*, Vol.273, No.18, pp. 10819-10822.

Dutta S. & Tan Y.-J. (2008). Structural and functional characterization of human SGT and its interaction with Vpu of the human immunodeficiency virus type 1. *Biochemistry*, Vol.47, No.38, pp. 10123-10131.

Echeverria P. C. & Picard D. (2010). Molecular chaperones, essential partners of steroid hormone receptors for activity and mobility. *Biochim Biophys Acta - Mol Cell Res*, Vol.1803, No.6, pp. 641-649.

Estébanez-Perpiñá E., Arnold L. A., Nguyen P., Rodrigues E. D., Mar E., Bateman R., Pallai P., Shokat K. M., Baxter J. D., Guy R. K., Webb P. & Fletterick R. J. (2007). A surface on the androgen receptor that allosterically regulates coactivator binding. *PNAS*, Vol.104, No.41, pp. 16074-16079.

Fang L., Ricketson D., Getubig L. & Darimont B. (2006). Unliganded and hormone-bound glucocorticoid receptors interact with distinct hydrophobic sites in the Hsp90 C-terminal domain. *PNAS*, Vol.103, No.49, pp. 18487-18492.

Febbo P. G., Lowenberg M., Thorner A. R., Brown M., Loda M. & Golub T. R. (2005). Androgen mediated regulation and functional implications of FKBP51 expression in prostate cancer. *J Urol*, Vol.173, No.5, pp. 1772-1777.

Felts S. J. & Toft D. O. (2003). p23, a simple protein with complex activities. *Cell Stress Chaperones*, Vol.8, No.2, pp. 108-113.

Fliss A. E., Fang Y., Boschelli F. & Caplan A. J. (1997). Differential *in vivo* regulation of steroid hormone receptor activation by Cdc37p. *Mol Biol Cell*, Vol.8, No.12, pp. 2501-2509.

Freeman B. C., Felts S. J., Toft D. O. & Yamamoto K. R. (2000). The p23 molecular chaperones act at a late step in intracellular receptor action to differentially affect ligand efficacies. *Genes Dev*, Vol.14, 422-434.

Freeman B. C. & Yamamoto K. R. (2001). Continuous recycling: a mechanism for modulatory signal transduction. *Trends Biochem Sci*, Vol.26, No.5, pp. 285-290.

Freeman B. C. & Yamamoto K. R. (2002). Disassembly of transcriptional regulatory complexes by molecular chaperones. *Science*, Vol.296, No.5576, pp. 2232-2235.

Frydman J. & Höhfeld J. (1997). Chaperones get in touch: the Hip-Hop connection. *Trends Biochem Sci*, Vol.22, No.3, pp. 87-92.

Fuller P. J., Smith B. J. & Rogerson F. M. (2004). Cortisol resistance in the New World revisited. *Trends Endocrinol Metab*, Vol.15, No.7, pp. 296-299.

Gallo L. I., Ghini A. A., Piwien Pilipuk G. & Galigniana M. D. (2007). Differential recruitment of tetratricopeptide repeat domain immunophilins to the mineralocorticoid receptor influences both heat-shock protein 90-dependent retrotransport and hormone-dependent transcriptional activity. *Biochemistry*, Vol.46, No.49, pp. 14044-14057.

Gebauer M., Zeiner M. & Gehring U. (1997). Proteins interacting with the molecular chaperone Hsp70/Hsc70: physical associations and effects on refolding activity. *FEBS Lett*, Vol.417, No.1, pp. 109-113.

Giannoukos G., Silverstein A. M., Pratt W. B. & Simons Jr S. S. (1999). The seven amino acids (547-553) of rat glucocorticoid receptor required for steroid and Hsp90 binding contain a functionally independent LXXLL motif that is critical for steroid binding. *J Biol Chem*, Vol.274, No.51, pp. 36527-36536.

Grad I., McKee T. A., Ludwig S. M., Hoyle G. W., Ruiz P., Wurst W., Floss T., Miller C. A., III & Picard D. (2006). The Hsp90 cochaperone p23 is essential for perinatal survival. *Mol Cell Biol*, Vol.26, No.23, pp. 8976-8983.

Gray P. J., Prince T., Cheng J., Stevenson M. A. & Calderwood S. K. (2008). Targeting the oncogene and kinome chaperone CDC37. *Nat Rev Cancer*, Vol.8, No.7, pp. 491-495.

Gray P. J., Stevenson M. A. & Calderwood S. K. (2007). Targeting Cdc37 inhibits multiple signaling pathways and induces growth arrest in prostate cancer cells. *Cancer Res*, Vol.67, No.24, pp. 11942-11950.

Harris S. F., Shiau A. K. & Agard D. A. (2004). The crystal structure of the carboxy-terminal dimerization domain of htpG, the *Escherichia coli* Hsp90, reveals a potential substrate binding site. *Structure*, Vol.12, No.6, pp. 1087-1097.

He B., Bowen N. T., Minges J. T. & Wilson E. M. (2001). Androgen-induced NH_2- and COOH-terminal interaction inhibits p160 coactivator recruitment by activation function 2. *J Biol Chem*, Vol.276, No.45, pp. 42293-42301.

He B., Gampe Jr R. T., Kole A. J., Hnat A. T., Stanley T. B., An G., Stewart E. L., Kalman R. I., Minges J. T. & Wilson E. M. (2004). Structural basis for androgen receptor interdomain and coactivator interactions suggests a transition in nuclear receptor activation function dominance. *Mol Cell*, Vol.16, No.3, pp. 425-438.

Hinds Jr T. D. & Sánchez E. R. (2008). Protein phosphatase 5. *Int J Biochem Cell Biol*, Vol.40, No.11, pp. 2358-2362.

Höhfeld J. & Jentsch S. (1997). GrpE-like regulation of the Hsc70 chaperone by the anti-apoptotic protein BAG-1. *EMBO J*, Vol.16, No.20, pp. 6209-6216.

Höhfeld J., Minami Y. & Hartl F.-U. (1995). Hip, a novel cochaperone involved in the eukaryotic Hsc70/Hsp40 reaction cycle. *Cell*, Vol.83, No.4, pp. 589-598.

Ikeda K., Ogawa S., Tsukui T., Horie-Inoue K., Ouchi Y., Kato S., Muramatsu M. & Inoue S. (2004). Protein phosphatase 5 is a negative regulator of estrogen receptor-mediated transcription. *Mol Endocrinol*, Vol.18, No.5, pp. 1131-1143.

Jackson S. E., Queitsch C. & Toft D. (2004). Hsp90: from structure to phenotype. *Nat Struct Mol Biol*, Vol.11, No.12, pp. 1152-1155.

Jiang J., Ballinger C. A., Wu Y., Dai Q., Cyr D. M., Höhfeld J. & Patterson C. (2001). CHIP is a U-box-dependent E3 ubiquitin ligase. Identification of Hsc70 as a target for ubiquitylation. *J Biol Chem*, Vol.276, 42938-42944.

Jinwal U. K., Koren J., Borysov S. I., Schmid A. B., Abisambra J. F., Blair L. J., Johnson A. G., Jones J. R., Shults C. L., O'Leary J. C., Jin Y., Buchner J., Cox M. B. & Dickey C. A. (2010). The Hsp90 cochaperone, FKBP51, increases Tau stability and polymerizes microtubules. *J Neurosci*, Vol.30, No.2, pp. 591-599.

Joseph J. D., Wittmann B. M., Dwyer M. A., Cui H., Dye D. A., McDonnell D. P. & Norris J. D. (2009). Inhibition of prostate cancer cell growth by second-site androgen receptor antagonists. *PNAS*, Vol.106, No.29, pp. 12178-12183.

Kajander T., Sachs J. N., Goldman A. & Regan L. (2009). Electrostatic interactions of Hsp-organizing protein tetratricopeptide domains with Hsp70 and Hsp90: computational analysis and protein engineering. *J Biol Chem*, Vol.284, 25364-25374.

Kallen J., Mikol V., Taylor P. & D.Walkinshaw M. (1998). X-ray structures and analysis of 11 cyclosporin derivatives complexed with cyclophilin A. *J Mol Biol*, Vol.283, No.2, pp. 435-449.

Karagöz G. E., Duarte A. M. S., Ippel H., Uetrecht C., Sinnige T., van Rosmalen M., Hausmann J., Heck A. J. R., Boelens R. & Rüdiger S. G. D. (2010). N-terminal domain of human Hsp90 triggers binding to the cochaperone p23. *PNAS*, Vol.108, No.2, pp. 580-585.

Kauppi B., Jakob C., Farnegardh M., Yang J., Ahola H., Alarcon M., Calles K., Engstrom O., Harlan J., Muchmore S., Ramqvist A.-K., Thorell S., Ohman L., Greer J., Gustafsson J.-A., Carlstedt-Duke J. & Carlquist M. (2003). The three-dimensional structures of antagonistic and agonistic forms of the glucocorticoid receptor ligand-binding domain: RU-486 induces a transconformation that leads to active antagonism. *J Biol Chem*, Vol.278, No.25, pp. 22748-22754.

Kim Y. S., Alarcon S. V., Lee S., Lee M. J., Giaccone G., Neckers L. & Trepel J. B. (2009). Update on Hsp90 inhibitors in clinical trial. *Curr Top Med Chem*, Vol.9, No.15, pp. 1479-1492.

Knoblauch R. & Garabedian M. J. (1999). Role for Hsp90-associated cochaperone p23 in estrogen receptor signal transduction. *Mol Cell Biol*, Vol.19, No.5, pp. 3748-3759.

Kosano H., Stensgard B., Charlesworth M. C., McMahon N. & Toft D. (1998). The assembly of progesterone receptor-Hsp90 complexes using purified proteins. *J Biol Chem*, Vol.273, No.49, pp. 32973-32979.

Lassle M., Blatch G. L., Kundra V., Takatori T. & Zetter B. R. (1997). Stress-inducible, murine protein mSTI1. Characterization of binding domains for heat shock proteins and in vitro phosphorylation by different kinases. *J Biol Chem*, Vol.272, No.3, pp. 1876-1884.

Li J., Richter K. & Buchner J. (2011). Mixed Hsp90-cochaperone complexes are important for the progression of the reaction cycle. *Nat Struct Mol Biol*, Vol.18, No.1, pp. 61-66.

Liou S.-T. & Wang C. (2005). Small glutamine-rich tetratricopeptide repeat-containing protein is composed of three structural units with distinct functions. *Arch Biochem Biophys*, Vol.435, No.2, pp. 253-263.

Louvion J.-F., Warth R. & Picard D. (1996). Two eukaryote-specific regions of Hsp82 are dispensable for its viability and signal transduction functions in yeast. *PNAS*, Vol.93, No.24, pp. 13937-13942.

MacLean M. & Picard D. (2003). Cdc37 goes beyond Hsp90 and kinases. *Cell Stress Chaperones*, Vol.8, No.2, pp. 114-119.

Magee J., Chang L., Stormo G. & Milbrandt J. (2006). Direct, androgen receptor-mediated regulation of the FKBP5 gene via a distal enhancer element. *Endocrinology*, Vol.147, No.1, pp. 590-598.

Marcu M. G., Chadli A., Bouhouche I., Catelli M. & Neckers L. M. (2000). The heat shock protein 90 antagonist novobiocin interacts with a previously unrecognized ATP-binding domain in the carboxyl terminus of the chaperone. *J Biol Chem*, Vol.275, No.47, pp. 37181-37186.

Matias P. M., Donner P., Coelho R., Thomaz M., Peixoto C., Macedo S., Otto N., Joschko S., Scholz P., Wegg A., Bäsler S., Schäfer M., Egner U. & Carrondo M. A. (2000a). Structural evidence for ligand specificity in the binding domain of the human androgen receptor: implications for pathogenic gene mutations. *J. Biol. Chem.*, Vol.275, No.34, pp. 26164-26171.

Matias P. M., Donner P., Coelho R., Thomaz M., Peixoto C., Macedo S., Otto N., Joschko S., Scholz P., Wegg A., Bäsler S., Schäfer M., Egner U. & Carrondo M. A. (2000b). Structural evidence for ligand specificity in the binding domain of the human androgen receptor: implications for pathogenic gene mutations. *J Biol Chem*, Vol.275, No.34, pp. 26164-26171.

McLaughlin S. H., Smith H. W. & Jackson S. E. (2002). Stimulation of the weak ATPase activity of human Hsp90 by a client protein. *J Mol Biol*, Vol.315, No.4, pp. 787-798.

McLaughlin S. H., Sobott F., Yao Z.-p., Zhang W., Nielsen P. R., Grossmann J. G. n., Laue E. D., Robinson C. V. & Jackson S. E. (2006). The co-chaperone p23 arrests the Hsp90 ATPase cycle to trap client proteins. *J Mol Biol*, Vol.356, No.3, pp. 746-758.

Mok D., Allan R. K., Carrello A., Wangoo K., Walkinshaw M. D. & Ratajczak T. (2006). The chaperone function of cyclophilin 40 maps to a cleft between the prolyl isomerase and tetratricopeptide repeat domains. *FEBS Lett*, Vol.580, No.11, pp. 2761-2768.

Moore T. W., Mayne C. G. & Katzenellenbogen J. A. (2010). Minireview: Not picking pockets: nuclear receptor alternate-site modulators (NRAMs). *Mol Endocrinol*, Vol.24, No.4, pp. 683-695.

Mousses S., Wagner U., Chen Y., Kim J. W., Bubendorf L., Bittner M., Pretlow T., Elkahloun A. G., Trepel J. B., Kallioniemi O.-P. & (2001). Failure of hormone therapy in prostate cancer involves systematic restoration of androgen responsive genes and activation of rapamycin sensitive signaling. *Oncogene*, Vol.20, No.46, pp. 6718-6723.

Murata S., Minami Y., Minami M., Chiba T. & Tanaka K. (2001). CHIP is a chaperone-dependent E3 ligase that ubiquitylates unfolded protein. *EMBO Rep*, Vol.2, No.12, pp. 1133-1138.

Nair S. C., A.Rimerman R., Toran E. J., Chen S., Prapapanich V., Butts R. N. & Smith D. F. (1997). Molecular cloning of human FKBP51 and comparisons of immunophilin interactions with Hsp90 and progesterone receptor. *Mol Cell Biol*, Vol.17, No.2, pp. 594-603.

Nathan D. F. & Lindquist S. (1995). Mutational analysis of Hsp90 function: interactions with a steroid receptor and a protein kinase. *Mol Cell Biol*, Vol.15, No.7, pp. 3917-3925.

Nelson G. M., Huffman H. & Smith D. F. (2003). Comparison of the carboxy-terminal DP-repeat region in the co-chaperones Hop and Hip. *Cell Stress Chaperones*, Vol.8, No.2, pp. 125-133.

Nelson G. M., Prapapanich V., Carrigan P. E., Roberts P. J., Riggs D. L. & Smith D. F. (2004). The heat shock protein 70 cochaperone Hip enhances functional maturation of glucocorticoid receptor. *Mol Endocrinol*, Vol.18, No.7, pp. 1620-1630.

Ni L., Yang C.-S., Gioeli D., Frierson H., Toft D. O. & Paschal B. M. (2010). FKBP51 promotes assembly of the Hsp90 chaperone complex and regulates androgen receptor signaling in prostate cancer cells. *Mol Cell Biol*, Vol.30, No.5, pp. 1243-1253.

Odunuga O. O., Hornby J. A., Bies C., Zimmermann R., Pugh D. J. & Blatch G. L. (2003). Tetratricopeptide repeat motif-mediated Hsc70-mSTI1 interaction. *J Biol Chem*, Vol.278, No.9, pp. 6896-6904.

Onuoha S. C., Coulstock E. T., Grossmann J. G. & Jackson S. E. (2008). Structural studies on the co-chaperone Hop and its complexes with Hsp90. *J Mol Biol*, Vol.379, No.4, pp. 732-744.

Pearl L. H. (2005). Hsp90 and Cdc37 - a chaperone cancer conspiracy. *Curr Opin Genet Dev*, Vol.15, No.1, pp. 55-61.

Periyasamy S., Hinds T., Jr., Shemshedini L., Shou W. & Sanchez E. R. (2010). FKBP51 and Cyp40 are positive regulators of androgen-dependent prostate cancer cell growth and the targets of FK506 and cyclosporin A. *Oncogene*, Vol.29, No.11, pp. 1691-1701.

Picard D. (2006). Chaperoning steroid hormone action. *Trends Endocrinol Metab*, Vol.17, No.6, pp. 229-236.

Picard D., Khursheed B., Garabedian M. J., Fortin M. G., Lindquist S. & Yamamoto K. R. (1990). Reduced levels of Hsp90 compromise steroid receptor action *in vivo*. *Nature*, Vol.348, 166-168.

Pirkl F. & Buchner J. (2001). Functional analysis of the Hsp90-associated human peptidyl prolyl cis/trans isomerases FKBP51, FKBP52 and CyP40. *J Mol Biol*, Vol.308, No.4, pp. 795-806.

Prapapanich V., Chen S., Nair S. C., Rimerman R. A. & Smith D. F. (1996a). Molecular cloning of human p48, a transient component of progesterone receptor complexes and an Hsp70-binding protein. *Mol Endocrinol*, Vol.10, No.4, pp. 420-431.

Prapapanich V., Chen S. & Smith D. F. (1998). Mutation of Hip's carboxy-terminal region inhibits a transitional stage of progesterone receptor assembly. *Mol Cell Biol*, Vol.18, No.2, pp. 944-952.

Prapapanich V., Chen S., Toran E. J., Rimerman R. A. & Smith D. F. (1996b). Mutational analysis of the hsp70-interacting protein Hip. *Mol Cell Biol*, Vol.16, No.11, pp. 6200-6207.

Pratt W. B. & Toft D. O. (1997). Steroid receptor interactions with heat shock protein and immunophilin chaperones. *Endocr Rev*, Vol.18, No.3, pp. 306-360.

Pratt W. B. & Toft D. O. (2003). Regulation of signaling protein function and trafficking by the Hsp90/Hsp70-based chaperone machinery. *Exp Biol Med*, Vol.228, No.2, pp. 111-133.

Prodromou C., Panaretou B., Chohan S., Siligardi G., O'Brien R., Ladbury J. E., Roe S. M., Piper P. W. & Pearl L. H. (2000). The ATPase cycle of Hsp90 drives a molecular 'clamp' via transient dimerization of the N-terminal domains. *EMBO J*, Vol.19, No.16, pp. 4383-4392.

Prodromou C., Siligardi G., O'Brien R., Woolfson D. N., Regan L., Panaretou B., Ladbury J. E., Piper P. W. & Pearl L. H. (1999). Regulation of Hsp90 ATPase activity by tetratricopeptide repeat (TPR)-domain co-chaperones. *EMBO J*, Vol.18, No.3, pp. 754-762.

Radanyi C., Chambraud B. & Baulieu E. E. (1994). The ability of the immunophilin FKBP59-HBI to interact with the 90-kDa heat shock protein is encoded by its tetratricopeptide repeat domain. *PNAS*, Vol.91, 11197-11201.

Ramsey A. J. & Chinkers M. (2002). Identification of potential physiological activators of protein phosphatase 5. *Biochemistry*, Vol.41, No.17, pp. 5625-5632.

Ramsey A. J., Russell L. C. & Chinkers M. (2009). C-terminal sequences of Hsp70 and Hsp90 as non-specific anchors for tetratricopeptide repeat (TPR) proteins. *Biochem J*, Vol.423, No.3, pp. 411-419.

Rao J., Lee P., Benzeno S., Cardozo C., Albertus J., Robins D. M. & Caplan A. J. (2001). Functional interaction of human Cdc37 with the androgen receptor but not with the glucocorticoid receptor. *J Biol Chem*, Vol.276, No.8, pp. 5814-5820.

Ratajczak T. (2001). Protein coregulators that mediate estrogen receptor function. *Reprod Fertil Dev*, Vol.13, 221-229.

Ratajczak T. & Carrello A. (1996). Cyclophilin 40 (CyP-40), mapping of its Hsp90 binding domain and evidence that FKBP52 competes with CyP-40 for Hsp90 binding. *J Biol Chem*, Vol.271, No.6, pp. 2961-2965.

Ratajczak T., Carrello A., Mark P. J., Warner B. J., Simpson R. J., Moritz R. L. & House A. K. (1993). The cyclophilin component of the unactivated estrogen receptor contains a tetratricopeptide repeat domain and shares identity with p59 (FKBP59). *J Biol Chem*, Vol.268, No.18, pp. 13187-13192.

Ratajczak T., Hlaing J., Brockway M. J. & Hahnel R. (1990). Isolation of untransformed bovine estrogen receptor without molybdate stabilization. *J Steroid Biochem*, Vol.35, No.5, pp. 543-553.

Ratajczak T., Ward B. K. & Minchin R. F. (2003). Immunophilin chaperones in steroid receptor signalling. *Curr Top Med Chem*, Vol.3, No.12, pp. 1348-1357.

Reynolds P. D., Ruan Y., Smith D. F. & Scammell J. G. (1999). Glucocorticoid resistance in the squirrel monkey is associated with overexpression of the immunophilin FKBP51. *J Clin Endocrinol Metab*, Vol.84, No.2, pp. 663-669.

Richter K., Walter S. & Buchner J. (2004). The co-chaperone Sba1 connects the ATPase reaction of Hsp90 to the progression of the chaperone cycle. *J Mol Biol*, Vol.342, No.5, pp. 1403-1413.

Riggs D. L., Cox M. B., Cheung-Flynn J., Prapapanich V., Carrigan P. E. & Smith D. F. (2004). Functional specificity of co-chaperone interactions with Hsp90 client proteins. *Crit Rev Biochem Mol Biol*, Vol.39, 279-295.

Riggs D. L., Cox M. B., Tardif H. L., Hessling M., Buchner J. & Smith D. F. (2007). Noncatalytic role of the FKBP52 peptidyl-prolyl isomerase domain in the regulation of steroid hormone signaling. *Mol Cell Biol*, Vol.27, No.24, pp. 8658-8669.

Riggs D. L., Roberts P. J., Chirillo S. C., Cheung-Flynn J., Prapapanich V., Ratajczak T., Gaber R., Picard D. & Smith D. F. (2003). The Hsp90-binding peptidylprolyl isomerase FKBP52 potentiates glucocorticoid signaling *in vivo*. *EMBO J*, Vol.22, No.5, pp. 1158-1167.

Roe S. M., Ali M. M. U., Meyer P., Vaughan C. K., Panaretou B., Piper P. W., Prodromou C. & Pearl L. H. (2004). The mechanism of Hsp90 regulation by the protein kinase-specific cochaperone p50cdc37. *Cell*, Vol.116, No.1, pp. 87-98.

Russell L. C., Whitt S. R., Chen M.-S. & Chinkers M. (1999). Identification of conserved residues required for the binding of a tetratricopeptide repeat domain to heat shock protein 90. *J Biol Chem*, Vol.274, No.29, pp. 20060-20063.

Scammell J. G., Denny W. B., Valentine D. L. & Smith D. F. (2001). Overexpression of the FK506-binding immunophilin FKBP51 is the common cause of glucocorticoid resistance in three New World primates. *Gen Comp Endocrinol*, Vol.124, No.2, pp. 152-165.

Schaufele F., Carbonell X., Guerbadot M., Borngraeber S., Chapman M. S., Ma A. A. K., Miner J. N. & Diamond M. I. (2005). The structural basis of androgen receptor activation: intramolecular and intermolecular amino–carboxy interactions. *PNAS*, Vol.102, No.28, pp. 9802-9807.

Scheufler C., Brinker A., Bourenkov G., Pegoraro S., Moroder L., Bartunik H., Hartl F. U. & Moarefi I. (2000). Structure of TPR domain-peptide complexes: critical elements in the assembly of the Hsp70-Hsp90 multichaperone machine. *Cell*, Vol.101, No.2, pp. 199-210.

Siligardi G., Hu B., Panaretou B., Piper P. W., Pearl L. H. & Prodromou C. (2004). Co-chaperone regulation of conformational switching in the Hsp90 ATPase cycle. *J Biol Chem*, Vol.279, No.50, pp. 51989-51998.

Silverstein A. M., Galigniana M. D., Chen M.-S., Owens-Grillo J. K., Chinkers M. & Pratt W. B. (1997). Protein phosphatase 5 is a major component of glucocorticoid receptor Hsp90 complexes with properties of an FK506-binding immunophilin. *J Biol Chem*, Vol.272, No.26, pp. 16224-16230.

Silverstein A. M., Galigniana M. D., Kanelakis K. C., Radanyi C., Renoir J.-M. & Pratt W. B. (1999). Different regions of the immunophilin FKBP52 determine its association with the glucocorticoid receptor, Hsp90, and cytoplasmic dynein. *J Biol Chem*, Vol.274, No.52, pp. 36980-36986.

Sinars C. R., Cheung-Flynn J., Rimerman R. A., Scammell J. G., Smith D. F. & Clardy J. (2003). Structure of the large FK506-binding protein FKBP51, an Hsp90-binding protein and a component of steroid receptor complexes. *PNAS*, Vol.100, No.3, pp. 868-873.

Skinner J., Sinclair C., Romeo C., Armstrong D., Charbonneau H. & Rossie S. (1997). Purification of a fatty acid-stimulated protein-serine/threonine phosphatase from bovine brain and its identification as a homolog of protein phosphatase 5. *J Biol Chem*, Vol.272, No.36, pp. 22464-22471.

Smith D. F. (1993). Dynamics of heat shock protein 90-progesterone receptor binding and the disactivation loop model for steroid receptor complexes. *Mol Endocrinol*, Vol.7, No.11, pp. 1418-1429.

Smith D. F. (2004). Tetratricopeptide repeat cochaperones in steroid receptor complexes. *Cell Stress Chaperones*, Vol.9, No.2, pp. 109-121.

Smith D. F. & Toft D. O. (2008). Minireview. The intersection of steroid receptors with molecular chaperones: observations and questions. *Mol Endocrinol*, Vol.22, No.10, pp. 2229-2240.

Sullivan W. P. & Toft D. O. (1993). Mutational analysis of hsp90 binding to the progesterone receptor. *J Biol Chem*, Vol.268, No.27, pp. 20373-20379.

Tai P. K., Albers M. W., Chang H., Faber L. E. & Schreiber S. L. (1992). Association of a 59-kilodalton immunophilin with the glucocorticoid receptor complex. *Science*, Vol.256, No.5061, pp. 1315-1318.

Takayama S., Bimston D. N., Matsuzawa S.-i., Freeman B. C., Aime-Sempe C., Xie Z., Morimoto R. I. & Reed J. C. (1997). BAG-1 modulates the chaperone activity of Hsp70/Hsc70. *EMBO J*, Vol.16, No.16, pp. 4887-4896.

Taylor P., Dornan J., Carrello A., Minchin R. F., Ratajczak T. & Walkinshaw M. D. (2001). Two structures of cyclophilin 40: folding and fidelity in the TPR domains. *Structure*, Vol.9, No.5, pp. 431-438.

Tomlins S. A., Mehra R., Rhodes D. R., Cao X., Wang L., Dhanasekaran S. M., Kalyana-Sundaram S., Wei J. T., Rubin M. A., Pienta K. J., Shah R. B. & Chinnaiyan A. M. (2007). Integrative molecular concept modeling of prostate cancer progression. *Nat Genet*, Vol.39, No.1, pp. 41-51.

Tranguch S., Cheung-Flynn J., Daikoku T., Prapapanich V., Cox M. B., Xie H., Wang H., Das S. K., Smith D. F. & Dey S. K. (2005). Cochaperone immunophilin FKBP52 is critical to uterine receptivity for embryo implantation. *PNAS*, Vol.102, No.40, pp. 14326-14331.

Vaughan C. K., Gohlke U., Sobott F., Good V. M., Ali M. M. U., Prodromou C., Robinson C. V., Saibil H. R. & Pearl L. H. (2006). Structure of an Hsp90-Cdc37-Cdk4 complex. *Mol Cell*, Vol.23, No.5, pp. 697-707.

Vaughan C. K., Mollapour M., Smith J. R., Truman A., Hu B., Good V. M., Panaretou B., Neckers L., Clarke P. A., Workman P., Piper P. W., Prodromou C. & Pearl L. H. (2008). Hsp90-dependent activation of protein kinases is regulated by chaperone-targeted dephosphorylation of Cdc37. *Mol Cell*, Vol.31, No.6, pp. 886-895.

Wang Z., Chen W., Kono E., Dang T. & Garabedian M. J. (2007). Modulation of glucocorticoid receptor phosphorylation and transcriptional activity by a C-terminal-associated protein phosphatase. *Mol Endocrinol*, Vol.21, No.3, pp. 625-634.

Ward B. K., Allan R. K., Mok D., Temple S. E., Taylor P., Dornan J., Mark P. J., Shaw D. J., Kumar P., Walkinshaw M. D. & Ratajczak T. (2002). A structure-based mutational analysis of cyclophilin 40 identifies key residues in the core tetratricopeptide repeat domain that mediate binding to Hsp90. *J Biol Chem*, Vol.277, No.43, pp. 40799-40809.

Whitesell L. & Lindquist S. L. (2005). Hsp90 and the chaperoning of cancer. *Nat Rev Cancer*, Vol.5, No.10, pp. 761-772.

Wochnik G. M., Young J. C., Schmidt U., Holsboer F., Hartl F. U. & Rein T. (2004). Inhibition of GR-mediated transcription by p23 requires interaction with Hsp90. *FEBS Lett*, Vol.560, 35-38.

Wu B., Li P., Liu Y., Lou Z., Ding Y., Shu C., Ye S., Bartlam M., Shen B. & Rao Z. (2004). 3D structure of human FK506-binding protein 52: implications for the assembly of the glucocorticoid receptor/Hsp90/immunophilin heterocomplex. *PNAS*, Vol.101, No.22, pp. 8348-8353.

Xu M., Dittmar K. D., Giannoukos G., Pratt W. B. & Simons Jr S. S. (1998). Binding of hsp90 to the glucocorticoid receptor requires a specific 7-amino acid sequence at the amino terminus if the hormone-binding domain. *J Biol Chem*, Vol.273, No.22, pp. 13918-13924.

Yang J., Roe S. M., Cliff M. J., Williams M. A., Ladbury J. E., Cohen P. T. W. & Barford D. (2005). Molecular basis for TPR domain-mediated regulation of protein phosphatase 5. *EMBO J*, Vol.24, No.1, pp. 1-10.

Yang Z., Wolf I. M., Chen H., Periyasamy S., Chen Z., Yong W., Shi S., Zhao W., Xu J., Srivastava A., Sanchez E. R. & Shou W. (2006). FK506-binding protein 52 is essential to uterine reproductive physiology controlled by the progesterone receptor A isoform. *Mol Endocrinol*, Vol.20, No.11, pp. 2682-2694.

Young E. T., Saario J., Kacherovsky N., Chao A., Sloan J. S. & Dombek K. M. (1998). Characterization of a p53-related activation domain in Adr1p that is sufficient for ADR1-dependent gene expression. *J Biol Chem*, Vol.273, No.48, pp. 32080-32087.

Zhang M., Windheim M., Roe S. M., Peggie M., Cohen P., Prodromou C. & Pearl L. H. (2005). Chaperoned ubiquitylation--crystal structures of the CHIP U Box E3 ubiquitin ligase and a CHIP-Ubc13-Uev1a complex. *Mol Cell*, Vol.20, No.4, pp. 525-538.

7

Protein-Protein Interactions and Disease

Aditya Rao, Gopalakrishnan Bulusu,
Rajgopal Srinivasan and Thomas Joseph
Life Sciences Division, TCS Innovation Labs, Tata Consultancy Services, Hyderabad
India

1. Introduction

Protein-protein interactions (PPI), in which, two or more proteins associate with each other by various means, are key to understanding all biological processes that occur within as well as between cells. In effect, biological processes are essentially interactions between multiple proteins (Zhang et al., 2011) with PPI networks controlling the flow of information both within and between biological processes.

Disruptions in PPI networks have been shown to result in diseases. This includes monogenic diseases such as hemophilia where a particular biochemical pathway is disrupted, as well as more complex diseases such as cancer, which involve several signaling pathways (Sam et al., 2007). Conversely, disruption of a set of PPI can lead to a particular disease or, in the case where the set is shared among several networks, to several diseases. While there is a wealth of protein-disease associations in the published literature that have been incorporated in PPI repositories, the challenge is to link such PPI to human disease (Ideker & Sharan, 2008).

In this chapter we discuss several examples of diseases that are caused by disruptions of PPI networks. Our goal is to illustrate through examples how the role of PPI in disease can be studied using a variety of computational tools and data sources. While we discuss tools and data sources that are of general interest, we also discuss methods for studying specific diseases and methods aimed at large scale analysis of PPI data to identify classes of diseases. In each case we provide specific examples from the literature and a brief discussion of the tools used.

2. PPI and disease example

Let us consider cerebral malaria as an example to understand how an analysis of PPI could be used to elucidate the molecular basis of disease. Here, a wide range of experimental and predicted human–*Plasmodium* (host-parasite), human-human (host-host) and *Plasmodium-Plasmodium* (parasite-parasite) PPI are combined and analyzed in the context of key events and processes of cerebral malaria, a dangerous infectious disease (Rao et al., 2010).

Cerebral malaria is a severe form of malarial infection, characterized by cerebral complications, such as neuronal damage and coma (Moxon et al., 2009). The disease is characterized by processes such as sequestration of infected red blood cells to cerebral capillaries and venules, systemic inflammation, hemostasis dysfunction and neuronal damage (van der Heyde et al., 2006; Wilson et al., 2008). PPI datasets from different sources were first obtained, summarized in Table 1. Since each dataset uses a different nomenclature system for the human and parasite proteins, a crucial step was to normalize all datasets using common gene names. This enabled creation of a unified host-parasite PPI dataset.

No	Source	Reference
1	Davis dataset	Davis et al., 2007
2	Dyer dataset	Dyer et al., 2007
3	Krishnadev dataset	Krishnadev & Srinivasan, 2008
4	Vignali dataset	Vignali et al., 2008
5	Literature PPI data	In-house manual curation

Table 1. Protein-Protein interaction datasets used in the cerebral malaria example.

An automated literature retrieval module was developed using Entrez Programming Utilities (Sayers et al., 2010) to retrieve the list of full-text articles relevant to the malarial parasite. This article set was pruned using the Medical Subject Headings (MeSH) controlled vocabulary for articles relevant to cerebral malaria. The resultant set was augmented by articles retrieved from the Google Scholar database using appropriate disease-specific query terms such as systemic inflammation, hemostasis dysfunction etc. This article corpus had two main uses:

- For extracting biochemical and signaling events of relevance in cerebral malaria.
- Identifying pairs of interacting proteins within the host, within the parasite and between host and parasite.

Gene Ontology (GO) cellular component annotations from PlasmoDB (Aurrecoechea et al., 2009), a comprehensive *Plasmodium* resource, were used to prune the unified PPI dataset using the approach of Mahdavi & Lin (2007). In the case of PPI involving parasite proteins, only those proteins that were annotated to be present on the parasite surface or were reported to be released during the relevant stage of the parasite were considered (Lyon et al., 1986). For the human protein annotations, tissue-specific annotations from UniProt (Hubbard et al., 2009) were used in the pruning process.

The resultant PPI subset was then analyzed by mapping the PPI to key events that influence the processes of the disease, as identified from the key review articles. The analysis showed the potential significance of apolipoproteins and heat-shock proteins on efficient *Plasmodium falciparum* erythrocyte membrane protein 1 (PfEMP1) presentation, role of the merozoite surface protein (MSP-1) in platelet activation, the role of albumin in astrocyte dysfunction and the effect of parasite proteins in transforming growth factor (TGF)-β regulation. The linking of these PPI to molecular events associated with the disease pathogenesis provides a basis for further experiments to determine the molecular basis of this fatal disease.

3. Tools of the trade

From the example, it is clear that the underpinnings for mapping PPI to disease are: (a) access to various repositories of PPI and (b) ability to filter these PPI in the context of disease and (c) using different tools for visualizing and analyzing the PPI in the context of diseases. Let us consider each of these in detail.

3.1 PPI repositories

There are a host of repositories that house experimental and predicted PPI data. The cerebral malaria example above considered malaria-specific PPI datasets. However, generic datasets such as BIND, DIP, HPRD, MINT, MIPS and STRING usually have the necessary PPI coverage required for a variety of disease studies.

The Biomolecular Interaction Network Database (BIND), a constituent database of the Biomolecular Object Network Databank, makes available a comprehensive collection of information for specific molecules such as proteins and small molecules (Bader et al., 2003). BIND has been one of the major sources of curated biomolecular interactions, especially PPI. The Database of Interacting Proteins (DIP) contains experimentally determined PPI (Salwinski et al., 2004). It has been created using both manual curation and computational approaches. The Human Protein Reference Database (HPRD) provides a platform to visually depict and integrate information, which are manually curated, pertaining to domain architecture, post-translational modifications, PPI networks and disease association for each protein of the human proteome (Prasad et al., 2009).

The Molecular INTeraction database (MINT) contains experimentally verified PPI that have been manually curated from the scientific literature (Ceol et al., 2010). The Mammalian Protein-Protein Interaction (MIPS) database is a collection of manually curated high-quality PPI data collected from the scientific literature by expert curators (Pagel et al., 2005). STRING is a database of known and predicted protein interactions (Szklarczyk et al., 2011). The interactions include direct (physical) as well as indirect (functional) associations.

Composite PPI resources are also available that integrate PPI data from some of these databases into a single resource. APID (Agile Protein Interaction DataAnalyzer), for instance, is one such resource that integrates experimentally validated PPI from databases such as BIND, DIP, HPRD and MINT, amongst others (Prieto et al., 2006). Protein Interaction Network Analysis (PINA) platform is another example of a composite PPI resource that integrates interactions from MINT, DIP, HPRD and MIPS, amongst others (Wu et al., 2009).

3.2 Integration and filtering of PPI

Databases and tools such as Reactome, GO, MeSH and the Entrez Programming Utilities are crucial for filtering the large number of PPI to obtain a PPI network relevant to a specific disease.

Reactome is a database of biological pathways from various organisms, especially humans (Matthews et al., 2009). This is manually curated by experts. It contains various entities such as proteins, chemicals, localization data, etc. The information in Reactome is cross-referenced to various standard bioinformatics databases such as Entrez Gene, UniProt, Ensembl, etc. The GO project attempts to standardize the description of gene and gene products across species and databases (The Gene Ontology Consortium, 2000). It consists of three ontologies that describe genes and gene products in relation to biological processes, molecular functions and cellular components.

MeSH (http://www.nlm.nih.gov/mesh/) is a controlled vocabulary thesaurus maintained by the National Library of Medicine. It is made up of terms naming descriptors in a hierarchical structure that permits searching at various levels of specificity. It currently has 16 major tree headings including "Diseases" and "Chemicals and Drugs". MeSH terms are used in various methods and tools to filter articles/abstracts and other data. Online Mendelian Inheritance in Man (OMIM) is a database of known Mendelian disorders and their related genes (Hamosh et al., 2002). Currently, there are around 12,000 genes described in this database. OMIM provides information on genotype-phenotype relationships in human Mendelian diseases.

Specific tools are available to access some of these databases, such as the Entrez Programming Utilities (Sayers et al., 2010). These are a set of server-side programs enabling a stable interface to utilize the Entrez query and database system at the National Center for Biotechnology Information. There are currently 38 databases in the Entrez system with a wide variety of information on nucleotide and protein structure and sequences, 3D-molecular structures, disease information and biomedical literature etc.

3.3 Visualization tools

Important tools that could aid in mapping PPI to disease include:

- Cytoscape (Shannon et al., 2003) for visualizing PPI datasets with nodes representing biological entities and edges representing the relationships between these entities
- Cell Circuits (Mak et al., 2007) for comparison of hand-curated pathway models to hypothetical models derived from large-scale 'omic' data.

Cytoscape plugins such as APID2NET (Hernandez-Toro et al., 2007) and PRINCIPLE (Gottlieb et al., 2011) are also very pertinent. APID2NET retrieves PPI data from the APID server for further analysis within the Cytoscape environment. PRINCIPLE, discussed later in this chapter, is built specifically for exploring PPI-disease associations. Given any disease as a query term, it provides a list of top-ranking genes associated with this disease and a Cytoscape visualization of the sub-networks formed by these genes and their direct interacting neighbors.

IPA (Ingenuity Systems, www.ingenuity.com) is an example of a commercially available platform that enables visualization of dynamically constructed pathway and network models.

4. PPI from literature

What happens when the PPI repositories do not have adequate coverage of the organism or specific protein-set under study? One possibility is that although such repositories do not have these PPI, the PPI have actually been reported in literature. One just needs to go look for them!

Let us consider an example of using text-mining to extract such PPI. In the cerebral malaria study, a basic text-mining approach has been used (Rao et. al., 2010). The article corpus was first checked for article-level co-occurrence of pairs of proteins. Full-text articles, wherever available, are automatically downloaded from the respective journal websites as Portable Document Format (PDF) files and converted to text format using the XPDF conversion utility (The FooLabs, http://www.foolabs.com/xpdf). All parasite and host proteins that occur in the full-text of each article were identified using a dictionary lookup approach, with PlasmoDB and UniProt/Ensembl being used to create the parasite and human protein dictionaries respectively. Only those articles that had at least one protein pair (host-parasite, host-host or parasite-parasite) were considered for further analysis.

Özgür et al. (2008) propose a more detailed approach based on integrating automatic text mining and network analysis methods to extract known disease genes and to predict unknown disease genes. They started by collecting an initial set of seed genes known to be related to a disease from curated databases such as OMIM. A disease specific gene network was created using advanced natural language processing techniques that capture both gene names as well as the semantic associations between them.

5. PPI Networks and SNPs

Genome wide measurement technologies such as microarrays have provided an opportunity to identify genes that are mutated or differentially expressed. In particular, SNP-arrays have been very useful in such studies and have resulted in identification of several genes that are associated with disease-risk or poor prognosis (Karinen et al., 2011). Such genes typically affect cellular functions by altering signaling in regulatory PPI networks.

The mainstay of this approach is the fact that genes related to the same disease are also known to have protein products that physically interact (Navlakha & Kingsford, 2010). However, that by itself is only one crucial component. The other important component is that a genetic disease is associated with a linkage interval on the chromosome if SNPs in the interval are correlated with an increased susceptibility to the disease. These linkage intervals define a potential disease-causing gene set. The computational approach boils down to using both these sources of information — PPI networks and linkage intervals to predict relationships between genes and diseases.

Let us look at a method called CANGES to identify the genetic basis of disease (Karinen et al., 2011). The strength of the method lies in its ability to cohesively integrate many different pieces of information to arrive at testable hypotheses. Genome wide association studies have identified many variations that are possibly linked to one or more diseases. How does one go about prioritizing these variations to get to a set of genes that cause the disease? Clearly, one needs to bring in other known information to help arrive at a decision. The CANGES method combines pathway data, PPI data and genetic variation data with analytical tools to rapidly evaluate the disease causing potential of variations and thus focus attention on one or a few genes. Using this method, a set of 158 SNPs in the p53 gene were identified that plays a central role in cancers. These SNPs are likely to have pathogenic consequences. The same method has also been used, in conjunction with clinical patient data, to identify genes associated with glioblastoma multiforme. It is clear that in the future we will see many more such methods which bring together PPI with several other pieces of information and analytical tools to identify disease genes and gene networks.

Several computational methods can be used to identify causal genes central to gene-disease relationships from large PPI networks. The methods include network neighbors and neighborhood methods, unsupervised graph partitioning and Markov clustering, semi-supervised graph partitioning, random walks, network flow methods and several of their variants (Navlakha & Kingsford, 2010). Navlakha & Kingsford tested these on two large PPI networks: (a) one derived from the Human Protein Reference Database, consisting of 8776 proteins and 35,820 PPI and (b) the other derived from Online Predicted Human Interaction Database containing 9842 proteins and 73130 PPI. Annotations from OMIM were used to associate diseases with genes and linkage intervals. They observed that the performance of most methods showed a significant correlation with neighborhood homophily. Based on this, they suggest that homophily could be used to assess the quality of network-based predictions of disease-protein relationships. They also observed that the individual methods capture different kinds of structure in the network and these unique abilities can be used together in a consensus method to enhance prediction quality.

6. Structural significance of PPI

One important disease class in which a study of PPI could shed light is cancer. Let us look at a study that analyzed cancer proteins in human PPI networks (Kar et al., 2009). This study is important from a methodology perspective as it uses structural properties of the proteins present in the PPI network. Integrating three-dimensional protein structural information into PPI networks revealed important aspects about cancer-related proteins. Analysis of the structural properties of cancer-related interface proteins showed that the interfaces are, on an average, smaller in size, more planar, less tightly packed and more hydrophilic than those of non-cancer proteins. For instance, in a breast cancer network used in the study, there was significant accuracy in discriminating cancer-protein interfaces from the non-cancer interfaces. Thus, there seems to be a clear distinction between the interfaces.

In addition, they observed that cancer-related proteins tend to interact with their partners via multi-interface hubs, which comprise 56% of cancer-related proteins. Cancer protein networks are therefore more enriched in multi-interface proteins. Cancer proteins, in general, are longer and have larger surface areas. Thus, to participate in many PPI at the same time, these tend to be multi-interface hubs, with distinct interfaces interacting with different proteins.

The processes involved in obtaining relevant PPI with regard to a disease are shown in Figure 1.

7. PPI common across diseases

The hitherto discussed examples link diseases with their possible proteomic underpinnings. Research is also underway that focuses on bridging the gap between PPI and their association to different diseases. The goal is to bring out underlying PPI that are common amongst different sets of diseases. Diseases with overlapping clinical phenotypes are caused by mutations in functionally related genes. Since PPI are the strongest manifestation of a functional relationship between disease genes, applying a network model is an effective approach for revealing the associations among diseases (Zhang et al., 2011).

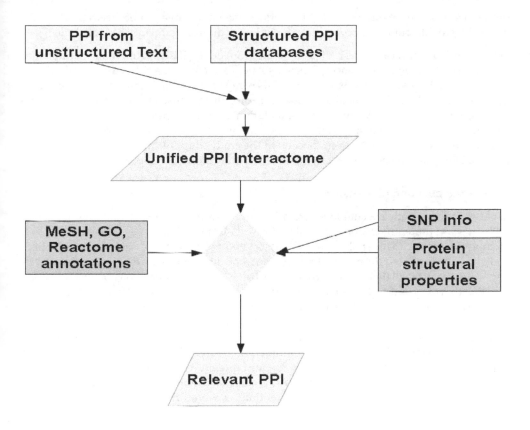

Fig. 1. Processes involved in obtaining relevant PPI with regard to a disease.

7.1 Background

Traditionally, diseases are defined as 'similar' mainly by their clinical appearance, with no correlation to underlying molecular processes. Conceptually, each monogenic disease has a collection of specific phenotypic features. This is true for about 2000 human single gene diseases with a defined genetic phenotype. Syndromes are defined in medicine as a set of phenotypes which, occurring together, serve to define a trait or disease. However, phenotypes very often overlap in the case of many syndromes. Recognition of this overlap brought about the concept of 'syndrome families' taking into account the common features shared between diseases (Sam et al., 2007).

The clustering of syndromes into these families in combination with genetic insights has led to the discovery that what were often thought as two different disorders were really variable expressions of the same disorder. On the other hand, it has long been known that mutations at different loci can lead to the same genetic disease. It has also been hypothesized that this genetic heterogeneity has its roots at the PPI level, suggesting that other genes associated with the phenotype also have some functional role. Therefore, it is plausible that functional properties of shared molecular networks reflect phenotypic overlap of diseases. Thus, PPI networks provide unique opportunities for exploring disease pathways (Sam et al., 2007).

Let us continue with cancer as a disease theme. Sam et al. (2007) highlight an example that links Fanconi's Anemia and cancer. Fanconi's Anemia is a hereditary DNA-repair deficiency disease characterized by defects in a set of DNA repair proteins, leading to, among others, hypersensitivity to DNA damaging agents. This disorder is caused by a mutation in any one of the genes in Fanconi's Anemia complementation group. Symptoms of the disease include anemia, several congenital malformations, etc. Importantly, patients suffering from it exhibit a strong predisposition to different cancers. In the study, this link was substantiated with 14 potential PPI common between Fanconi's Anemia and colorectal neoplasms.

7.2 PPI and common phenotypes

Let us consider another example where a PPI network has been systematically combined with disease-protein relationship data derived from mining GO annotations with phenotypic context (Sam et al., 2007). PPI associated with pairs of diseases were identified and the statistical significance of the occurrence of interactions in the protein interaction knowledgebase calculated. This study demonstrates that the associations between diseases are directly correlated to their underlying PPI networks. A subset of PhenoGO (Lussier et al., 2006; Sam et al., 2009) restricted to human diseases was examined to study the relationships between diseases according to the following criteria. Two basic types of relationships were considered, which determine whether two diseases share PPI networks: a) an identity relationship where common proteins are shared by two diseases, and b) direct interactions between protein A of one disease and protein B of another. A total of 10 pairs of diseases were identified that are significantly correlated due to their shared proteins and PPI. These pairs were analyzed based on mentions in literature, and their correlations were confirmed.

Xeroderma pigmentosum and Cockayne syndrome provide an example of how two diseases are correlated through their PPI networks. Xeroderma pigmentosum is a disorder causing susceptibility of the skin to ultraviolet radiation as a result of deficiencies in one of the XPA-XPG complementation group genes involved in nucleotide excision repair (NER). Cockayne syndrome results from deficiencies in transcription-coupled repair genes, like ERCC6 and ERCC8, leading to a number of conditions including abnormal sensitivity to sunlight (Sam et al., 2007; Spivak et al., 2004). There were 27 direct PPI and 5 common proteins shared amongst these two diseases. Majority of the proteins in the common networks between the two diseases are related to DNA repair processes - Global Genomic NER and Transcription-coupled NER. While the Global Genomic NER repairs lesions from non-transcribed regions of genome independent to transcription, the Transcription-coupled NER repairs UV induced damage in the transcribed strands of active genes. Both the diseases are seen to be associated with these processes, suggesting defects in the DNA damage repair processes are the cause of the diseases.

7.3 Of PRINCE and PRINCIPLEs

PRINCIPLE is very relevant tool specifically built for finding out common diseases based on PPI. It is a Cytoscape plugin implementation of the PRINCE algorithm (Vanunu, et al.,

2010). Given a query disease, it provides a list of top ranking genes associated with it and an additional visualization of the sub-networks formed by these top ranking genes and their direct interacting neighbors. The underlying logic is that genes causing similar diseases often lie close to one another in a PPI network (Oti & Brunner, 2007; Oti, et al., 2006). Given a disease as the query term, PRINCE (a) identifies a set of phenotypically similar diseases, (b) retrieves the known causal genes of these diseases based on their similarity to the query and (c) propagates the scores of the prior set of genes over a human PPI network to provide association scores for all genes. It uses a comprehensive set of weighted PPI compiled from disparate sources (Vanunu, et al., 2010), disease-disease similarity measures (van Driel, et al., 2006), and on the disease-gene associations present in OMIM.

7.4 Human disease network – The holy grail!

Zhang et al. (2011) constructed an expanded Human Disease Network by combining disease-gene information with PPI information. Work such as this is very important, since a network model to represent relationships between diseases is very useful in looking at relationships amongst diseases on a large scale. Analysis of the network's topological features and functional properties showed that the network was hierarchical. Most diseases in the network were connected to only a few diseases, while a small set of diseases were linked to many different diseases. Also, diseases in a specific disease category tended to cluster together, and genes associated with the same disease were functionally related. While this might intuitively sound obvious, it establishes a molecular basis for disease-disease associations.

The limitation of the network is that only known and available disease phenotypic data has been incorporated. However, as more data is made available in databases and in literature, this network provides an ideal template to analyze relationships amongst diseases from a PPI perspective.

8. The road ahead

This is a new field and there are many more approaches than what has been brought out in this chapter. For instance, Bandyopadhyay et al. (2006) use a network analysis of gene expression and PPI data to identify active pathways related to HIV pathogenesis. A functional analysis of the detected sub-networks provides useful insights into various stages of the HIV replication cycle. Chen et al. (2006) developed a framework to mine disease-related proteins from OMIM and PPI data. They demonstrate the power of their method by applying it to Alzheimer's disease. The key to their method is a scoring function that ranks proteins according to their relevance to a particular disease pathway.

Methods to arrive at high-precision predictions that are translatable to effective steps in disease prevention, diagnosis and prognosis should be the goal of PPI studies. The generated leads should be tested experimentally to determine their relevance.

9. References

Aurrecoechea C, Brestelli J, Brunk BP, Dommer J, Fischer S, Gajria B, Gao X, Gingle A, Grant G, Harb OS, Heiges M, Innamorato F, Iodice J, Kissinger JC, Kraemer E, Li W,

Miller JA, Nayak V, Pennington C, Pinney DF, Roos DS, Ross C, Stoeckert CJ Jr, Treatman C & Wang H. (2009). PlasmoDB: a functional genomic database for malaria parasites. *Nucleic Acids Research* D539-543.

Bader GD, Betel D & Hogue CW. (2003). BIND: the Biomolecular Interaction Network Database. *Nucleic Acids Research* 31:248-250.

Bandyopadhyay S, Kelley R & Ideker T. (2006). Discovering regulated networks during IV-1 latency and reactivation. *Pacific Symposium on Biocomputing* 354-366.

Ceol A, ChatrAryamontri A, Licata L, Peluso D, Briganti L, Perfetto L, Castagnoli L & Cesareni G. (2010). MINT, the molecular interaction database: 2009 update. *Nucleic Acids Research* (Database issue):D532-D539.

Chen JY, Shen C & Sivachenko AY. (2006). Mining Alzheimer disease relevant proteins from integrated protein interactome data. *Pacific Symposium on Biocomputing* 367-378.

Davis FP, Barkan DT, Eswar N, McKerrow JH & Sali A. (2007). Host pathogen protein interactions predicted by comparative modeling. *Protein Science* 16:2585-2596.

Dyer MD, Murali TM & Sobral BW. (2007). Computational prediction of host pathogen protein-protein interactions. *Bioinformatics* 23:59-66.

Gottlieb A., Magger O., Berman I., Ruppin E. & Sharan R. (2011). PRINCIPLE: A tool for associating genes with diseases via network propagation. *Bioinformatics.doi: 10.1093/bioinformatics/btr584.*

Hernandez-Toro J, Prieto C & De las Rivas J. (2007). APID2NET: unified interactome graphic analyzer. *Bioinformatics* 23:2495-2497.

Hamosh A, Scott AF, Amberger J, Bocchini C, Valle D & McKusick VA. (2002). Online Mendelian Inheritance in Man (OMIM), a knowledgebase of human genes and genetic disorders. *Nucleic Acids Research* 30:52-55.

Hubbard TJ, Aken BL, Ayling S, Ballester B, Beal K, Bragin E, Brent S, Chen Y, Clapham P, Clarke L, Coates G, Fairley S, Fitzgerald S, Fernandez-Banet J, Gordon L, Graf S, Haider S, Hammond M, Holland R, Howe K, Jenkinson A, Johnson N, Kahari A, Keefe D, Keenan S, Kinsella R, Kokocinski F, Kulesha E, Lawson D, Longden I, Megy K, Meidl P, Overduin B, Parker A, Pritchard B, Rios D, Schuster M, Slater G, Smedley D, Spooner W, Spudich G, Trevanion S, Vilella A, Vogel J, White S, Wilder S, Zadissa A, Birney E, Cunningham F, Curwen V, Durbin R, Fernandez-Suarez XM, Herrero J, Kasprzyk A, Proctor G, Smith J, Searle S & Flicek P. (2009). Ensembl 2009. *Nucleic Acids Research* 37:D690-697.

Ideker T & Sharan R. (2008). Protein networks in disease. *Genome Research* 18: 644-652.

Kann M, Ofran Y, Punta M & Radivojac P. (2006). Protein Interactions and disease. *Pacific Symposium on Biocomputing* 11:351-353.

Kar G, Gursoy A & Keskin O. (2009). Human cancer protein-protein interaction network: a structural perspective. *PLoS Computational Biology* 5:e1000601.

Karinen S, Heikkinen T, Nevanlinna H & Hautaniemi S. (2011). Data integration workflow for search of disease driving genes and genetic variants. *PLoS One* 6:e18636.

Krishnadev O & Srinivasan N. (2008). A data integration approach to predict host- pathogen protein-protein interactions: application to recognize protein interactions between human and a malarial parasite. *In Silico Biology* 8:235-250.

Lussier Y, Borlawsky T, Rappaport D, Liu Y, Friedman C. (2006). PhenoGO: assigning phenotypic context to gene ontology annotations with natural language processing. *Pacific Symposium on Biocomputing* 64-75.

Lyon JA, Haynes JD, Diggs CL, Chulay JD & Pratt-Rossiter JM. (1986). *Plasmodium falciparum* antigens synthesized by schizonts and stabilized at the merozoite surface by antibodies when schizonts mature in the presence of growth inhibitory immune serum. *Journal of Immunology* 136:2252-2258.

Mahdavi MA & Lin YH. (2007). False positive reduction in protein-protein interaction predictions using gene ontology annotations. *BMC Bioinformatics* 8:262.

Mak HC, Daly M, Gruebel B & Ideker T. (2007). CellCircuits: a database of protein network models. *Nucleic Acids Research* 35:D538-545.

Matthews L, Gopinath G, Gillespie M, Caudy M, Croft D, de Bono B, Garapati P, Hemish J, Hermjakob H, Jassal B, Kanapin A, Lewis S, Mahajan S, May B, Schmidt E, Vastrik I, Wu G, Birney E, Stein L & D'Eustachio P. (2009). Reactome knowledgebase of human biological pathways and processes. *Nucleic Acids Research* (Database issue): D619-622.

Moxon CA, Heyderman RS & Wassmer SC. (2009). Dysregulation of coagulation in cerebral malaria. *Molecular and Biochemical Parasitology* 166:99-108.

Navlakha S & Kingsford C. (2010). The power of protein interaction networks for associating genes with diseases. *Bioinformatics* 26:1057-1063.

Oti M & Brunner HG. (2007). The modular nature of genetic diseases. *Clinical Genetics* 71(1):1-11.

Oti M, Snel B, Huynen MA & Brunner HG. (2006). Predicting disease genes using protein-protein interactions. *Journal of Medical Genetics* 43(8):691-698.

Özgür A, Vu T, Erkan G & Radev DR. (2008). Identifying gene-disease associations using centrality on a literature mined gene-interaction network. *Bioinformatics* 24, i277-i1285.

Pagel P, Kovac S, Oesterheld M, Brauner B, Dunger-Kaltenbach I, Frishman G, Montrone C, Mark P, Stümpflen V, Mewes HW, Ruepp A, Frishman D (2005). The MIPS mammalian protein-protein interaction database. *Bioinformatics* 21(6):832-834.

Prasad TSK, Goel R, Kandasamy K, Keerthikumar S, Kumar S, Mathivanan S, Telikicherla D, Raju R, Shafreen B, Venugopal A, Balakrishnan L, Marimuthu A, Banerjee S, Somanathan DS, Sebastian A, Rani S, Ray S, Harrys Kishore CJ, Kanth S, Ahmed M, Kashyap MK, Mohmood R, Ramachandra YL, Krishna V, Rahiman BA, Mohan S, Ranganathan P, Ramabadran S, Chaerkady R & Pandey A. (2009). Human Protein Reference Database - 2009 Update. *Nucleic Acids Research* 37: D767-72.

Prieto C.& De Las Rivas J. (2006). APID: Agile Protein Interaction DataAnalyzer.Nucl. Acids Research 34:W298-W302.

Rao A, Kumar MK, Joseph T & Bulusu G. (2010). Cerebral malaria: Insights from host-parasite protein protein interactions. *Malaria Journal* 9:155.

Salwinski L, Miller CS, Smith AJ, Pettit FK, Bowie JU & Eisenberg D. (2004). TheDatabase of Interacting Proteins: 2004 update. *Nucleic Acids Research* 32:D449-451.

Sam L, Liu Y, Li J, Friedman C & Lussier YA. (2007). Discovery of protein interaction networks shared by diseases. *Pacific Symposium on Biocomputing* 76-87.

Sam LT, Mendonça EA, Li J, Blake J, Friedman C & Lussier YA. (2009). PhenoGO: an integrated resource for the multiscale mining of clinical and biological data. *BMC Bioinformatics* 10:S8.

Sayers EW, Barrett T, Benson DA, Bolton E, Bryant SH, Canese K, Chetvernin V, Church DM, Dicuccio M, Federhen S, Feolo M, Geer LY, Helmberg W, Kapustin Y, Landsman D, Lipman DJ, Lu Z, Madden TL, Madej T, Maglott DR, Marchler-Bauer A, Miller V,

Mizrachi I, Ostell J, Panchenko A, Pruitt KD, Schuler GD, Sequeira E, Sherry ST, Shumway M, Sirotkin K, Slotta D, Souvorov A, Starchenko G, Tatusova TA, Wagner L, Wang Y, John Wilbur W, Yaschenko E & Ye J. (2010). Database resources of the National Center for Biotechnology Information. *Nucleic Acids Research* 38:D5-16.

Shannon P, Markiel A, Ozier O, Baliga NS, Wang JT, Ramage D, Amin N, Schwikowski B & Ideker T. (2003). Cytoscape: a software environment for integrated models of biomolecular interaction networks. *Genome Research* 13:2498-2504.

Spivak G. (2004). The many faces of Cockayne syndrome. *Proceedings of the ational Academy of Sciences of the United States of America* 101(43):15273-15274.

Szklarczyk D, Franceschini A, Kuhn M, Simonovic M, Roth A, Minguez P, Doerks T, Stark M, Muller J, Bork P, Jensen LJ, von Mering C. (2011). The STRING database in 2011: functional interaction networks of proteins, globally integrated and scored. *Nucleic Acids Research* 39(Database issue):D561-568.

The Gene Ontology Consortium. (2000). Gene ontology: tool for the unification of biology. *Nature Genetics* 25(1):25-29.

The UniProt Consortium. (2009). The Universal Protein Resource (UniProt). *Nucleic Acids Research* 37:D169-D174.

van der Heyde HC, Nolan J, Combes V, Gramaglia I & Grau GE. (2006). A unified hypothesis for the genesis of cerebral malaria: sequestration, inflammation   and hemostasis leading to microcirculatory dysfunction. *Trends in Parasitology* 22:503-508.

van Driel MA, Bruggeman J, Vriend G, Brunner HG & Leunissen JA. (2006). A text-mining analysis of the human phenome. *European Journal of Human Genetics* 14(5):535-542.

Vanunu, O., Magger, O., Ruppin, E., Shlomi, T. & Sharan R. Associating genes and protein complexes with disease via network propagation. (2010). *PLoS Computational Biology* 6, e1000641.

Vignali M, McKinlay A, LaCount DJ, Chettier R, Bell R, Sahasrabudhe S, Hughes RE & Fields S. (2008). Interaction of an atypical *Plasmodium falciparum* ETRAMP with human apolipoproteins. *Malaria Journal* 7:211.

Wilson NO, Huang MB, Anderson W, Bond V, Powell M, Thompson WE, Armah HB, Adjei AA, Gyasi R, Tettey Y & Stiles JK. (2008). Soluble factors from *Plasmodium falciparum*-infected erythrocytes induce apoptosis in human brain vascular endothelial and neuroglia cells. *Molecular and Biochemical Parasitology* 162:172-176.

Wu J, Vallenius T, Ovaska K, Westermarck J, Mäkelä TP, Hautaniemi S. (2009). Integrated network analysis platform for protein-protein interactions *Nature Methods* 6(1):75-77.

Zhang X, Zhang R, Jiang Y, Sun P, Tang G, Wang X, Lv H & Li X. (2011). The expanded human disease network combining protein-protein interaction information. *European Journal of Human Genetics* 19:783-788.

8

Direct Visualization of Single-Molecule DNA-Binding Proteins Along DNA to Understand DNA–Protein Interactions

Hiroaki Yokota
Institute for Integrated Cell-Material Sciences,
Kyoto University
Japan

1. Introduction

Single-molecule fluorescence imaging has recently developed into a powerful method for studying biophysical and biochemical phenomena (Moerner, 2007), including DNA metabolism (Ha, 2004; Hilario & Kowalczykowski, 2010; Zlatanova & van Holde, 2006). While classical biochemical methods yield parameters that are ensemble averaged, single-molecule fluorescence imaging can be used to observe real-time behavior of individual biomolecules, allowing us to study their dynamic characteristics in great detail. Among the various biomolecules, protein molecules, which play central roles in many biological functions, are the prime targets for single-molecule imaging. Direct observations of single-molecule DNA-binding proteins acting on their DNA targets in real time have provided new insights into DNA metabolism. In this chapter, I focus primarily on recent advances in direct visualization of single-molecule DNA-binding proteins *in vitro*, especially on the key techniques employed for this visualization.

2. Fluorophores and fluorescence labeling methods

Direct visualization of protein dynamics by single-molecule fluorescence imaging requires labeling of target protein molecules with a fluorophore. To be used in single-molecule imaging, fluorophores must be (1) bright (have high extinction coefficients and high quantum yield), (2) photostable, and (3) relatively small so as not to perturb the functions of target protein molecules. With regard to photostability, fluorophores in single-molecule imaging tend to undergo photobleaching and blinking (repetitive fluorescence turning-on and -off) because of the high-power excitation. Agents that minimize such photophysical events have been reported (Aitken et al., 2008; Dave et al., 2009; Harada et al., 1990; Rasnik et al., 2006). In general, two classes of fluorophores are used for single-molecule fluorescence imaging: organic small-molecule fluorophores and quantum dots (Qdots) (Table 1). Despite the advantage of the labeling capability with genetic engineering, fluorescent proteins are not popularly applied to single-molecule fluorescence imaging because of their lower intensity and instability of their fluorescence emission. This section briefly describes fluorescence properties of dyes and Qdots and their fluorescence labeling methods.

2.1 Organic small-molecule fluorophores (dyes)

Organic small-molecule fluorophores or dyes with a molecular weight < 1 kDa are mainly used for covalent labeling of protein molecules. Frequently used dyes in single-molecule imaging are Cy3, Cy5, and Alexa derivatives. They emit sufficeint photons to be detected, but subsequently photobleach in approximately 1 minute, and often exhibit blinking (Table 1).

The most popular site-specific labeling method with dyes involves labeling the sulfhydryl group of the cysteine residues with a dye containing a maleimide group. The double bond of the maleimide group may undergo an alkylation reaction with the sulfhydryl group to form a stable thioester bond. The reaction is specific for thiols in the physiological pH range of 6.5–7.5. At pH 7.0, the reaction proceeds 1,000 times faster than its reaction with amines (Hermanson, 2008). Another popular labeling method, which is less site-specific, involves labeling the amine group of amino acids in protein molecules with a dye containing an N-hydroxysuccinimide ester (NHS). NHS ester reacts principally with the ε–amines of the lysine side chains and α–amines at the N-terminals (Hermanson, 2008).

2.2 Quantum dots (Qdots)

Qdots are inorganic semiconductor nanocrystals, typically composed of a cadmium selenide (CdSe) core and a zinc sulphide (ZnS) shell measuring 10–20 nm in diameter (Michalet et al., 2005). They are commonly used in single-molecule imaging owing to their resistance to photobleaching and extreme brightness (Table 1). They are characterized by broad absorption profiles, high extinction coefficients, and narrow and spectrally tunable emission profiles, depending on their sizes. Although they are resistant to photobleaching, they often exhibit blinking, which seems to be related to charging of the nanocrystal upon excitation (Table 1). However, recently, the problem of blinking was reported to be overcome by the use of a nanocrystalline CdZnSe core capped with a ZnSe semiconductor shell, in which the transition between ZnSe and CdSe is not abrupt, but radially graded (Wang et al., 2009).

Qdots are usually labeled to protein molecules via immunolabeling. The most popular method is based on an avidin-biotin interaction, an antigen-antibody reaction with the highest affinity. In this method, target protein molecules, which are conjugated with a biotin by chemical crosslinking or genetic engineering, can be labeled with avidin-coated Qdots that are commercially available.

2.3 Advances in site-specific labeling methods

Recent advances in tagging technologies based on genetic have engineering enabled site-specific targeting of fluorophores to protein molecules. One approach is to fuse the target protein molecule to a peptide or a protein recognition sequence, which then recruits a fluorophore coated with the proteins that have affinity for the specific peptide or the protein recognition sequence. For labeling of a Qdot used in single-molecule visualization, HA tag (Kad et al., 2010), Flag tag and HA tag (Gorman et al., 2010), and His tag (Dunn et al., 2011) have been used. Enzyme-mediated covalent protein labeling is also used, in which a recognition peptide is fused to the protein of interest and a natural or engineered enzyme ligates the small-molecule probe to the recognition peptide. This approach can confer highly specific and rapid labeling, with the benefit of a small directing peptide sequence. This approach involved the following pairs: Halo tag and dehalogenase, biotin ligase acceptor peptide and biotin ligase, and SNAP tag and O^6-alkylguanine-DNA alkyltransferase (Fernández-Suárez & Ting, 2008).

Fluorophore	λ_{em} (nm)	M. W. or size	ε ($M^{-1}cm^{-1}$)	Brightness	Resistance to photobleaching	Resistance to blinking
Dye						
Cy3 maleimide	570	791.0 Da	150,000 (550 nm)	++	++	++
Cy5 maleimide	670	817.0 Da	250,000 (649 nm)	++	+	+
Quantum dot						
Qdot655	~655	~15 to 20 nm	2,100,000 (532 nm)	+++	+++	+
Fluorescent protein						
EGFP	507	32.7 kDa	56,000 (484 nm)	+	+	+

Table 1. Size (or molecular weight) and fluorescence properties of commonly used dyes and a Qdot compared to those of a fluorescence protein.

3. Surface-coating methods

Minimization of the background noise arising from non-specific adsorption of the fluorescently labeled protein molecules on glass substrates is essential for their single-molecule fluorescence visualization. The non-specifically adsorbed protein molecules severely interfere with the visualization because the fluorescence associated with these molecules confounds the fluorescence signal from the target protein molecules. Poly (ethylene glycol) (PEG) and lipid are commonly used for reducing the non-specific protein adsorption on the glass substrate.

3.1 Poly (Ethylene Glycol) (PEG)

PEG is a biocompatible polymer that exhibits protein and cell resistance when immobilized onto metal (Prime & Whitesides, 1991), plastic (Ito et al., 2007), and glass surfaces (Cuvelier et al., 2003), which serves many applications especially in biosensors and medical devices. The feature of PEG comes from its high hydrophilicity and appreciable chain flexibility that induce an effective exclusion volume effect (Harris, 1992). To my knowledge, Ha et al. first used PEG coated glass substrates for single-molecule fluorescence imaging (Ha et al., 2002). The key factors affecting the suppression of nonspecific adsorption by PEG are its length and density (Harris, 1992), both of which are trade-offs. An increase in the chain length of PEG to construct a defined tethered chain layer results in a decrease in the density of the PEG chain due to the exclusion volume effect. PEG with a molecular weight of 5,000 appears to find a compromise between the trade-offs to maximize the suppression of non-specific adsorption of protein molecules on a glass surface (Heyes et al., 2004; Malmstena et al., 1998; McNamee et al., 2007; Pasche et al., 2003; Yang et al., 1999). A general approach to the immobilization of PEG onto a glass surface involves coupling of PEG to amine groups that have been conjugated on the surface by silanization (Figure 1a). Another approach involves coupling of PEG through poly (L-lysine) adsorption on the surface (Figure 1b). The inclusion of a small fraction of biotinylated PEG provides anchor points for tethering DNA (Section 4).

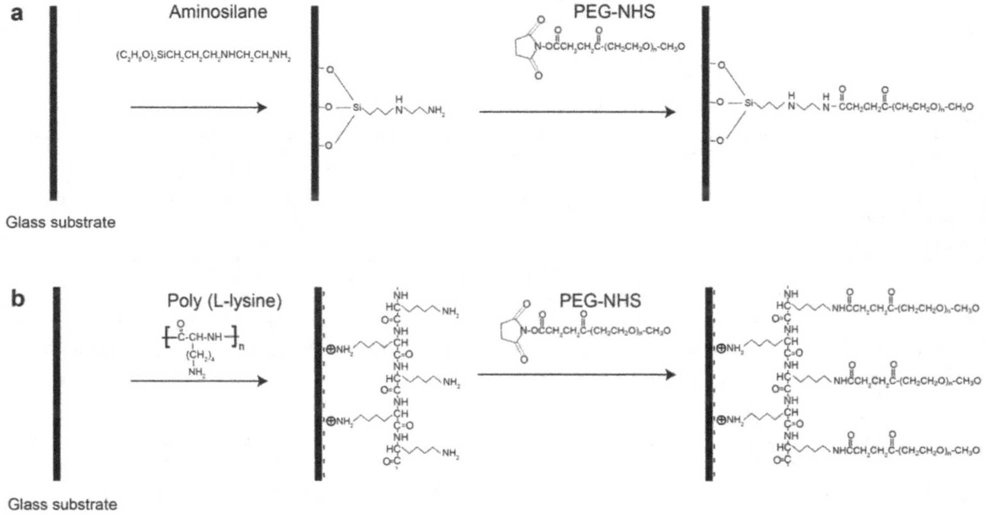

Fig. 1. PEG coating methods of glass substrates through either (a) silanization or (b) poly (L-lysine) adsorption.

3.1.1 PEGylation through aminosilanization

The standard PEG coating method on quartz slides and silicate coverslips for single-molecule fluorescence imaging is performed by silanization with N-2-(aminoethyl)-3-aminopropyl-triethoxysilane in methanol containing acetic acid for few hours (Figure 1 (a)) (Joo et al., 2006). The aminosilanization treatment yields a surface that is densely coated with exposed primary amines. A covalently attached PEG layer is formed on the surfaces by a PEGylation reaction with the amine reactive N-hydroxy-succinimidyl (NHS)-PEG (M.W. = 5,000 Da) dissolved commonly in freshly prepared 0.1 M sodium bicarbonate buffer (pH 8.3) for a couple of hours.

3.1.2 Improvement of non-specific adsorption suppression capability on silicate coverslips

I found that the standard PEG coating method mentioned above did not suppress non-specific adsorption as effectively on silicate coverslips as it did on quartz (Yokota et al., 2009). Therefore, an improved method for efficient PEG coating on silicates is required to reduce their non-specific adsorption. Then, I found that performing the PEGylation reaction in 50 mM 3-(N-morpholino)propanesulfonic acid (MOPS, pH 7.5) instead of the usual 0.1 M sodium bicarbonate (pH 8.3) reduced non-specific adsorption on silicate surfaces by up to an order of magnitude. Figure 2 shows the single-molecule fluorescence images of a cyanine dye-labeled protein (the *E. coli* helicase UvrD (Lohman et al., 2008) labeled with Cy5: Cy5-UvrD) non-specifically adsorbed on a silicate coverslip that was PEGylated in MOPS buffer as compared to one with no treatment and one coated following the standard method. The improvement is crucial since silicate coverslips are extensively used for single-molecule fluorescence microscopy with epi-illumination (Funatsu et al., 1995), highly inclined thin illumination (Tokunaga et al., 2008), and total-internal reflection fluorescence (Tokunaga et al., 1997) microscopies (sometimes combined with other techniques such as optical tweezers (Hohng et al., 2007; Lang et al., 2004; Zhou et al., 2011)). The improvement in non-specific adsorption using the new buffer condition for PEGylation was also observed on quartz, though to a lesser extent (Table 2) and with a Qdot-labeled protein molecule (UvrD labeled with Qdot655:Qdot655-UvrD) (Table 3). Figure 3 shows the single-molecule fluorescence images of a Qdot-labeled protein (Qdot-UvrD) non-specifically adsorbed on a silicate coverslip that was PEGylated in MOPS buffer as compared to one with no treatment. While PEGylation is required to reduce the non-specific adsorption of Qdot-labeled protein on both silicate and quartz surfaces, PEGylation in MOPS buffer reduced non-specific adsorption of Qdot-UvrD on glass substrates by a factor of approximately 2 as compared to PEG coating in sodium bicarbonate buffer (Table 3). The key issue determining the non-adsorption properties of PEG-coated glass surfaces is their coverage by the polymer (Harris, 1992), which is improved with the new method. PEG coating at pH 7.5 may be enhanced with respect to coating at higher pH owing to better stability of the NHS ester. Its lifetime is in the order of hours at physiological pH and decreases steeply at higher pH owing to increased hydrolysis (Hermanson, 2008).

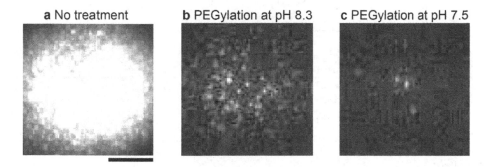

a No treatment   **b** PEGylation at pH 8.3   **c** PEGylation at pH 7.5

Fig. 2. Enhancement of protein-non-adsorption capability on silicate coated with PEG dissolved in 50 mM MOPS (pH 7.5). Fluorescence images of Cy5-UvrD non-specifically adsorbed on silicate with (a) no treatment, coated PEG dissolved in (b) 0.1 M sodium bicarbonate (pH 8.3) and (c) 50 mM MOPS (pH 7.5). The Cy5-UvrD concentration used was 2 nM. Scale bar, 10 μm.

	Silicate			Quartz		
	No treatment	PEG (pH 8.3)	PEG (pH 7.5)	No treatment	PEG (pH 8.3)	PEG (pH 7.5)
Mean number of Cy5-UvrD non-specifically adsorbed (/nM/1,000µm²)	1.6×10^3 $\pm 0.5 \times 10^3$ (21)	39 ±4 (9)	3.9 ±1.4 (30)	2.2×10^3 $\pm 0.3 \times 10^3$ (6)	1.4 ±0.2 (9)	0.78 ±0.20 (13)
Ratio to the number for PEG (pH 7.5)	4.1×10^2	10	1	2.8×10^3	1.8	1

The mean numbers of Cy5-UvrD non-specifically adsorbed per 1,000 µm² are normalized by the UvrD concentrations (nM) used for the experiments. Values represent mean ± standard deviation (n). (Reproduced from (Yokota et al., 2009) , Copyright (2009), with permission from The Chemical Society of Japan).

Table 2. Protein-non-adsorption capability on silicate coverslips and quartz slides (Cy5-UvrD).

a No treatment   b PEGylation at pH 7.5

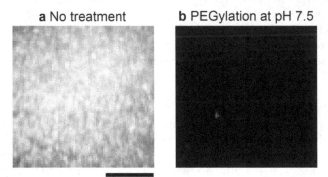

Fig. 3. Reduction of the non-specific adsorption of Qdot-labeled protein on glass substrates coated with PEG dissolved in 50 mM MOPS (pH 7.5). (b) Fluorescence images of Qdot-UvrD non-specifically adsorbed on silicate with no treatmemt (a) and coated with PEG dissolved in 50 mM MOPS (pH 7.5) (b). The Qdot-UvrD concentration used was 1 nM. Scale bar, 10 µm.

	Silicate			Quartz		
	No treatment	PEG (pH 8.3)	PEG (pH 7.5)	No treatment	PEG (pH 8.3)	PEG (pH 7.5)
Mean number of Qdot-UvrD non-specifically adsorbed (/nM/1,000 µm²)	6.8×10^3 $\pm 0.5 \times 10^3$ (8)	1.9 ±0.8 (23)	1.3 ±0.3 (22)	1.6×10^4 $\pm 0.3 \times 10^4$ (9)	1.7 ±0.2 (9)	0.51 ±0.20 (9)
Ratio to the number for PEG (pH 7.5)	5.2×10^3	1.5	1	3.1×10^4	3.3	1

The mean numbers of Qdot-UvrD non-specifically adsorbed per 1,000 µm² are normalized by the UvrD concentrations (nM) used for the experiments. Values represent mean ± standard deviation (n). (Reproduced from (Yokota et al., 2009) , Copyright (2009), with permission from The Chemical Society of Japan).

Table 3. Protein-non-adsorption capability on silicate coverslips and quartz slides (Qdot-UvrD).

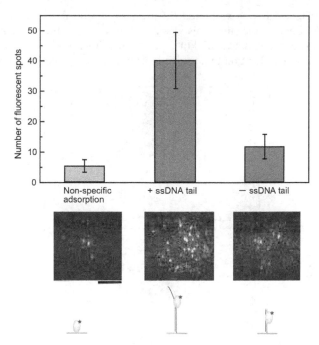

Fig. 4. Comparison of the numbers of Cy5-UvrD interacting with 18-bp dsDNA with or without an ssDNA tail immobilized on a PEG-coated silicate coverslip. Below are the single-molecule fluorescence images. The Cy5-UvrD concentration used was 2 nM. Scale bar, 10 μm. (Reproduced from (Yokota et al., 2009), Copyright (2009), with permission from The Chemical Society of Japan).

The results on the reduction of non-specific protein adsorption on glass surfaces indicate that our PEGylation method facilitates direct visualization of single-molecule interactions between fluorescently labeled protein molecules and DNA. Indeed, we could visualize how Cy5-UvrD interacts with a 18-bp dsDNA with or without an ssDNA tail immobilized on PEG-coated surfaces. The number of Cy5-UvrD fluorescent spots observed can be compared in Figure 4; this figure clearly shows that non-specific adsorption of Cy5-UvrD on the surface was effectively suppressed enough to conclude that Cy5-UvrD has higher affinity for the dsDNA with an ssDNA tail than for that without one; this finding is in agreement with that of a previous report (Maluf et al., 2003).

3.1.3 Single-molecule visualization of binding mode of helicase to DNA on silicate coverslips

Using microscopy, we could compare the number of UvrD molecules bound to 18-nt ssDNA, 4.7-kbp dsDNA, and 4.7-knt ssDNA immobilized on a PEG-coated silicate coverslip. Figure 5 shows a comparison of fluorescence intensity distributions of Cy5-UvrD bound to 18-nt ssDNA, 4.7-kbp dsDNA, or 4.7-knt ssDNA. In agreement with the results in Figure 4, Cy5-UvrD had low affinity for dsDNA, and thus, the fluorescence intensity distribution for 4.7-kbp dsDNA peaked at a fluorescence intensity that corresponds to that of single Cy5-UvrD, which was validated by the observation that most of the fluorescent spots photobleached in a single step. This was also the case with the experiment using 18-nt ssDNA. For the experiment using 4.7-knt ssDNA, the fluorescence intensity distribution shifted to a larger value, indicating that multiple Cy5-UvrD attached to the ssDNA.

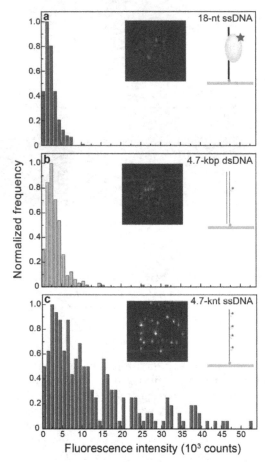

Fig. 5. Comparison of fluorescence intensity distributions of Cy5-UvrD bound to (a) 18-nt ssDNA, (b) 4.7-kbp dsDNA, or (c) 4.7-knt ssDNA immobilized on a PEG-coated silicate coverslip. The insets show the single-molecule fluorescence images. The Cy5-UvrD concentration used was 2 nM. Scale bar, 10 μm. (Reproduced from (Yokota et al., 2009) , Copyright (2009), with permission from The Chemical Society of Japan).

3.1.4 Stability of PEG-coated surfaces

The long-term stability of PEG has been examined in several studies. On examination of a PEG-coated coverslip surface for up to 1 month, the surface, immersed in 0.1 M sodium phosphate buffer (pH 7.4), has been found to degrade after 25 days (Branch et al., 2001). Another study examined the stability of PEG-modified silicon substrates that were incubated in PBS (37 °C, pH 7.4, 5% CO_2) for different periods of time up to 28 days. This study also concluded that PEG-modified surfaces retain their protein and cell repulsive nature for the period of investigation, i.e., 28 days (Sharma et al., 2004). With regard to PEG stability in the air, no degradation was reported after 75 days (Anderson et al., 2008). In contrast to these macroscopic experimental studies, PEG-coated surfaces used for single-molecule imaging imposes several challenges. The protein-non-adsorption capability of PEG-coated glass surfaces sharply deteriorated with time (incubation > 24 h) when incubated in a buffer (Yokota et al., 2009). Moreover, PEG-coated coverslips and quartz slides exhibited increasing non-specific adsorption with time when stored, exposed to the air, at room temperature (Yokota et al., 2009). PEG is known to degrade by oxidation (Han et al., 1997), and thus, to prevent the oxidative degradation, PEG-coated slides/coverslips can be stored in the dark under dry conditions at –20 °C until use (Joo et al., 2006).

3.1.5 PEGylation through poly(L-lysine) adsorption

PEGylation is also feasible through poly(L-lysine) adsorption on the glass surface (Figure 1b). The amino groups of the side chain of poly(L-lysine) are positively charged at pH values below 10, and are therefore easily attached to the negatively charged glass surface (Iler, 1979) through an electrostatic interaction at physiological pH. The protein non-adsorption capability is comparable to that of PEG-coated surfaces through aminosilanization (unpublished data). A similar strategy for PEGylation, which is not used for single-molecule fluorescence imaging, involves using a poly(L-lysine) grafted with PEG (PLL-g-PEG), a polycationic copolymer that is positively charged at neutral pH. The compound spontaneously adsorbs from aqueous solution onto negatively charged surfaces, resulting in the formation of stable polymeric monolayers and rendering the surfaces protein-resistant to a degree relative to the PEG surface density (Huang et al., 2001; Pasche et al., 2003).

3.2 Lipid bilayer

Artificial lipid membranes deposited on solid supports have proven to be useful for many types of biochemical studies (Chan & Boxer, 2007). The membranes provide a surface environment that is similar to that inside a living cell and can prevent any non-specific interactions of biomolecules on the surface. Lipid bilayers that are spontaneously formed on quartz surfaces have been used for single-molecule fluorescence imaging. The bilayers can be modified through the incorporation of lipids with various functional groups such as biotin or PEG. A mixture of 1,2-dioleoyl-*sn*-glycerophosphocholine dioleoylphosphatidylcholine (DOPC) with a small amount of 1,2-dioleoyl-*sn*-glycero-3-phosphoethanolamine-N-[methoxy(polyethylene glycol)-550] (mPEG 550-DOPE) was reported to help minimize non-specific binding of Qdot-tagged proteins to the lipid bilayer (Gorman et al., 2010). For experiments that require tethering DNA that is biotinylated at the end(s), avidin is attached either via non-specific interaction with the glass surface or via interaction with biotin head groups of the lipids present.

4. Platforms

A common feature among direct visualization of single-molecule DNA-binding protein molecules is the need for DNA to be tethered and extended. A pioneering platform published in 1993 employs dielectrophoresis between aluminium electrodes to extend λDNA (Fig. 6g) (Kabata et al., 1993). λDNA is a bacteriophage DNA that consists of 48,490 base pairs of double-stranded linear DNA with 12-nucleotide single-stranded segments at both 5' ends (approximately 16.5 μm in total length). The platform can form "DNA belts" and has uncovered RNA polymerase dynamics along DNA, including sliding and jumping. Another pioneering platform published in 1998 employs optical tweezers, which was incorporated into a single-molecule fluorescence microscope, to extend λDNA (Fig. 6d) (Harada et al., 1999), which revealed strain-dependent DNA-binding kinetics of RNA polymerase. At that time, casein, a conventional blocking agent that is found in milk in large quantities, was used to block non-specific protein adsorption.

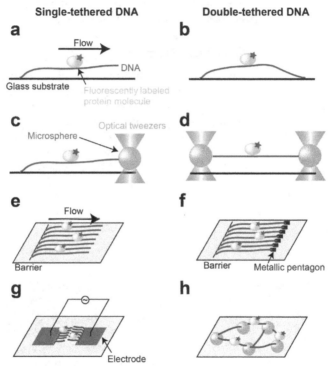

Fig. 6. Platforms used for direct visualization of single-molecule DNA-binding protein molecules along DNA. (a) DNA tethered at one end on a glass surface with buffer flow. (b) DNA tethered at both ends on a glass surface. (c) DNA tethered on a glass surface at one end and with optical tweezers at the other end. (d) DNA tethered with optical tweezers at both ends. (e) DNA curtains. (f) DNA racks. (g) DNA belts. (f) Tightropes.

To minimize non-specific protein adsorption, most of the visualization has been performed on glass surfaces modified by the coating methods discussed in the previous section. Many platforms capable of real-time visualization of single-molecule DNA-binding protein molecules along tethered DNA, most of which are used in combination with total-internal fluorescence microscopy (Funatsu et al., 1995; Tokunaga et al., 1997), have been developed to understand protein dynamics. Among many methods for DNA tethering by non-specific adsorption, or specific attachments mediated by a non-covalent interaction, the most popular DNA tethering method used for the platforms is tethering biotinylated λDNA on glass surfaces via an avidin-biotin interaction. The two single-stranded segments at both 5′ ends of λDNA, termed the *cos* site, are the sticky ends and can be easily biotinylated via ligation with corresponding the complimentary biotinylated oligos. The biotinylated λDNA can be tethered on glass surfaces at either one or both end(s) via an avidin-biotin interaction. In the following sections, I briefly review some of the modern platforms and present their target DNA protein molecules.

4.1 Platforms with continuous buffer flow

Tethering DNA at one end on the surface via an avidin-biotin interaction is the simplest method for tethering DNA. However, owing to the flexible nature of DNA, tethering alone cannot make the DNA extend sufficiently enough for use in single-molecule visualization with high spatial resolution. To stretch such DNA, some investigators use shear force generated by hydrodynamic flow (Fig. 6a) (Brewer & Bianco, 2008; Graneli et al., 2006; Greene & Mizuuchi, 2002; S. Kim et al., 2007; Lee et al., 2006; van Oijen et al., 2003). In this platform, buffer is continuously input into a flow cell made of sandwiched glass substrates. The glass substrates are coated with PEG (Blainey et al., 2006; Blainey et al., 2009; S. Kim et al., 2007) or a lipid bilayer (Graneli et al., 2006) to reduce non-specific adsorption of protein molecules. With this platform, dynamics of many single-molecule proteins have been uncovered. These include dynamics of single-protein polymers along DNA (Greene & Mizuuchi, 2002; Han & Mizuuchi, 2010; Tan et al., 2007) and one-dimensional diffusion of many DNA-binding protein molecules, for example, a base-excision DNA-repair protein along DNA (Blainey et al., 2006; Etson et al., 2010; Kochaniak et al., 2009; Komazin-Meredith et al., 2008; Tafvizi et al., 2011). A comparison of one-dimensional diffusion constants as a function of protein size with theoretical predictions indicates that DNA-binding proteins undergo rotation-coupled sliding along the DNA helix (Blainey et al., 2009).

4.2 Tethering DNA at multiple points on glass surfaces

As mentioned above, tethering DNA at one end on the surface alone is not sufficient for the single-molecule visualization. Several papers report methods of DNA tethering at many points via non-specific interaction or at both ends (Fig. 6b) on the surface via an avidin-biotin interaction. These include λDNA immobilized on a polystyrene-coated coverslip via a non-specific interaction (Kim & Larson, 2007) or on a lipid bilayer-deposited fused silica slide via an avidin-biotin interaction (Graneli et al., 2006), and T7 bacteriophage DNA immobilized on a silanized coverslip via an avidin-biotin interaction (Bonnet et al., 2008). Using these platforms, one-dimensional diffusion and transcription of single T7 RNA polymerases along DNA (Kim & Larson, 2007) and sliding and jumping of EcoRV restriction enzyme (Bonnet et al., 2008) along DNA were visualized.

4.3 DNA curtains

4.3.1 Single-tethered DNA curtains

To tether many DNA strands on the surface for a high-throughput single-molecule analysis, the Greene group has developed a new platform referred to as "DNA curtains" (Fig. 6e) (Fazio et al., 2008; Graneli et al., 2006; Visnapuu et al., 2008). This assay allows simultaneous study of up to hundreds of individual DNA strands anchored to a lipid bilayer and aligned with respect to one another within a single field-of-view with TIRFM. The DNA curtains were constructed in the following manner: (i) the surface of a glass slide was first mechanically etched to form lipid-diffusion barriers perpendicular to the direction of buffer flow. (ii) The flow cell was coated with a lipid bilayer by injecting lipid vesicles comprising DOPC, biotinylated lipids, and PEGylated lipids into the sample chamber (discussed in 3.2). (iii) After removing the excess vesicles by a buffer flush, neutravidin was injected into the flow cell. (iv) λDNA that was biotinylated at one end was injected into the flow cell. The PEGylated lipid enhanced protein non-specific adsorption suppression capability on the surface. The barriers formed on the flow cell could not be traversed by the lipids, and thus, buffer flow extended the lipid-tethered DNA in the flow direction and confined the DNA within the evanescent field generated by total-internal reflection. This platform has been used to study a broad range of DNA-binding proteins that function in homologous recombination (Prasad et al., 2006; Prasad et al., 2007) and mismatch repair (Gorman et al., 2007).

4.3.2 Double-tethered DNA curtains

The DNA curtains require buffer flow to extend DNA for single-molecule visualization of protein dynamics along the DNA. The hydrodynamic force exerted by the buffer flow can potentially influence the behavior of protein molecules; this influence is expected to be proportional to the hydrodynamic radius of the protein molecules under observation (Gorman & Greene, 2008; Tafvizi et al., 2008). To circumvent such an issue, the Greene group has developed double-tethered DNA curtains without the need of buffer flow during the visualization (Fig. 6f) (Gorman et al., 2010). The group used electron beam lithography to fabricate diffusion barriers with nanoscale dimensions, which allows for much more precise control over both the location and lateral distribution of the DNA within the curtains. These patterns, which the group termed "DNA racks," comprise linear barriers to lipid diffusion along with arrays of metallic pentagons, which are used for the scaffold of antibodies to anchor DNA at one end. The lipid-tethered DNA is first aligned along the linear barriers; subsequently, the DNA is anchored on the pentagons positioned at a defined distance downstream from the linear barriers. Once aligned and anchored, the "double-tethered" DNA curtains maintain their extended form. This sophisticated platform has been applied to biological systems such as nucleosomes and chromatin remodeling (Fig. 7) (Finkelstein et al., 2010; Gorman et al., 2010).

4.4 DNA tightropes

Another platform that did not require buffer flow was developed for single-molecule visualization of protein dynamics along extended DNA. In this platform, λDNA is suspended between poly (L-lysine)-coated microspheres non-specifically attached on a PEGylated surface to form "DNA tightropes" in a flow cell, allowing the DNA to maintain its extended form during the visualization (Fig. 6h) (Kad et al., 2010). Since DNA is negatively charged at physiological pH, DNA can attach to the positively charged poly (L-lysine) that is coated on the microspheres via an electrostatic interaction. Injection of high concentration of DNA (1.6 nM) forms many DNA tightropes, allowing a high-throughput single-molecule analysis by simultaneous observation of multiple DNA-binding protein

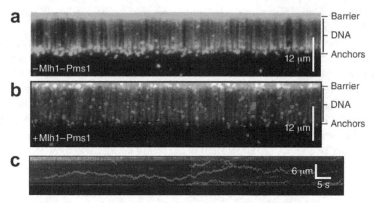

Fig. 7. Nanofabricated racks of DNA for visualizing one-dimensional diffusion of a postreplicative mismatch repair protein complex (Mlh1–Pms1). (a) YOYO1-stained λDNA curtains (green; 48,502 base pairs), (b) YOYO1-stained λDNA curtains with a Qdot-labeled protein complex (magenta), and (c) Kymogram illustrating the motion of the protein complex. (Modified from *Nature Structural & Molecular Biology* (Gorman et al., 2010) , Copyright (2010), with permission from Macmillan Publishers Ltd).

molecules. Since the DNA tightropes formed on the microspheres are off the glass surface, there is no interaction between the DNA and the glass surface, which may not interfere with the interactions with the protein molecules. Furthermore, the no buffer-flow condition enables the detection of various dynamic mechanisms of DNA repair proteins, including not only diffusion along DNA (Dunn et al., 2011) but also jumping from one DNA to another DNA (Fig. 8) (Kad et al., 2010).

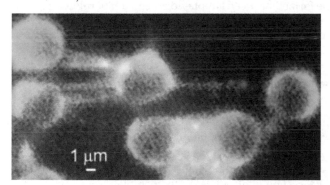

Fig. 8. Fluorescence image of Qdot-labeled single protein molecules (UvrA, red spots) bound to DNA tightropes (green). (Reproduced from (Kad et al., 2010), Copyright (2010), with permission from Elsevier).

4.4.1 Minimization of protein non-specific adsorption on microspheres by PEGylaton

One of the drawbacks of the DNA tightrope platform with poly (L-lysine)-coated microspheres is that many protein molecules are non-specifically adsorbed on the microspheres. Fluorescence from the Qdots that are labeled to the adsorbed protein molecules interferes with single-molecule fluorescence imaging. Kad et al. circumvented the interference by manually masking the bead portions on screen using an image analysis software before analysis (Kad et al., 2010). Here, I describe another tightrope platform developed to overcome the issue by PEGylation, in which the glass surface as well as microspheres are coated with biotinylated PEG. The excellent resistance of PEG to protein adsorption (discussed in 3.1) efficiently suppresses the non-specific adsorption of Qdot-labeled protein molecules on the microspheres. The biotin of the PEG allows the microspheres to be immobilized on the glass surface and biotinylated DNA to be tethered between the microspheres via an avidin-biotin interaction (Fig. 9). With this improved platform, dynamics of a DNA repair protein is currently being investigated.

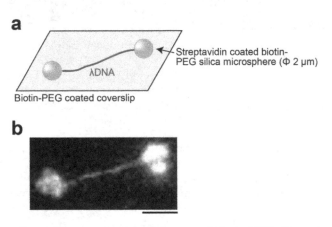

Fig. 9. A tightrope formed between streptavidin coated biotin-PEG silica microspheres immobilized on a biotin-PEG coated coverslip. (a) Schematic drawing of this tightrope platform. (b) Fluorescence image of YOYO1-labeled λDNA tethered between microspheres. Scale bar, 5 μm.

4.5 DNA extended by single-molecule DNA manipulation

Technical advances in combining single-molecule fluorescence visualization with single-molecule manipulation provide deeper insights into DNA–protein interactions. Some groups incorporated optical tweezers into a single-molecule fluorescence set-up (Fig. 6c,d) for studying DNA dynamics or DNA–protein interactions (Amitani et al., 2006; Comstock et al., 2011; Galletto et al., 2006; Harada et al., 1999; Hohng et al., 2007; Lang et al., 2004; van Mameren et al., 2009; Zhou et al., 2011). With such platforms, translocation (Amitani et al., 2006), association and dissociation (Harada et al., 1999; Zhou et al., 2011), filament assembly (Galletto et al., 2006), and disassembly (van Mameren et al., 2009) of protein molecules along DNA were visualized. Some of these studies investigated the effect of DNA tension on DNA–protein interactions. Recently, my colleagues and I have incorporated magnetic tweezers (Allemand et al., 2007), another apparatus for single-molecule DNA manipulation, into the single-molecule fluorescence set-up through collaboration with my research colleagues, which has allowed us to visualize single-molecule DNA helicase interacting with single dsDNA that is tethered by the magnetic tweezers.

5. Outlook

I have provided an overview recent of advances in direct visualization of single-molecule DNA-binding proteins, especially of key techniques used for the visualization. Fluorescence labeling methods as well as modern platforms employing biocompatible surface-coating methods described here have allowed the dynamics of single-molecule DNA-binding proteins to be elucidated. Frequently used fluorophores, dyes and Qdots have allowed visualization to a certain extent. However, the fluorophores have not satisfied all the requirements related to optimum size and photophysical properties, for example, dyes are small but short lived whereas Qdots are bright and resistant to photobleaching, but relatively large in size. Thus, it is highly desirable that new all-round fluorophores be synthesized or extracted from materials. One such promising fluorophore is the nitrogen-vacancy center, a crystal defect in diamonds (Aharonovich et al., 2011). The defect that emits near-infrared fluorescence exhibits neither blinking nor photobleaching, which is distinct from the common fluorophores.

Technical advances in combining single-molecule fluorescence visualization with single-molecule manipulation (discussed in 4.5) will provide deeper insights into DNA–protein interactions. Zero-mode waveguides (Levene et al., 2003), a nano-hole (<100 nm in diameter) array, is another promising microscopy for visualizing DNA–protein interactions. The waveguides attract increasing attention because compared to TIRFM, zero-mode waveguides make it possible to perform single-molecule fluorescence imaging under higher ligand concentration (μM range) conditions due to its extremely small excitation volume. And thus, a few biological processes were visualized at single-molecule resolution by zero-mode waveguides (Miyake et al., 2008; Sameshima et al., 2010; Uemura et al., 2010).

In a recent study (Kad et al., 2010), single-molecule interaction of two different DNA protein molecules along DNA was visualized by labeling them with different fluorophores, although most studies typically focused on the dynamics of single DNA binding-proteins. In the future, simultaneous visualization of different DNA binding protein molecules may be possible by labeling the molecules with different fluorophores. This will help elucidate fundamental processes of DNA metabolism. In fact, biological processes in DNA metabolism involve various protein molecules as a form of macromolecular complexes. Moreover, the processes can be associated with chromatin for eukaryotes. To approach the protein dynamics in such an environment, a recent study investigated single-molecule interaction between a protein complex and chromatin (Gorman et al., 2010). Future developments in various fields, including platforms, fluorophores, surface coating, and fluorescence labelling, will provide fruitful information on complex DNA–protein interactions that cannot otherwise be explored.

6. Acknowledgments

I would like to thank the Harada group at Kyoto University and the Bensimon & Croquette group at École Normale Supérieure for supporting my research. The presented work performed in the groups was partly funded by JSPS and JST.

7. References

Aharonovich, I., Greentree, A. D. & Prawer, S. (2011). Diamond photonics. *Nature Photonics,* 5(7), 397-405.

Aitken, C. E., Marshall, R. A. & Puglisi, J. D. (2008). An oxygen scavenging system for improvement of dye stability in single-molecule fluorescence experiments. *Biophysical Journal*, 94(5), 1826-1835.

Allemand, J.-F., Bensimon, D., Charvin, G., Croquette, V., Lia, G., Lionnet, T., Neuman, K. C., Saleh, O. A. & Yokota, H. (2007). Studies of DNA-Protein Interactions at the single molecule level with magnetic tweezers. *Lecture Notes on Physics*, 711, 123-140.

Amitani, I., Baskin, R. J. & Kowalczykowski, S. C. (2006). Visualization of Rad54, a chromatin remodeling protein, translocating on single DNA molecules. *Molecular Cell*, 23(1), 143-148.

Anderson, A. S., Dattelbaum, A. M., Montano, G. A., Price, D. N., Schmidt, J. G., Martinez, J. S., Grace, W. K., Grace, K. M. & Swanson, B. I. (2008). Functional PEG-modified thin films for biological detection. *Langmuir*, 24(5), 2240-2247.

Blainey, P. C., van Oijen, A. M., Banerjee, A., Verdine, G. L. & Xie, X. S. (2006). A base-excision DNA-repair protein finds intrahelical lesion bases by fast sliding in contact with DNA. *Proceedings of the National Academy of Sciences of the United States of America*, 103(15), 5752-5757.

Blainey, P. C., Luo, G., Kou, S. C., Mangel, W. F., Verdine, G. L., Bagchi, B. & Xie, X. S. (2009). Nonspecifically bound proteins spin while diffusing along DNA. *Nature Structural & Molecular Biology*, 16(12), 1224-1229.

Bonnet, I., Biebricher, A., Porte, P. L., Loverdo, C., Benichou, O., Voituriez, R., Escude, C., Wende, W., Pingoud, A. & Desbiolles, P. (2008). Sliding and jumping of single EcoRV restriction enzymes on non-cognate DNA. *Nucleic Acids Research*, 36(12), 4118-4127.

Branch, D. W., Wheeler, B. C., Brewer, G. J. & Leckband, D. E. (2001). Long-term stability of grafted polyethylene glycol surfaces for use with microstamped substrates in neuronal cell culture. *Biomaterials*, 22(10), 1035-1047.

Brewer, L. R. & Bianco, P. R. (2008). Laminar flow cells for single-molecule studies of DNA-protein interactions. *Nature Methods*, 5(6), 517-525.

Chan, Y. H. & Boxer, S. G. (2007). Model membrane systems and their applications. *Current Opinion in Chemical Biology*, 11(6), 581-587.

Comstock, M. J., Ha, T. & Chemla, Y. R. (2011). Ultrahigh-resolution optical trap with single-fluorophore sensitivity. *Nature Methods*, 8(4), 335-340.

Cuvelier, D., Rossier, O., Bassereau, P. & Nassoy, P. (2003). Micropatterned "adherent/repellent" glass surfaces for studying the spreading kinetics of individual red blood cells onto protein-decorated substrates. *European Biophysics Journal*, 32(4), 342-354.

Dave, R., Terry, D. S., Munro, J. B. & Blanchard, S. C. (2009). Mitigating unwanted photophysical processes for improved single-molecule fluorescence imaging. *Biophysical Journal*, 96(6), 2371-2381.

Dunn, A. R., Kad, N. M., Nelson, S. R., Warshaw, D. M. & Wallace, S. S. (2011). Single Qdot-labeled glycosylase molecules use a wedge amino acid to probe for lesions while scanning along DNA. *Nucleic Acids Research*, 39(17), 7487-7498.

Etson, C. M., Hamdan, S. M., Richardson, C. C. & van Oijen, A. M. (2010). Thioredoxin suppresses microscopic hopping of T7 DNA polymerase on duplex DNA. *Proceedings of the National Academy of Sciences of the United States of America*, 107(5), 1900-1905.

Fazio, T., Visnapuu, M. L., Wind, S. & Greene, E. C. (2008). DNA curtains and nanoscale curtain rods: high-throughput tools for single molecule imaging. *Langmuir*, 24(18), 10524-10531.

Fernández-Suárez, M. & Ting, A. Y. (2008). Fluorescent probes for super-resolution imaging in living cells. *Nature Reviews Molecular Cell Biology*, 9(12), 929-943.

Finkelstein, I. J., Visnapuu, M. L. & Greene, E. C. (2010). Single-molecule imaging reveals mechanisms of protein disruption by a DNA translocase. *Nature*, 468(7326), 983-987.

Funatsu, T., Harada, Y., Tokunaga, M., Saito, K. & Yanagida, T. (1995). Imaging of single fluorescent molecules and individual ATP turnovers by single myosin molecules in aqueous solution. *Nature*, 374(6522), 555-559.

Galletto, R., Amitani, I., Baskin, R. J. & Kowalczykowski, S. C. (2006). Direct observation of individual RecA filaments assembling on single DNA molecules. *Nature*, 443(7113), 875-878.

Gorman, J., Chowdhury, A., Surtees, J. A., Shimada, J., Reichman, D. R., Alani, E. & Greene, E. C. (2007). Dynamic basis for one-dimensional DNA scanning by the mismatch repair complex Msh2-Msh6. *Molecular Cell*, 28(3), 359-370.

Gorman, J. & Greene, E. C. (2008). Visualizing one-dimensional diffusion of proteins along DNA. *Nature Structural & Molecular Biology*, 15(8), 768-774.

Gorman, J., Fazio, T., Wang, F., Wind, S. & Greene, E. C. (2010). Nanofabricated racks of aligned and anchored DNA substrates for single-molecule imaging. *Langmuir*, 26(2), 1372-1379.

Gorman, J., Plys, A. J., Visnapuu, M. L., Alani, E. & Greene, E. C. (2010). Visualizing one-dimensional diffusion of eukaryotic DNA repair factors along a chromatin lattice. *Nature Structural & Molecular Biology*, 17(8), 932-938.

Graneli, A., Yeykal, C. C., Prasad, T. K. & Greene, E. C. (2006). Organized arrays of individual DNA molecules tethered to supported lipid bilayers. *Langmuir*, 22(1), 292-299.

Greene, E. C. & Mizuuchi, K. (2002). Direct observation of single MuB polymers: evidence for a DNA-dependent conformational change for generating an active target complex. *Molecular Cell*, 9(5), 1079-1089.

Ha, T., Rasnik, I., Cheng, W., Babcock, H. P., Gauss, G. H., Lohman, T. M. & Chu, S. (2002). Initiation and re-initiation of DNA unwinding by the *Escherichia coli* Rep helicase. *Nature*, 419(6907), 638-641.

Ha, T. (2004). Structural dynamics and processing of nucleic acids revealed by single-molecule spectroscopy. *Biochemistry*, 43(14), 4055-4063.

Han, S., Kim, C. & Kwon, D. (1997). Thermal/oxidative degradation and stabilization of polyethylene glycol. *Polymer*, 38, 317-323.

Han, Y. W. & Mizuuchi, K. (2010). Phage Mu transposition immunity: protein pattern formation along DNA by a diffusion-ratchet mechanism. *Molecular Cell*, 39(1), 48-58.

Harada, Y., Sakurada, K., Aoki, T., Thomas, D. D. & Yanagida, T. (1990). Mechanochemical coupling in actomyosin energy transduction studied by *in vitro* movement assay. *Journal of Molecular Biology*, 216(1), 49-68.

Harada, Y., Funatsu, T., Murakami, K., Nonoyama, Y., Ishihama, A. & Yanagida, T. (1999). Single-molecule imaging of RNA polymerase-DNA interactions in real time. *Biophysical Journal*, 76(2), 709-715.

Harris, J. M. (1992). *Poly(Ethylene Glycol) Chemistry (Ethylene Glycol Chemistry : Biotechnical and Biomedical Applications)*. New York: Plenum Press.

Hermanson, G. T. (2008). *Bioconjugate Techniques (2nd Edition)*: Academic Press.

Heyes, C. D., Kobitski, A. Y., Amirgoulova, E. V. & Nienhaus, G. U. (2004). Biocompatible surfaces for specific tethering of individual protein molecules. *Journal of Physical Chemistry B,* 108(35), 13387-13394.

Hilario, J. & Kowalczykowski, S. C. (2010). Visualizing protein-DNA interactions at the single-molecule level. *Current Opinion in Chemical Biology,* 14(1), 15-22.

Hohng, S., Zhou, R., Nahas, M. K., Yu, J., Schulten, K., Lilley, D. M. & Ha, T. (2007). Fluorescence-force spectroscopy maps two-dimensional reaction landscape of the holliday junction. *Science,* 318(5848), 279-283.

Huang, N.-P., Michel, R., Voros, J., Textor, M., Hofer, R., Rossi, A., Elbert, D. L., Hubbell, J. A. & Spencer, N. D. (2001). Poly(L-lysine)-g-poly(ethylene glycol) layers on metal oxide surfaces: surface-analytical characterization and resistance to serum and fibrinogen adsorption. *Langmuir,* 17(2), 489-498.

Iler, R. K. (1979). *The Chemistry of Silica.* New York: Wiley-Interscience.

Ito, Y., Hasuda, H., Sakuragi, M. & Tsuzuki, S. (2007). Surface modification of plastic, glass and titanium by photoimmobilization of polyethylene glycol for antibiofouling. *Acta Biomaterialia,* 3(6), 1024-1032.

Joo, C., McKinney, S. A., Nakamura, M., Rasnik, I., Myong, S. & Ha, T. (2006). Real-time observation of RecA filament dynamics with single monomer resolution. *Cell,* 126(3), 515-527.

Kabata, H., Kurosawa, O., Arai, I., Washizu, M., Margarson, S. A., Glass, R. E. & Shimamoto, N. (1993). Visualization of single molecules of RNA polymerase sliding along DNA. *Science,* 262(5139), 1561-1563.

Kad, N. M., Wang, H., Kennedy, G. G., Warshaw, D. M. & Van Houten, B. (2010). Collaborative dynamic DNA scanning by nucleotide excision repair proteins investigated by single-molecule imaging of quantum-dot-labeled proteins. *Molecular Cell,* 37(5), 702-713.

Kim, J. H. & Larson, R. G. (2007). Single-molecule analysis of 1D diffusion and transcription elongation of T7 RNA polymerase along individual stretched DNA molecules. *Nucleic Acids Research,* 35(11), 3848-3858.

Kim, S., Blainey, P. C., Schroeder, C. M. & Xie, X. S. (2007). Multiplexed single-molecule assay for enzymatic activity on flow-stretched DNA. *Nature Methods,* 4(5), 397-399.

Kochaniak, A. B., Habuchi, S., Loparo, J. J., Chang, D. J., Cimprich, K. A., Walter, J. C. & van Oijen, A. M. (2009). Proliferating cell nuclear antigen uses two distinct modes to move along DNA. *Journal of Biological Chemistry,* 284(26), 17700-17710.

Komazin-Meredith, G., Mirchev, R., Golan, D. E., van Oijen, A. M. & Coen, D. M. (2008). Hopping of a processivity factor on DNA revealed by single-molecule assays of diffusion. *Proceedings of the National Academy of Sciences of the United States of America,* 105(31), 10721-10726.

Lang, M. J., Fordyce, P. M., Engh, A. M., Neuman, K. C. & Block, S. M. (2004). Simultaneous, coincident optical trapping and single-molecule fluorescence. *Nature Methods,* 1(2), 133-139.

Lee, J. B., Hite, R. K., Hamdan, S. M., Xie, X. S., Richardson, C. C. & van Oijen, A. M. (2006). DNA primase acts as a molecular brake in DNA replication. *Nature,* 439(7076), 621-624.

Levene, M. J., Korlach, J., Turner, S. W., Foquet, M., Craighead, H. G. & Webb, W. W. (2003). Zero-mode waveguides for single-molecule analysis at high concentrations. *Science,* 299(5607), 682-686.

Lohman, T. M., Tomko, E. J. & Wu, C. G. (2008). Non-hexameric DNA helicases and translocases: mechanisms and regulation. *Nature Reviews Molecular Cell Biology,* 9(5), 391-401.

Malmstena, M., Emoto, K. & Alstine, J. M. V. (1998). Effect of chain density on inhibition of protein adsorption by poly(ethylene glycol) Based Coatings. *Journal of Colloid Interface Science,* 202(2), 507-517.

Maluf, N. K., Fischer, C. J. & Lohman, T. M. (2003). A dimer of *Escherichia coli* UvrD is the active form of the helicase in vitro. *Journal of Molecular Biology,* 325(5), 913-935.

McNamee, C. E., Yamamoto, S. & Higashitani, K. (2007). Preparation and characterization of pure and mixed monolayers of poly(ethylene glycol) brushes chemically adsorbed to silica surfaces. *Langmuir,* 23(8), 4389-4399.

Michalet, X., Pinaud, F. F., Bentolila, L. A., Tsay, J. M., Doose, S., Li, J. J., Sundaresan, G., Wu, A. M., Gambhir, S. S. & Weiss, S. (2005). Quantum dots for live cells, *in vivo* imaging, and diagnostics. *Science,* 307(5709), 538-544.

Miyake, T., Tanii, T., Sonobe, H., Akahori, R., Shimamoto, N., Ueno, T., Funatsu, T. & Ohdomari, I. (2008). Real-time imaging of single-molecule fluorescence with a zero-mode waveguide for the analysis of protein-protein interaction. *Analytical Chemistry,* 80(15), 6018-6022.

Moerner, W. E. (2007). Single-molecule Chemistry and Biology Special Feature: New directions in single-molecule imaging and analysis. *Proceedings of the National Academy of Sciences of the United States of America,* 104(31), 12596-12602.

Pasche, S., Paul, S. M. D., Vörös, J., Spencer, N. D. & Textor, M. (2003). Poly(L-lysine)-graft-poly(ethyleneglycol) assembled monolayers on niobium oxide surfaces: a quantitative study of the influence of polymer interfacial architecture on resistance to protein adsorption by ToF-SIMS and *in situ* OWLS. *Langmuir,* 19, 9216-9235.

Prasad, T. K., Yeykal, C. C. & Greene, E. C. (2006). Visualizing the assembly of human Rad51 filaments on double-stranded DNA. *Journal of Molecular Biology,* 363(3), 713-728.

Prasad, T. K., Robertson, R. B., Visnapuu, M. L., Chi, P., Sung, P. & Greene, E. C. (2007). A DNA-translocating Snf2 molecular motor: *Saccharomyces cerevisiae* Rdh54 displays processive translocation and extrudes DNA loops. *Journal of Molecular Biology,* 369(4), 940-953.

Prime, K. L. & Whitesides, G. M. (1991). Self-assembled organic monolayers: model systems for studying adsorption of proteins at surfaces. *Science,* 252(5010), 1164-1167.

Rasnik, I., McKinney, S. A. & Ha, T. (2006). Nonblinking and long-lasting single-molecule fluorescence imaging. *Nature Methods,* 3(11), 891-893.

Sameshima, T., Iizuka, R., Ueno, T., Wada, J., Aoki, M., Shimamoto, N., Ohdomari, I., Tanii, T. & Funatsu, T. (2010). Single-molecule study on the decay process of the football-shaped GroEL-GroES complex using zero-mode waveguides. *Journal of Biological Chemistry,* 285(30), 23159-23164.

Sharma, S., Johnson, R. W. & Desai, T. A. (2004). Evaluation of the stability of nonfouling ultrathin poly(ethylene glycol) films for silicon-based microdevices. *Langmuir,* 20(2), 348-356.

Tafvizi, A., Huang, F., Leith, J. S., Fersht, A. R., Mirny, L. A. & van Oijen, A. M. (2008). Tumor suppressor p53 slides on DNA with low friction and high stability. *Biophysical Journal*, 95(1), L01-03.

Tafvizi, A., Huang, F., Fersht, A. R., Mirny, L. A. & van Oijen, A. M. (2011). A single-molecule characterization of p53 search on DNA. *Proceedings of the National Academy of Sciences of the United States of America*, 108(2), 563-568.

Tan, X., Mizuuchi, M. & Mizuuchi, K. (2007). DNA transposition target immunity and the determinants of the MuB distribution patterns on DNA. *Proceedings of the National Academy of Sciences of the United States of America*, 104(35), 13925-13929.

Tokunaga, M., Kitamura, K., Saito, K., Iwane, A. H. & Yanagida, T. (1997). Single molecule imaging of fluorophores and enzymatic reactions achieved by objective-type total internal reflection fluorescence microscopy. *Biochemical and Biophysical Research Communications*, 235(1), 47-53.

Tokunaga, M., Imamoto, N. & Sakata-Sogawa, K. (2008). Highly inclined thin illumination enables clear single-molecule imaging in cells. *Nature Methods*, 5(2), 159-161.

Uemura, S., Aitken, C. E., Korlach, J., Flusberg, B. A., Turner, S. W. & Puglisi, J. D. (2010). Real-time tRNA transit on single translating ribosomes at codon resolution. *Nature*, 464(7291), 1012-1017.

van Mameren, J., Modesti, M., Kanaar, R., Wyman, C., Peterman, E. J. & Wuite, G. J. (2009). Counting RAD51 proteins disassembling from nucleoprotein filaments under tension. *Nature*, 457(7230), 745-748.

van Oijen, A. M., Blainey, P. C., Crampton, D. J., Richardson, C. C., Ellenberger, T. & Xie, X. S. (2003). Single-molecule kinetics of lambda exonuclease reveal base dependence and dynamic disorder. *Science*, 301(5637), 1235-1238.

Visnapuu, M. L., Duzdevich, D. & Greene, E. C. (2008). The importance of surfaces in single-molecule bioscience. *Molecular Biosystems*, 4(5), 394-403.

Wang, X., Ren, X., Kahen, K., Hahn, M. A., Rajeswaran, M., Maccagnano-Zacher, S., Silcox, J., Cragg, G. E., Efros, A. L. & Krauss, T. D. (2009). Non-blinking semiconductor nanocrystals. *Nature*, 459(7247), 686-689.

Yang, Z., Galloway, J. A. & Yu, H. (1999). Protein interactions with poly(ethylene glycol) self-assembled monolayers on glass substrates: diffusion and adsorption *Langmuir*, 15(24), 8405-8411.

Yokota, H., Han, Y. W., Allemand, J.-F., Xi, X. G., Bensimon, D., Croquette, V. & Harada, Y. (2009). Single-molecule visualization of binding modes of helicase to DNA on PEGylated surfaces. *Chemistry Letters*, 38, 308-309.

Zhou, R., Kozlov, A. G., Roy, R., Zhang, J., Korolev, S., Lohman, T. M. & Ha, T. (2011). SSB functions as a sliding platform that migrates on DNA via reptation. *Cell*, 146(2), 222-232.

Zlatanova, J. & van Holde, K. (2006). Single-molecule biology: what is it and how does it work? *Molecular Cell*, 24(3), 317-329.

9

Defining the Cellular Interactome of Disease-Linked Proteins in Neurodegeneration

Verena Arndt[1,*] and Ina Vorberg[1,2],
[1]German Center for Neurodegenerative Diseases (DZNE), Bonn,
[2]Department of Neurology, Rheinische Friedrich-Wilhelms-University Bonn,
Germany

1. Introduction

Age-related neurodegenerative diseases are recognized as a major health issue worldwide. Due to our aging society, the number of people suffering from dementia is drastically increasing, creating serious challenges for society and the public health system. Our current paucity of effective treatments and total lack of cures, when coupled with this increasing prevalence, makes the exploration of novel strategies for therapeutic interventions of the utmost importance. Although many of the disease-causing proteins have been identified, the molecular mechanisms that underlie disease pathogenesis are still not fully understood. One common feature of neurodegenerative diseases is the accumulation of misfolded proteins into toxic oligomers and aggregates. Gaining extensive knowledge regarding the formation of these cytotoxic species, the cellular machineries that guarantee their persistence or clearance and the basis of their toxicity is essential for the development of both preventive and therapeutic interventions. A powerful approach to increase our knowledge on these disease processes is the use of proteomic tools to define the interaction networks of disease-related proteins. Significant technical progress has been made in the last decade that now allows high-throughput screening for protein-protein interactions on a proteome level. In this book chapter, we review the diverse proteomic approaches that have been used to define the interactomes of disease-linked proteins and the impact of these findings on the understanding of pathogenic processes.

1.1 Cellular mechanisms of neurodegeneration

The intracellular and extracellular aggregation of mutated and/or misfolded proteins into highly ordered β-sheet rich aggregates, termed amyloid, is a common hallmark of many neurodegenerative diseases such as Alzheimer's (AD), Parkinson's (PD) and Huntington's (HD) disease (Fig.1).

AD is the most common progressive neurodegenerative disease, affecting millions of people worldwide. AD patients show neuronal loss, the deposition of extracellular amyloid plaques consisting of amyloid β peptides (Aβ, (Aβ$_{40}$ and Aβ$_{42}$)) and intracellular neurofibrillary

---

* Corresponding Author

tangles consisting of hyperphosphorylated and cleaved Tau. Although various models have been proposed to explain the pathogenic processes underlying the disease, the exact mechanisms leading to AD are not fully understood (Mudher & Lovestone, 2002).

The amyloid cascade hypothesis proposes that aberrant cleavage of the amyloid precursor protein (APP) by two different proteases (β/γ-secretases) leads to the accumulation of aggregation-prone Aβ peptides that eventually cause disease through multiple mechanisms, including microglial infiltration, generation of reactive oxygen species and synaptic damage (Hardy & Selkoe, 2002). According to this model, early intracellular Aβ accumulation induces the aggregation of the microtubule associated protein Tau. In familial forms of AD, genetic mutations within the APP or γ-sectretase cause extensive formation of Aβ protofibrils, which leads to neurotoxicity.

The Tau or tangle hypothesis, however, claims that it is the disruption of microtubule binding and aggregation of Tau by phosphorylation or genetic mutations that initiates the disease cascade (Maccioni et al., 2010). The formation of neurofibrillary tangles subsequently leads to disintegration of the neuronal cytoskeleton, which causes the disruption of neuronal transport and cell death.

Although Tau and Aβ became the focus of extensive research, the exact disease mechanisms remain unclear. The complexity of the disease is reflected in the fact that many other susceptibility genes or proteins involved in development and progression of this disease have been identified in the last decade.

After AD, PD represents the second most prevalent neurodegenerative disease. It is characterized by the loss of dopaminergic neurons in the substantia nigra and formation of intracellular inclusions (Lewy bodies) consisting of α-synuclein, a presynaptic protein of unknown function (Martin, I. et al., 2011). Importantly, in idiopathic Parkinson's disease, α-synuclein inclusions first localize to defined brain areas and pathology appears to progress in a topographically predictable manner (Braak et al., 2004). The majority of cases are sporadic and age-related. However the identification of several mutated genes in the small number of early onset familial forms of PD implicate various proteins contributing to disease progression. So far, genetic mutations in α-synuclein (SNCA), parkin (PARK2), ubiquitin carboxyl-terminal esterase L1 (UCHL1), parkinson protein 7 (PARK7, DJ-1), PTEN-induced putative kinase 1 (PINK1), leucine-rich repeat kinase 2 (LRRK2), α-synuclein interacting protein (SNCAIP) and glucosidase, beta, acid (GBA) have been linked to familial forms of PD (Martin, I. et al., 2011).

Polyglutamine expansions within unrelated proteins are the underlying cause of nine different neurodegenerative pathologies, including HD, spinobulbar muscular atrophy and spinocerebellar ataxias (SCAs) (Hands & Wyttenbach, 2010). HD, a dominantly inherited neurodegenerative disease that is caused by an expansion of the polyglutamine tract in the Huntingtin (HTT) protein, is the most widely studied of these diseases. It manifests in movement disorder, psychological disturbances and cognitive dysfunction. Hallmarks of HD are cytoplasmic and nuclear inclusions that consist of N-terminal fragments of expanded HTT. Mutant HTT causes cellular dysfunction and neurodegeneration, probably through a combination of toxic gain-of-function and loss-of-function mechanisms. Many proteins have been found to localize to HTT inclusions and it has been postulated that mutant HTT interferes with the function of a diverse variety of cellular proteins, leading to toxic alterations of many pathways.

Although the disease-related proteins in AD, PD and HD have different identities, common cellular mechanisms underlie the formation of oligomeric complexes and amyloidogenic aggregates. Whether neurotoxicity is caused by a toxic loss-of-function of the disease-associated proteins or a toxic gain-of-function of the built up amyloids is still under debate. Gaining more insights into the cellular and molecular pathways leading to aggregation and neurotoxicity could, thus, help to identify general and specific targets for therapeutic intervention in a variety of neurodegenerative diseases.

1.2 Protein aggregation and neurotoxicity

Maintaining the functionality of the proteome in a highly dynamic cellular environment constantly exposed to physical, metabolic and environmental stresses is essential for cell survival. Protein homeostasis, or "proteostasis", is controlled by a highly interconnected network of different protein quality control pathways that balance protein folding, degradation and aggregation (Kettern et al., 2010). Molecular chaperones are central players in this network. According to the current concept of chaperone function, chaperones act as surveillance factors that scan the cell and recognize non-native proteins. Upon binding to a substrate protein, they prevent its aggregation by facilitating folding or disposal through the proteasomal (CAP – chaperone assisted proteasomal degradation) or autophagic (CMA – chaperone mediated autophagy, CASA – chaperone assisted selective autophagy) machineries. The association of regulatory cochaperones determines the function of the chaperone as a "folding" or "degradation" factor. Inefficient or unsuccessful folding of a substrate protein thereby enhances the probability that degradation-inducing cochaperones will associate and initiate ubiquitination of the client protein. The different protein quality control pathways do not operate independently from each other and their coordinated action allows the cell to adjust to different alterations that endanger the integrity of the proteome.

Aggregation of disease-linked proteins probably represents a second line of defense against cytotoxic effects of misfolded proteins. At this stage, aggregate formation is probably a cytoprotective mechanism by which the cell sequesters misfolded proteins or oligomers that cannot be degraded by the proteasome. It has been shown that aggregate formation is not an uncontrolled process, but can be defined as part of the cellular protein quality control (Tyedmers et al., 2010). Usually, protein aggregates can be cleared by macroautophagy (Martinez-Vicente & Cuervo, 2007). However, this second line of defense is also overtaxed in neurodegenerative diseases, as the enhanced production of misfolded proteins is cumulative. Thus, the continuous presence of large or numerous aggregates at later disease stages probably exerts a negative effect on many cellular functions and enhances toxicity. The inhibition of cell function may be due to either sterical hindrance of cellular processes, such as axonal transport, or to coaggregation of other proteins that are then depleted from the cell (Olzscha et al., 2011). Taken together, understanding the mechanisms that lead to aggregation of disease-linked proteins is crucial for the identification of processes that cause toxicity in neurodegeneration.

Sporadic neurodegenerative diseases are usually age-related and reflect the overtaxing of a large variety of cellular processes that normally control protein homeostasis (Douglas & Dillin, 2011). In contrast hereditary early-onset dementias are caused by a set of genetic mutations that lead to the constant production of misfolded and aggregation-prone proteins. Mutations causing familial neurodegenerative diseases have been shown to increase the aggregation tendency of several disease-associated proteins in different ways.

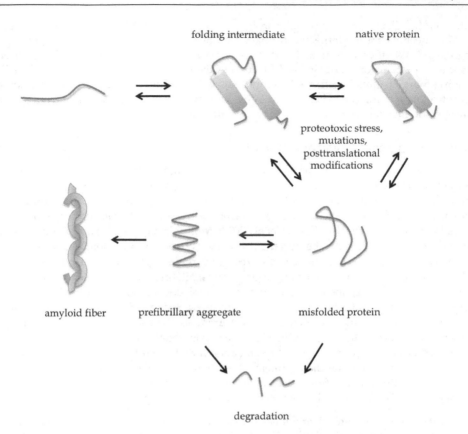

Fig. 1. Protein aggregation in neurodegenerative disease. To be functional, a protein has to fold into an appropriate three-dimensional structure. Aggregation of unfolded cellular proteins is usually prevented by the quality control machinery, which supports the folding process and ensures the removal of unfolded or misfolded proteins. Mutations that enhance the tendency of disease-linked proteins to aggregate, overexpression of aggregation-prone proteins or proteotoxic stress during aging may, however, overwhelm the cellular folding and degradation machineries. Once the quality control machinery is overwhelmed, protein aggregation may be a second line of defense to prevent cytotoxic effects of misfolded proteins or prefibrillary aggregates. Most of the disease-linked proteins in neurodegenerative diseases form highly ordered β-sheet-rich amyloid fibrils. The continued production of aggregation-prone proteins concomitant with decreased protein degradation leads to increased aggregation, likely leading to toxicity at later stages.

Often, mutations in the coding sequence of the disease-causing gene affect the propensity of the protein to misfold and, eventually, aggregate. In HD and several spinocerebellar ataxias (SCAs), a clear correlation between the lengths of the polyglutamine stretch in the disease-causing proteins and the tendency to aggregate has been observed (Martindale et al., 1998). Furthermore, three missense mutations in the α-synuclein gene that are associated with early-onset forms of PD (A30P, E46K and A53T) have been found to enhance the propensity of α-synuclein to aggregate (Li et al., 2001, Greenbaum et al., 2005). Several mutations in Tau have been linked to various neurodegenerative diseases; while Tau mutations affect different aspects of the protein's function, all result in increased aggregation (Wolfe, 2009).

Importantly, altered gene dosage is sometimes sufficient to cause disease. A genomic duplication or triplication involving the α-synuclein locus has been found to cause some forms of familial early-onset PD (Chartier-Harlin et al., 2004, Singleton et al., 2003). In addition, the higher risk for dementias with neuropathological features of AD in Down's syndrome has been attributed to the triplication of the APP gene in these patients (Rumble et al., 1989).

Mutations can, however, also alter posttranslational modifications of the disease-causing proteins. Proteolytic cleavage of disease-related proteins often precedes amyloid deposition and mutations in the coding region can also affect proteolytic processing of the disease-related protein. Generation of the neurotoxic Aβ peptide by β- and γ-secretase mediated sequential proteolysis of APP plays a central role in AD (O' Brien & Wong, 2011). Pathogenic APP mutations that cause early-onset familial AD are clustered around the α-, β-, γ-secretase cleavage sites and affect the ratio of $A\beta_{40/42}$, the latter having an increased propensity to form amyloid plaques (van Dam & De Deyn, 2006). The role of proteolytic cleavage of Tau in neurodegeneration is less well understood, but proteolysis of Tau can be clearly linked to aggregation and neurotoxicity (Wang et al., 2010). Proteolytic cleavage of several polyglutamine disease-linked proteins liberates toxic protein fragments that can form aggregates (Shao & Diamond, 2007). Finally, hyperphosphorylation of Tau and phosphorylation of α-synuclein and ataxin-1 has been shown to enhance their aggregation. Recently, other posttranslational modifications such as oxidation, sumoylation, ubiquitination or nitration have also been implicated in the aggregation of Tau, α-synuclein and polyglutamine-rich proteins (Beyer, 2006, Martin, L. et al., 2011, Pennuto et al., 2009). In addition, glycosylation affects the processing of APP (Georgopoulou et al., 2001).

2. Interactome mapping in diseases

In the last decades, significant progress has been made uncovering a large number of genetic mutations that are associated with a variety of genetically inherited disorders, summarized in the Online Mendelian Inheritance in Man database (OMIM) (Amberger et al., 2011). Clinical symptomatology, however, is less dependent on single mutations than on how whole organisms and systems are altered.

The pathology of one single gene mutation mapped to a specific disease is rarely caused by the malfunction of just one mutated gene product, but rather reflects perturbations of the whole interaction network in which the altered protein is embedded. This concept is in line with the finding that cellular proteins linked to the same disease exhibit a high tendency to interact with each other (Barabasi et al., 2011). In the interactome, the corresponding disease module, thus, consists of a group of proteins that function together in a cellular pathway or process whose breakdown results in a specific pathophenotype. To understand local network perturbations underlying the diseases phenotype, it is necessary to systematically explore the complex interaction network in which the disease-associated proteins are interconnected. Systems-based approaches to human diseases can, thus, lead to the identification of new disease genes and pathways. As drug discovery starts to concentrate on network-based targets rather than single gene targets, a deeper understanding of the disease module is crucial for the development of treatments (Morphy & Rankovic, 2007). Moreover the identification of disease modules can help to uncover subnetworks of interacting proteins that are shared between diseases with similar pathologies, such as neurodegeneration. A powerful approach to detect disease modules in neurodegenerative diseases is the interactome mapping of the different disease-linked proteins in combination with the generation of a complete map of the human interactome. The integration of the interactome of a diseases module with gene expression data or structural proteomic data will significantly advance our understanding of the pathophysiology of the disease in search for effective treatments.

2.1 The human interactome

To date, most attempts to map protein-protein interaction networks have been made using model organisms such as the yeast *Saccharomyces cerevisiae*, the nematode *Caenorhabditis elegans* and the fruitfly *Drosophila melanogaster*. This model organism usage is due to the better and earlier annotation of their genomes. In the last decade, several attempts to map the human interactome have been made, representing a first crucial step towards the understanding of cellular interconnectivity and the identification of networks that play a key role in human diseases (Vidal et al., 2011). In addition, the combination of high-throughput datasets and literature-based protein-protein interactions into databases helped to extend existing interactome maps and made information obtained from high-throughput screens more accessible (Tab. 1).

Database	Description	Webadress
BioGRID	*Biological General Repository for Interaction Datasets*	http://thebiogrid.org/
BIND	*Biomolecular Interaction Network Database:* component database of BOND (*Biomolecular Object Network Databank*) integrates a range of component databases including Genbank and BIND, the Biomolecular Interaction Network Database, resource for cross database searches	http://bond. unleashedinformatics. Com/
DIP	*Database of Interacting Proteins*	http://dip.doe-mbi.ucla.edu/dip/Main.cgi
GWIDD	*Genome Wide Docking Database:* integrated resource for structural studies of protein-protein interactions on genome-wide scale	http://gwidd. bioinformatics.ku.edu/

Database	Description	Webadress
Human Protein Interaction Database	interactions of HIV-1 proteins with host cell proteins, other HIV-1 proteins, or proteins from disease organisms associated with HIV / AIDS	http://www.ncbi.nlm.nih.gov/RefSeq/HIVInteractions/index.html
HPID	*Human Protein Interaction Database*	http://wilab.inha.ac.kr/hpid/
HPRD	*Human Protein Reference Database:* integrates information on domain architecture, post-translational modifications, interaction networks and disease association for each protein in the human proteome	http://www.hprd.org/
HUGE ppi	*Human Unidentified Gene-Encoded large proteins*	http://www.kazusa.or.jp/huge/ppi/
IBIS	*Inferred Biomolecular Interactions Server*	http://www.ncbi.nlm.nih.gov/Structure/ibis/ibis.cgi
IntAct	curated from published protein-protein interaction data	http://www.ebi.ac.uk/intact/
MINT	*Molecular Interaction Database*	http://mint.bio.uniroma2.it/mint/Welcome.do
MIPS	*Mammalian Protein-Protein Interaction Database*	http://mips.helmholtz-muenchen.de/proj/ppi/
NetPro	Comprehensive database of protein-protein and protein-small molecules interaction	http://www.molecularconnections.com/home/en/home/products/NetPro
POINeT	prediction of the human protein-protein interactome based on available orthologous interactome datasets	http://point.bioinformatics.tw/main.do?file=architecture.txt&selectMenu=environment.ini

Table 1. A selection of protein-protein interaction databases.

In the past years, improvement of techniques for high-throughput interaction screening has helped to generate high-quality protein interaction data for many organisms, including humans. In addition, an empirical framework has been proposed to evaluate the quality of data generated by high-throughput mapping approaches and to estimate the size of interactome networks (Venkatesan et al., 2009). Based on empirical sizing, the authors predict the human interactome to consist of ~130,000 interactions. Previous estimates of human interactome size range from 150,000 – 650,000 interactions (Venkatesan et al., 2009). To date ~23000 human protein-protein interactions have been reported. With the authors' estimation that 42% of these reported interactions represent true positives, only ~8% of the full interactome size has been identified so far (Venkatesan et al., 2009). Comprehensive mapping of the human interactome will require further development of complementary, systematic, unbiased and cost-effective high-throughput mapping approaches.

2.2 Experimental strategies for mapping protein-protein interactions

A large variety of experimental and computational methods can be used to identify protein-protein interactions. Two complementary methods are primarily applied for the large-scale mapping of protein-protein interactions. Mapping of binary interactions was first accomplished by a high-throughput adaption of the yeast two-hybrid (Y2H) system, originally developed by Fields and Song (Fields & Song, 1989). Mapping of indirect protein associations within protein complexes can be carried out by a combination of affinity purification and mass spectrometry (AP/MS) (Rigaut et al., 1999). Unfortunately both techniques have their limitations in terms of quality and coverage. Quality assessment of datasets from different interaction studies has shown that Y2H and AP/MS data provide the same quality but represent different subpopulations of the interactome, resulting in networks with different topologies and biological functions (Seebacher & Gavin, 2011). Binary maps from Y2H screens were enriched for transient signalling interactions and interactions between different protein complexes, whereas data generated from AP/MS experiments preferentially detected interaction within a protein complex. The combinatorial application of both techniques is, therefore, recommended to obtain more complete protein interaction maps. The lack of a complete interactome map can be overcome to a certain extent by using data-mining strategies to identify sub-networks from different incomplete interactome maps.

2.3 The yeast two-hybrid system (Y2H)

Originally developed by Field and Song (Field & Song, 1989) to detect the interactions of two proteins, Y2H has been further adapted to high-throughput screening (Koegl & Uetz, 2007). The Y2H is based on the finding that many transcription factors can be divided into a DNA-binding domain (*BD*) and an activation domain (*AD*) that maintain their functionality when separated and recombined. In the two-hybrid approach the BD (e.g. from the yeast Gal4 or the *E. coli* LexA protein) is fused to a protein of interest to generate the bait (Fig. 2A). The prey is constructed by the fusion of the *AD* (e.g. from the yeast Gal4 *or* the heterologous B42 peptide) with a set of open reading frames (ORFs). Bait and prey fusions are then coexpressed in yeast. When the bait and prey proteins interact, a functional transcription factor is reconstituted. Reconstitution of the transcription factor is detected by measuring the activity of a reporter gene. Several reporter genes have been used so far. In the "classic" two-hybrid, auxotrophic markers such as HIS3 and LEU2 allow selection by growth on selective media lacking histidine or leucine. Another commonly used reporter is the bacterial β-galactosidase. Recently, GFP has been successfully used as a reporter gene and many others are under investigation. Although Y2H has proved valuable in the confirmation and identification of many protein-protein interactions, there are experimental limitations of this system.

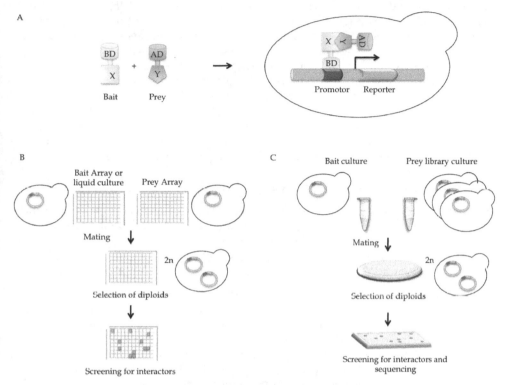

Fig. 2. The yeast two-hybrid system and high-throughput adaptions.
A Principle of the yeast two-hybrid system. To test for a direct interaction, proteins X and Y are coexpressed as fusions with the binding domain (BD), or bait, and the activation domain (AD), or prey, of a transcription factor (e.g. Gal4) in yeast. If protein X and Y interact directly a functional transcription factor is reconstituted which induces transcription of a reporter gene that allows detection of X-Y interaction (e.g. GFP) or selection for X-Y interaction (e.g. HIS3).
B Workflow of the matrix or array approach for high-throughput interaction screening. A prey array is generated by dispensing yeast clones that each express a different AD fusion in a multiwell plate. In an automated step, the prey array is then pinned on a multiwell plate containing yeast clones that express the bait (BD fusion). Prey and bait clones are allowed to mate and diploids are selected based on the expression of selection genes. If the bait and prey proteins interact directly, then the expression of a reporter gene that allows screening for interactors of the bait protein is induced.
C Principle of the exhaustive or random library screen for high-throughput screening. In this screen, a bait fusion is screened against a pooled prey library consisting of ORFs or ORF fragments. Diploids and positive interactors are selected based on growth of selection plates. In contrast to the array approach each positive clone has to be picked and sequenced after selection to identify the prey protein.

To obtain worthwhile Y2H results, the bait and prey fusion proteins must properly fold and not be hindered from proper interaction. Their interaction must be stable and not require posttranslational modification. Furthermore, the interaction in the Y2H takes place in the nucleus, though this restriction can be circumvented by the use of protein fragments that fold more efficiently and are able to translocate to the nucleus. AD and BD fusions alone can also sometimes auto-activate reporter gene expression. This is excluded by performing a self-activation assay of the bait and prey constructs. In this test, a control plasmid coding for an unrelated BD or AD fusion-protein is added along with the construct of interest. The auto-activation background of a HIS3 reporter can be reduced by the addition of the HIS3 inhibitor, 3-AT. The usage of the opposite fusion construct or a different protein fragment is also often effective in suppressing auto-activation. In addition, several alternative genetic screening techniques have been developed to allow the detection of interactions for transcription activators and membrane proteins (Auerbach et al., 2002).

2.3.1 Large-scale yeast two-hybrid screens

The most powerful application of the Y2H system, one that can generate comprehensive protein-protein interaction maps, is the unbiased screening of whole libraries. Two adaptions of the "classical" Y2H for high-throughput screening are the "matrix approach" or "array approach" and the "exhaustive library screening" or "random library screening" approach (Fig. 2B and 2C) (Koegl & Uetz, 2007).

In the array approach, a set of defined prey proteins is tested for interaction with a bait protein (Fig. 2B). Bait and prey constructs are individually transformed into isogenic reporter strains of opposite mating types. Yeast clones expressing a single AD fusion are dispensed into single wells of a multi-well plate, generating a matrix of AD fusions. The array of AD fusions is then spotted onto another multi-well plate containing one yeast clone expressing a single BD fusion for mating. In this way, all BD fusions are mated to AD arrays. A positive interaction is then detected by the ability of a diploid cell to activate the reporter gene (e.g. growth on selective media). As multiple assays can be performed with the same system under identical conditions, they can later be compared. To exclude false positives, experiments are usually performed in duplicates and only interactions found in both experiments are considered to be true. As whole genomes are available as ordered clone sets, each component of the array has a known identity and no sequencing is required after identifying a positive interaction. To accelerate the screening procedure, a pooling strategy can be applied. In the pooled array screen, AD fusions of known identity are tested as pools of AD fusions against the BD strains. This means that the identification of interaction partners from a positive pool requires retesting of all members. It is also possible to use pools of baits against an AD array.

A second approach to analyse the interactome of whole proteomes is the "exhaustive library screening" or "random library screening" approach (Fig. 2C). In this assay, a BD fusion is screened against a library of AD fusions of full Length ORFs or ORF fragments. For this approach, it is not necessary to know the sequence of the whole genome, as random prey libraries can be generated from randomly cut genomic DNA. In addition a large number of cDNA libraries is commercially available. Like the array screen, the bait and prey constructs are individually transferred into isogenic yeast strains of opposing mating types. However, in contrast to the matrix approach, the different AD fusions are not separated on an array.

Instead each *BD* fusion strain is mated with pooled *AD* fusion strains. After mating of the two strains, the diploid yeast cells are plated on selective media to screen for interactions. The identity of the *AD* fusion has to be determined by yeast colony PCR of positive colonies followed by sequencing.

In terms of quality assessment, rigorous evaluation and filtering of the raw data will enhance the quality and reliability of results. Routine testing of positive interactions is performed in duplicate and protein interactions that are not reproducible are discarded. From interactome studies, it is estimated that the coverage of array-based two-hybrid screens is only ~20% (Rajagopala et al., 2011). This high false negative rate is probably caused by the technical limitations of the system. Furthermore, different attempts to map the yeast interactome show very little overlap with each other and with annotated protein-protein interactions, probably due to the usage of different Y2H systems (Uetz et al., 2000, Ito et al., 2001). Recent attempts to generate a high quality dataset of the yeast interactome have revealed over 1000 new interactions (Yu et al., 2008). Further technical development and improvement is needed to increase the coverage of interactome screens. However, as the Y2H only covers a subset of interactions, a complementary AP/MS approach is necessary to generate a comprehensive map of the interactome (Yu et al., 2008).

2.4 Affinity purification coupled to mass spectrometry

Most proteins function in cellular processes as multi-subunit protein complexes. A well-established method to identify protein co-complexes is based on affinity purification followed by mass spectrometry (Fig.3) (Bauer & Kuster, 2003). The classical coimmunoprecipitation protocol is commonly used to detect whether two proteins interact in cellular systems, but can be also used to identify new interaction partners. In this approach, one protein is affinity captured along with its associated proteins by a specific antibody immobilized on Protein A or G sepharose. Experimental details like the affinity tag, lysis conditions, incubation time and washing conditions have a significant impact on the output and need to be optimized depending on the protein complex stability and localization. To preserve protein-protein interactions and native conformations, relatively mild conditions should be used during lysis. TritonX-100 or NP-40 are widely used in cell lysis buffers as non-ionic detergents and efficiently lyse membranes but are mild enough to preserve protein-protein interactions. Other variables that can influence the outcome of the affinity purification are salt concentrations, divalent cation concentrations and pH. The purified complex is rinsed several times to remove unspecifically bound proteins and complexes are eluted from the resin by low or high pH, high salt concentrations, by competition with a counter ligand or by adding Laemmli buffer to the beads. After purification, the isolated protein complex is usually separated on 1D or 2 D gels and protein lanes or spots are digested by trypsin and subsequently analyzed by mass spectrometry. It is also possible to precipitate eluted protein complexes and trypsinize the pellet. Comparison of the proteins detected in the sample with a negative control leads to the identification of specific interaction partners. Although low background binding is favoured, the stringency of washing steps needs to be adjusted to preserve more dynamic or transient interactions. Nonspecific binding can be reduced by adding low levels of detergent or by adjusting the salt concentration.

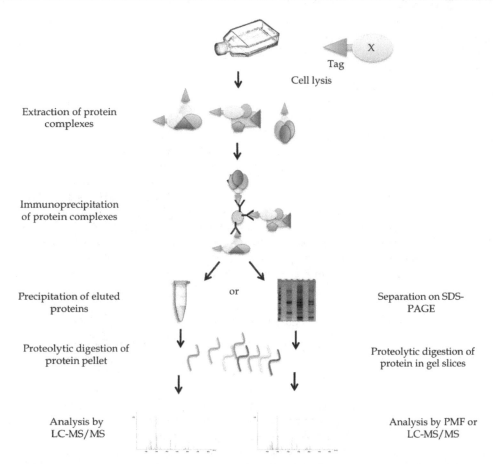

Fig. 3. Workflow of AP/MS for high throughput screening. Cells that express an epitope-tagged form of the protein of interest are lysed to extract protein complexes. The protein complexes are then isolated using a tag- specific antibody. The antibody-bound protein complex is immobilized on Protein A or Protein G sepharose and non-specific interactors are eliminated by several wash steps. As a control, the immobilized antibody is incubated with an untagged cell extract or a cell extract containing a tagged control protein (e.g. GFP). After purification, the protein complexes can be either separated on a gel or eluted and precipitated. The gel slices or the protein pellet are then digested to generate peptide fragments that can be analyzed by mass spectrometry to identify co-complexed proteins.

2.4.1 Large scale AP/MS screens

Proteome wide interaction studies need to meet certain criteria, such as reproducibility and low background binding. To circumvent the need of a specific antibody, the bait protein can be fused to an epitope tag that allows the pull down of different protein complexes using the same antibody under standardized conditions. A large variety of different epitopes has been used in the past and are commercially available (e.g. HA, FLAG, MYC). One disadvantage of this technique, however, is that the epitope-tagged proteins are overexpressed in the cell. Cytotoxicity induced by overexpression can be prevented by the usage of inducible expression systems. The tag might also interfere with protein folding and localization. Despite these limitations, coimmunoprecipitation via an epitope tag has been successfully used in proteome wide screens (Gavin et al., 2002, Ho et al., 2002).

Another coimmunoprecipitation protocol that is frequently used for high-throughput experiments is the tandem affinity purification (TAP) developed by Rigaut (Rigaut et al., 1999). The basic concept is similar to the coimmunoprecipitation of epitope tagged proteins. The main difference, however, is the use of two tags. One commonly used TAP tag consists of an IgG binding protein (Protein A) linked to a calmodulin-binding domain (CBD) via a TEV protease recognition site. Protein complexes are purified by the incubation of lysates with IgG sepharose beads that capture the Protein A tag. After washing the immobilized complex is eluted by TEV cleavage. Eluted protein can then be bound to a calmodulin sepharose column via their CBD domain. As the binding to calmodulin is calcium dependent, immobilized protein complexes can be eluted by addition of EDTA. The major advantage of this technique over one-step purification is increased specificity, reducing background binding to levels suitable for large-scale analysis (Gavin et al., 2002, Krogan et al., 2006). However, the long purification protocol preserves only stable interactions. For this reason, one-step purification with an epitope-tagged protein of interest is preferable if a broader range of interactions should be captured. Although the stringency of the purification protocol needs to be adjusted, the high sensitivity of mass spectrometry requires a minimum of non-specific binding to avoid laborious post-experimental filtering, especially in the case of large-scale experiments. An alternative approach to eliminate non-specific interactors without loosing low abundant or transiently interacting proteins is the use of quantitative mass spectrometry (Kaake et al., 2010).

The analysis of the isolated protein complex typically starts from separation of the complex components on 1D SDS-PAGE gels based on their molecular weight or 2D gels based on their charge and molecular weight followed by protease digestion. Usually, trypsin or Lys-C are employed to cleave the protein in peptide fragments because they generate peptides that have basic amino acids at their C-termini, which is favourable for detection and sequencing by mass spectrometry. In large-scale experiments, however, all purified proteins are precipitated and digested together to avoid time-consuming separation steps by gel electrophoresis and instead protein samples are fractionated by liquid chromatography (LC) after tryptic digest. Depending on the complexity of the sample, different approaches for protein identification by mass spectrometry can be used (Bauer and Kuster 2003). Complex peptide mixtures are analyzed by liquid chromatography in combination with tandem mass spectrometry (LC-MS/MS), whereas protein samples that have been separated on 1D SDS-PAGE or 2D gels and have therefore a lower complexity are analyzed by peptide mass fingerprinting (PMF).

PMF is an analytical method in which the absolute masses of peptides from a tryptic digest can be measured by matrix-assisted laser desorption/ionization time-of-flight mass spectrometry (MALDI-TOF). The measured masses of the sample are then looked up in a database that contains predicted peptide masses that have been generated by *in silico* digests for every protein. The results are statistically analyzed to find a significant overlap between the experimentally generated and predicted peptide masses. The identification method is based on the assumption that it is very unlikely that the exact same combination of peptide masses is found in more than one protein. Although this method can be applied to protein mixtures, the presence of different proteins complicates the analysis. Therefore, PMF is usually applied to identify proteins or protein mixtures isolated from 1 or 2D gels. Mixtures of more than five proteins require the additional use of tandem mass spectrometry for reliable protein identification.

Tandem mass spectrometry, uses a combination of sequence and mass information for the identification of proteins (MS/MS). In a first step, the masses of peptides from the tryptic digest are measured. Single peptides are then isolated from the mixture in the mass spectrometer and collided with inert gas molecules, which leads to fragmentation of the peptide backbone. Peptide fragments generated by this method differ in length by one amino acid. Measurement in the mass spectrometer allows determination of mass differences between two adjacent fragments that indicate a particular amino acid residue. The peptide mass, fragment mass and sequence is then compared against one or more databases to identify the proteins in the sample. Each peptide is individually identified and attributed to a protein in the mixture. For the analysis of complex samples, tandem mass spectrometry is generally combined with liquid chromatography to fractionate the peptide mixture before analysis by tandem mass spectrometry (LC-MS/MS).

Due to the sensitivity of the method and depending on the stringency of the purification protocol, post-experimental filtering is necessary to eliminate non-specific interactors. The ultimate goal of post-experimental analysis is the maximum reduction of false positives while maintaining the maximum coverage. Usually proteins that coimmunoprecipitate with a protein of interest are compared to proteins that are detected in a control sample to define unspecific interactions. Each of the experiments is commonly done at least in duplicate to identify high confidence interacting proteins. Whereas reproducibility seems to be a good indicator for true positive interactions, the quality assessment by comparisons with existing datasets depends on the selected gold standard. An ideal gold standard reference data set would be confirmed by other sources and should be generated by a comparable method to that being applied. Yu et al. (Yu et al., 2008) used an approach in which they tested several reference databases using various techniques to select a good gold standard.

2.5 Quality assessment of interactome mapping datasets

The identification of disease-modules and the interactome mapping of disease-linked proteins relies on the quality of the available reference datasets and the filtering of the experimentally generated interaction data. In early studies the precision of interactome maps was estimated by the integration of other biological attributes, such as gene ontology, or comparison with literature curated datasets. However, these methods suffer limitations in the estimation of data quality, as they need to be complete and unbiased. Recent efforts to establish an empirical framework for protein interaction maps will improve the estimation of accuracy and sensitivity for interaction maps generated by high-throughput interaction screens (Venkatesan et al., 2009). This empirical framework evaluates four different quality parameters of the currently used methods to estimate quality:

1. *Screening completeness*: the fraction of overall possible ORF pairs that are tested to generate the interaction map.
2. *Assay sensitivity*: the fraction of all interactions that could be identified by the applied experimental method.
3. *Sampling sensitivity*: the fraction of all detectable interactions that are identified in one trial of an assay under specific experimental conditions.
4. *Precision:* the fraction of known true positive pairs in the dataset.

Estimation of these parameters offers a quantitative idea of the coverage and accuracy of an interaction map and, when used in a standardized way, enables the comparison of different datasets.

3. Interactome mapping of disease-linked proteins in neurodegenerative diseases

Although great efforts have been made to uncover specific and shared pathways underlying neurodegeneration, many cellular mechanisms involved remain to be uncovered. As many disease-linked proteins function in multisubunit complexes and are functionally embedded in large cellular networks, gaining extensive knowledge of the architecture and composition of these complexes and networks is crucial to identify disease-linked pathways and to establish new disease markers. Based on this knowledge, diagnosis of early-onset dementias could be significantly improved and new therapeutic strategies could be developed. For age-related dementias, the discovery and characterization of shared pathways could pave the way towards common therapeutic interventions. To uncover the networks and complexes in which disease-linked proteins are embedded, interaction studies of many disease linked proteins, such as AD–associated (Chen et al., 2006, Krauthammer et al., 2004, Liu et al., 2006, Norstrom et al., 2010, Perreau et al., 2010, Soler-Lopez e a., 2010, Tamayev et al., 2009), PD-associated (Engelender et al., 1999, Meixner et al., 2011, Schnack et al., 2008, Suzuki, 2006, Woods et al., 2007, Zheng et al., 2008), HD-associated (Goehler et al., 2004, Kaltenbach et al., 2007) and ataxia-associated proteins (Kahle et al., 2011, Lim et al., 2006) as well as of prion protein (PrP) (Nieznanski, 2010) have been conducted. In addition, an interactome study that utilizes artificially designed amyloid-like fibrils has been performed and focused more on the general mechanisms underlying the toxic gain-of-function of β-sheet rich proteins (Olzscha et al., 2011).

3.1 Interactome mapping of disease- linked genes by large-scale Y2H

Several Y2H studies have been conducted to identify interaction partners of disease-related proteins in neurodegenerative diseases. In the following section, we highlight several studies that have contributed significant insight into disease networks by using new approaches or developing new ideas.

3.1.1 Generating the AD interaction network

An interesting approach to generate a complete interaction map for AD has been employed by Soler-Lopez et al. (Soler-Lopez et al., 2010). The authors elegantly combined different strategies to establish the most complete AD interaction network to date by exploring the interactomes of all known AD-linked proteins. The integration of their data with literature-based information provides new insights into the molecular interplay of different functional modules within the AD network and has helped to identify new candidate genes. Whereas most of the other interactome mapping approaches only focused on one prominent disease-linked protein, this study underlines the need to study the complete set of disease-associated genes in order to get the whole picture of a complex disorder like AD.

Although extensively studied, the cellular mechanisms that underlie the neurophathological changes associated with AD remain elusive. Despite the central role of APP and Tau, AD is a genetically complex disease and several genetic risk factors have been identified in the last decades. As it has been shown that the susceptibility/causative genes for many diseases are often interconnected within the same biological module, Soler-Lopez et al. applied different network biology strategies in their recently published study to build the most complete AD related interactome (Soler-Lopez et al., 2010).

The authors used available information from the Online Mendelian Inheritance in Man database to assemble a set of 12 well established AD related genes that they termed "seed". To be classified as a seed gene at least one mutation of the gene must be associated with AD in the OMIM Morbid Map. Interestingly, quantification of the degree of connectivity by computing the minimal path length between seed genes shows that the seed genes are connected by three links on average, whereas control sets (randomly picked genes and randomly picked disease causing genes from different disorders) were connected by more then four links on average. Starting from their defined AD seed genes, the authors developed a strategy to discover new disease genes. For this purpose they aimed to identify proteins that match at least one of the following criteria: 1.) the encoded proteins must directly interact with the seed genes, 2.) the gene must locate to a known genetic susceptibility locus and 3.) the gene must have altered expression in AD.

To define the interaction network of their selected seed proteins, they performed Y2H interaction screens. Baits were constructed for nine genes from the seed (three ORFs were not available) and yeast clones for the individual baits were transformed with an adult brain cDNA prey library. Each experiment was done in five replicates. 72 positive interactions resulted from this screen and were retested by pairwise cotransformant Y2H arrays. 32 high confidence interactors for the seed proteins could be validated.

In a second approach, the authors performed a pairwise candidate screen based on published genome wide association studies that identified four chromosomal regions (7q36, 10q24, 19q13.2, 20p) containing unknown AD susceptibility genes. They focused on 185 genes that are located in chromosomal regions linked to AD. Of those, 44 candidate genes known to be coexpressed with AD genes and suitable for Y2H screening were tested for their interaction with the 9 seed proteins. A systematic matrix-based Y2H was performed to identify interactors of the AD seed among the candidates extracted from the AD linked chromosomal regions. Two different technical approaches, a mating and a cotransformation screen, were used to perform the Y2H. With the identified interactions from the two Y2H, the authors generated a high-confidence interaction core (HC). The interactors identified in the library screen had to be validated to be included in the HC and the interactors from the pairwise candidate screen had to activate at least two reporter genes to be defined as high-confidence interaction. The final HC contained 8 seed genes (no HC interactors were identified for one seed gene) and 66 interactors, 27 from the library screen and a further 39 from the pairwise candidate screen. Interestingly, the different interactor set showed no overlap and only a low overlap with other studies, highlighting the importance of utilizing multiple approaches to increase coverage.

The HC interactions were validated by complementary strategies, such as GST pull down assays, coimmunoprecipitations and colocalization studies. Due to the high stringency applied for the identification of HC interactors, 87% of the interactors could be validated. The analysis of the HC identified four novel direct interactions between well-established AD related proteins (APP, A2M, APOE, PSEN1, PSEN2), implicating a possible link between plaque formation and inflammatory processes and providing insights into the regulation of APP cleavage. The assembled network showed an enrichment of 3 biological processes (oxidation reduction, regulation of apoptosis and negative regulation of cell motion), 5 molecular functions (protein binding, mono-oxygenase activity, oxygen binding, actin binding and integrin binding) and 6 cellular compartment terms (cytoplasm, pseudopodium, platelet alpha granule lumen, cytosol, cytoskeleton and internal side of plasma membrane) as well as altered expression of 17 out of 66 interactors, based on microarray data (Blalock et al., 2004). Furthermore, 6 out of 58 direct interactors were listed in the AlzGene database, whereas none of the non-interacting proteins were found in the database. Network analysis suggests a role for the programmed cell death 4 protein (PDCD4) in regulating neuronal death. PCDC4 was found to bind to the seed proteins PSEN2 and APOE and is located in a functionally homogenous network module enriched for translational elongation. As PCDC4 is upregulated in AD brain tissue, the authors suggest that the protein plays a role in Aβ toxicity. Another interesting member in the AD interactome network (AD-PIN) is ESCIT (evolutionary conserved signalling intermediate in Toll pathway), which links the redox signalling and immune response modules. Based on published data, the authors hypothesize that ESCIT may represent a molecular link between mitochondrial processes and AD lesions. Taken together, newly identified genes in the HC are likely related to AD onset or progression.

To generate a complete AD-PIN, the interaction partners of the HC were retrieved from literature-curated databases. The AD-PIN was then used to identify functional modules within the network that help to link processes potentially involved in AD and to identify the relationship between new candidate genes.

Taken together, this integrative approach combined stringent interaction screening and extensive validation by complementary strategies to build a comprehensive disease interaction map and highlight the importance of a network view of disease and the necessity of data integration from different sources when exploring disease interactomes.

3.1.2 The ataxia-ome

Disorders having a common clinical presentation likely also have common altered pathways. Examples of such diseases are familial spinocerebellar ataxias. A multitude of disease-associated mutations have been discovered, each leading to gain or loss of normal protein function. These mutations all inevitably lead to loss of Purkinje cells through an as yet unknown common pathway. An interesting study by Lim et al. used an Y2H approach to explore the interaction network of 54 proteins involved in 23 ataxias (Lim et al., 2006).

This study nicely showed that interactome mapping of different disease related genes can help to identify common pathways shared by diseases with similar presentation, such as neurodegenerative disorders.

Similar to the AD interactome study, the authors first defined a set of 23 ataxia-causing genes (11 recessive and 12 dominant, including the polyglutamine-mediated SCAs) whose mutations are linked to ataxias in humans or mice. Paralogs and an additional 31 directly interacting proteins were grouped together as ataxia-associated proteins. Bait and prey constructs for each of the genes were constructed to perform matrix-based mating type reciprocal screens against the human ORFeome. To minimize false positives, they used a stringent version of the Y2H system that expresses fusion proteins at low levels. Yeast clones containing single ataxia baits were mated with yeast clones of the opposite mating type containing 188 different ORF minilibrary pools (Rual et al., 2005). In a second screen, reciprocal interactions were tested between human ORFeome baits and ataxia preys to exclude effects caused by misfolding of the fusion protein and to include autoactivation baits. The overlap of the two screens comprises only 5.2% of the observed interactions, which is typical for reciprocal studies (Rual et al., 2005). To include a tissue specific library, the authors screened a human brain cDNA library using the same experimental setup. 29 interactions were overlapping in both screens, indicating that screening different types of libraries can enhance the coverage of the interactome map, as splice forms of a single gene are often spread across cDNA libraries. 83% of the interactions were validated experimentally by GST pull down assays from HEK293 cells with randomly picked interaction pairs. Analysis of the interaction sets found that 72% of those with compartment annotations colocalize together and 98% of the interactors share a GO branch, demonstrating the high quality of the Y2H generated interaction network.

A large interconnected network between the 36 ataxia genes and 541 preys was revealed for ataxia processes. In addition, 13 ataxia causing baits were linked directly or through common interacting proteins, indicating that the proteins in this interconnected network are functionally linked.

To build the ataxia network, Y2H data were integrated with interaction data from available databases. The network was then divided into a direct ataxia network, which contains first-order protein interactions, and the expanded ataxia network, which contains additional second- or third-order interactions. Further comparison of this ataxia network with networks generated from unrelated disease proteins showed that ataxia network nodes have a shorter path length, with a higher number of single hub interconnections. Surprisingly, 18 out of 23 ataxia causing proteins interacted directly or indirectly.

The ataxia interactome assembled by Lim et al. shows that for different ataxias, the similar pathophenotype is indeed caused by the alteration of shared pathways and processes. Main hubs in this ataxia network are involved in DNA repair and maintenance, transcription, RNA processing and protein quality control. Taken together, this kind of approach shows that genes and proteins that are involved in the same or related disorders are highly interconnected, often operating in the same pathways. In this way, interactome studies can prove valuable in identifying the shared pathways underlying related phenotypes.

3.2 Interactome mapping of disease-linked genes by AP/MS

Several AP/MS approaches have been successfully applied to identify interaction partners of disease-linked proteins in different neurodegenerative diseases. The following section summarizes two quantitative approaches that aim to identify distinct disease-relevant interaction networks.

3.2.1 QUICK screen to identify a PD-associated interaction network

Although PD represents a complex disorder for which several genetic risk factors have been identified, no complementary PD network has yet been generated, unlike for AD (Soler-Lopez et al., 2010). To explore the physiological function of the disease associated leucine-rich repeat kinase 2 (LRRK2), Meixner et al. used a new AP/MS protocol to identify the interaction network under stoichiometric constraints, called QUICK (quantitative immunoprecipitation combined with knockdown) (Meixner et al., 2011, Selbach & Mann, 2006). The QUICK screen aims to detect protein-protein interactions at endogenous levels and in their normal cellular environment by using a combination of SILAC, RNA interference, coimmunoprecititation and quantitative MS. For the QUICK approach, NIH3T3 cells were transduced with a lentiviral shRNA construct to knock down endogenous LRRK2. WT and LRRK2 knock down cells were then grown in SILAC medium containing either normal heavy isotope labeled lysine or arginine. For coimmunoprecipitation, an LRRK2 specific antibody was crosslinked to Protein G sepharose. Cells were lysed in buffer containing 1% NP-40 and phosphatase inhibitors. Equal numbers of heavy and light SILAC labeled cells were incubated with LRRK2 sepharose and then pooled. Purified LRRK2 complexes were eluted after several washing steps in Laemmli buffer. Separation of the proteins was achieved by SDS-PAGE gel electrophoresis, followed by tryptic digestion of gel slices. Interaction partners were identified by subsequent LC-MS/MS. Identified interaction partners were verified using the same approach, without prior SILAC labeling. Using this protocol, 36 proteins were identified as high confidence LRRK2 interactors. Bioinformatic analysis and integration of interaction data from different databases showed that these proteins are mainly members of the actin family and actin-associated proteins, pointing to an important role for LRRK2 in actin cytoskeleton based processes. Additional experiments demonstrated that LRRK2 binds to F-Actin *in vitro* and regulates its assembly. Knockdown of LRRK2 leads to morphological alterations and shortened neurite processes in primary neurons, further indicating a physiological role for LRRK2 in cytoskeletal organization. As the experimental strategy of the QUICK approach aims to identify interactors at physiological conditions (endogenous expression of the target protein) with specific antibodies, this method is well suited to explore the physiological function of other disease-linked proteins. Moreover, using knock-down or knock-out cells for control immunoprecipitations with a specific antibody proved as a suitable control. So far, this approach is not suited for large-scale applications involving several candidate proteins, but is of interest for generating interactome maps for one protein.

3.2.2 Interactome mapping of amyloid-like aggregates

Ordered proteinacious aggregates with high ß-sheet content, termed amyloid, are a characteristic of many neurodegenerative diseases. Whether this aggregation is cytoprotective or cytotoxic is still under debate. One hypothesis for aggregate toxicity is that these aggregates sequester cellular proteins, thereby leading to functional impairment (Chiti & Dobson, 2006).

To uncover the gain-of-function toxicity of amyloid-like fibrils in general, Olzscha et al. defined the interactome of artificial β-sheet proteins designed to form amyloids (West et al., 1999). The advantage of this method is that none of the artificial proteins have endogenous interaction partners, which allows for the identification of common amyloid interacting proteins independent of the identity of the aggregating protein. β4, 17 and 23, which differ in their β-sheet propensity, with β23 having the highest tendency to form β-sheets, were selected for interactome mapping. As a control, the authors used an α-helical protein with a similar amino acid composition (α-S284). Interaction analysis of the different β-sheet proteins was performed using SILAC followed by LC-MS/MS. Constructs coding for MYC-labelled proteins were transfected into HEK293 cells and proteins were labelled with light, medium or heavy arginine and lysine isotopes. Different experimental setups were chosen: (1) empty MYC tag vector, MYC α-S284, MYC β23; (2) MYC α-S284, MYC β4, MYC β17; (3) MYC β4, MYC β17, MYC β23 for quantitative MS. Lysates from heavy, medium and light labelled cells were mixed in a 1:1:1 ratio and amyloidogenic aggregates were isolated using anti-MYC coupled magnetic beads to avoid loss of the protein aggregates due to centrifugation. After purification the bound proteins were eluted and processed for LC-MS/MS analysis. Experiments were performed in triplicate. Proteins were defined as interactors if they were enriched relative to the α-S284 control or one of the β proteins with >95% confidence in two of the three repeats. Interactors were validated by coimmunoprecipitation and western blotting or immunofluorescence analysis. Bioinformatic analysis of the amyloid interactomes revealed that aggregates sequester a large amount of large, unstructured proteins that occupy essential hub position in housekeeping functions such as transcription and translation, chromatin regulation, vesicular transport, cell motility and morphology, as well as protein quality control. The authors hypothesize that amyloidogenic aggregation coaggregates a metastable subproteome, thereby causing perturbations in essential cellular networks leading to toxicity.

3.3 Interactome mapping by combinatorial approaches: The HD interactome

As Y2H and AP/MS approaches identify different groups of interactors, the combination of the two experimental strategies may prove valuable for the generation of comprehensive interaction maps of disease-linked proteins. In a study on HD, Kaltenbach et al. used these two approaches to generate a comprehensive HD interactome, which was then tested for modulators of polyglutamine toxicity. Importantly, the two approaches revealed distinct but biologically relevant interactions (Kaltenbach et al., 2007).

In this study the authors performed Y2H and AP/MS approaches in parallel to generate a comprehensive HTT interactome. Assuming that HTT and its interactors are functionally linked and regulate each other, the HTT interactome should be enriched for genetic modifiers of neurotoxicity. To test this hypothesis, the authors generated a *Drosophila* model of polyglutamine toxicity. For the AP/MS, different recombinant TAP-tagged HTT fragment baits were constructed (HTT 1-90: Q23, Q48, Q75; HTT 443-1100 and HTT-2758-3114) and purified from bacteria for pull down assays with mouse or human brain tissue and mouse muscle tissue. Purified complexes were then analyzed by MS. HTT fragments 1-90 were also used to probe HeLa, HEK293 and M17 neuroblastoma cell lysates. The interaction lists from the different pull downs were filtered by excluding proteins observed in a control pull down with the TAP tag alone and subjected to statistical analysis. The interaction lists were compared to a database of high-scoring peptides from 88 other unrelated pull downs. By using this approach, the authors generated a list of 145 mouse and human specific interactors with WT and expanded HTT fragments. In a complementary assay, a set of additional HTT interactors was identified by Y2H with HTT fragment baits (WT: 23Q and mutant: >45Q) against prey libraries from 17 different human tissue cDNA sources. Again, only baits generated from N-terminal fragments gave reproducible results. Filtering of the data was performed to generate a high confidence interaction dataset. Only interactions that were observed at least three times in the Y2H were integrated into the dataset; interactors were further compared to a database of Y2H interaction screens such that promiscuous interactors could be excluded. Finally, all positive prey constructs were retested by cotransformation into yeast strains expressing bait constructs. Previously published HTT interactors were included regardless of whether they matched the quality assessment criteria. A total of 104 interactors were identified by this Y2H approach.

The interactome was tested for possible genetic interaction in a fly model of polyglutamine toxicity. In this model, the N-terminal fragment of human HTT cDNA, including a 128Q expansion in exon 1, is expressed in the eye. The resulting neurodegenerative phenotype is visible by examining the retinal histology. Interactors that have orthologues in the fly and for which suitable fly stocks for overexpression or partial loss-of-function were available were selected from both high-confidence datasets. 60 interactors divided equally between interactors from the Y2H and AP/MS approach were tested as modifiers of polyglutamine mediated toxicity. 80% of the tested interactors altered the readout of more than one allele in a single background or a single allele in multiple backgrounds. The high confidence interaction dataset was further validated by coimmunoprecipitation studies using lysates from WT and HD mice. Data from this study strongly suggest that Y2H and AP/MS are of comparable quality in identifying biologically relevant interactions. Gene ontology analysis demonstrated that modifiers cluster into different groups, such as cytoskeletal organization, signal transduction, synaptic transmission, proteolysis and regulation of transcription and translation. Based on their experimentally generated interaction datasets, the authors built an interaction map of HD by integrating interaction data from different databases.

Interacting proteins of the HD network that were confirmed as modifiers in the fly model were shown to have diverse biological functions, such as synaptic transmission, cytoskeletal organization, signal transduction and transcription. In addition the authors revealed an unknown association between the HTT fragment and components of the vesicle secretion apparatus, suggesting that modulation of SNARE-mediated neurotransmitter secretion may be a physiological function of HTT. The involvement of HTT in these processes suggests a model in which mutant HTT interferes with different cellular pathways, leading to pathology.

4. Summary

Proteins do not operate alone, but instead function as a large interconnected network. Disease-related mutations, consequently, disrupt not only the function of individual proteins, but also the larger network in which these proteins function, only thereby leading to clinically relevant pathology. While many complex diseases are caused by an array of mutations in any of a multitude of genes, these all lead to a similar pathology. Focusing not just on individual proteins but, rather, on the network of proteins altered in diseases will prove essential to both identifying new candidate genes for this and related disease, as well as to provide new targets for the development of therapeutic interventions. A combination of different experimental and bioinformatic approaches has proved valuable in the network analysis of neurodegenerative diseases. Further studies are necessary to generate high quality, comprehensive datasets, which can be used to identify shared disease pathways.

5. Acknowledgements

We thank Devon Ryan, Daniele Bano, Donato A. Di Monte and Moritz Hettich for providing a critical reading of the manuscript and for helpful suggestions.

6. References

Amberger, J., Bocchini, C. & Hamosh, A. (2011). A new face and new challenges for Online Mendelian Inheritance in Man (OMIM®). *Human Mutation*, Vol. 32, No. 5, (May2011), pp.(564-567), ISSN 1059-7794

Auerbach, D., Thaminy, S., Hottiger, M.O. & Stagljar, I. (2002). The post-genomic era of interactive proteomics: facts and perspectives. *Proteomics*, Vol. 2, No. 6, (June 2002), pp. (611-623), ISSN 1615-9861

Barabási, A.L., Gulbahce, N. & Loscalzo, J. (2011). Network medicine: a network-based approach to human disease. *Nature Reviews Genetics,* Vol. 12, No. 1, (January 2011), pp. (56-68), ISSN 1471-0064

Bauer, A. & Kuster, B. (2003). Affinity purification-mass spectrometry. Powerful tools for the characterization of protein complexes. *European Journal of Biochemistry*, Vol. 270, No. 4, pp. (570-578), ISSN 1432-1033

Braak, H., Ghebremedhin, E., Rüb, U., Bratzke, H. & Del Tredici, K. (2004). Stages in the development of Parkinson's disease-related pathology. *Cell and Tissue Research,* Vol. 318, No. 1, (October 2004), pp. (121-134), ISSN 1432-0878

Beyer, K. (2006). Alpha-synuclein structure, posttranslational modification and alternative splicing as aggregation enhancers. *Acta Neuropathologica*, Vol. 112, No. 3, (September 2006), pp. (237-251), ISSN 1432-0533

Blalock, E.M., Geddes, J.W., Chen, K.C., Porter, N.M., Markesbery, W.R. & Landfield, P.W. (2004). Incipient Alzheimer's disease: Microarray correlation analyses reveal major transcriptional and tumor suppressor responses. *Proceedings of the National Academy of Sciences of the United States of America*, Vol. 101, No. 7, (February 2004), pp. (2173–2178), ISSN 1091-6490

Chartier-Harlin M.C., Kachergus J., Roumier C., Mouroux V., Douay X., Lincoln S., Levecque C., Larvor L., Andrieux J., Hulihan M., Waucquier N., Defebvre L., Amouyel P., Farrer M. & Destée A. (2004). Alpha-synuclein locus duplication as a cause of familial Parkinson's disease. *Lancet*, Vol. 364, No. 9440, (October 2004), pp. (1167-1169), ISSN 0140-6736

Chen, J.Y., Shen, C. & Sivachenko, A.Y. (2006). Mining Alzheimer disease relevant proteins from integrated protein interactome data. *Pacific Symposium on Biocomputing*, Vol.11, pp. (367-378), ISSN 1793-5091

Chiti, F. & Dobson, C.M. (2006). Protein misfolding, functional amyloid, and human disease. *Annual Review of Biochemistry*, Vol. 75, (2006), pp. (333-366), ISSN 0066-4154

Douglas, P.M. & Dillin, A. (2011). Protein homeostasis and aging in neurodegeneration. *The Journal of Cell Biology*, Vol. 190, No. 5, (September 2010), pp. (719-729), ISSN 1540-8140

Engelender, S., Kaminsky, Z., Guo, X., Sharp, A.H., Amaravi, R.K., Kleiderlein, J.J., Margolis, R.L., Troncoso J.C., Lanahan, A.A., Worley, P.F., Dawson, V.L., Dawson, T.M. & Ross, C.A. (1999). Synphilin-1 associates with alpha-synuclein and promotes the formation of cytosolic inclusions. *Nature Genetics*, Vol. 22, No. 1, (May 1999), pp. (110-114), ISSN 1546-1718

Fields, S. & Song, O. (1989). A novel genetic system to detect protein-protein interactions. *Nature*, Vol. 340, No. 6230, (July 1989), pp. (245-246), ISSN 0028-0836

Gavin, A.C., Bösche, M., Krause, R., Grandi, P., Marzioch, M., Bauer, A., Schultz, J., Rick, J.M., Michon, A.M., Cruciat, C.M., Remor, M., Höfert, C., Schelder, M., Brajenovic, M., Ruffner, H., Merino, A., Klein, K., Hudak, M., Dickson, D., Rudi, T., Gnau, V., Bauch, A., Bastuck, S., Huhse, B., Leutwein, C., Heurtier, M.A., Copley, R.R., Edelmann, A., Querfurth, E., Rybin, V., Drewes, G., Raida, M., Bouwmeester, T., Bork, P., Seraphin, B., Kuster, B., Neubauer, G. & Superti-Furga, G. (2002). Functional organization of the yeast proteome by systematic analysis of protein complexes. *Nature*, Vol. 415, No. 6868, (January 2002), pp. (141-147), ISSN 0028-0836

Georgopoulou, N., McLaughlin, M., McFarlane, I. & Breen, K.C. (2001). The role of post-translational modification in beta-amyloid precursor protein processing. *Biochemical Society Symposium*, Vol. 67, (2001), pp. (23-36), ISSN 0067-8694

Goehler, H., Lalowski, M., Stelzl, U., Waelter, S., Stroedicke, M., Worm, U., Droege, A., Lindenberg, K.S., Knoblich, M., Haenig, C., Herbst, M., Suopanki, J., Scherzinger, E., Abraham, C., Bauer, B., Hasenbank, R., Fritzsche, A., Ludewig, A.H., Büssow, K., Coleman, S.H., Gutekunst, C.A., Landwehrmeyer, B.G., Lehrach, H. & Wanker, E.E. (2004). A protein interaction network links GIT1, an enhancer of huntingtin aggregation, to Huntington's disease. *Molecular Cell*, Vol. 15, No. 6, (September 2004), pp. (853-865), ISSN 1097-2765

Greenbaum, E.A., Graves, C.L., Mishizen-Eberz, A.J., Lupoli, M.A., Lynchm D.R., Englander, S.W., Axelsen, P.H. & Giasson, B.I. (2005). The E46K mutation in alpha-synuclein increases amyloid fibril formation. *The Journal of Biological Chemistry*, Vol. 280, No. 9, (March 2005), pp. (7800-7807), ISSN 1083-351X

Hands, S.L. & Wyttenbach, A. (2010). Neurotoxic protein oligomerisation associated with polyglutamine diseases. *Acta Neuropathologica*, Vol. 120, No. 4, (October 2010), ISSN 1432-0533

Hardy, J. & Selkoe, D.J. (2002). The amyloid hypothesis of Alzheimer's disease: progress and problems on the road to therapeutics. *Science*, Vol. 297, No. 5580, (July 2002), pp. (353-356), ISSN 1095-9203

Ho, Y., Gruhler, A., Heilbut, A., Bader, G.D., Moore, L., Adams, S.L., Millar, A., Taylor, P., Bennett, K. & Boutilier, K., et al. (2002). Systematic identification of protein complexes in Saccharomyces cerevisiae by mass spectrometry. *Nature*, Vol. 415, No. 6868, pp. (180-183), ISSN 0028-0836

Ito, T., Chiba, T., Ozawa, R., Yoshida, M., Hattori & M., Sakaki, Y. (2001). A comprehensive two-hybrid analysis to explore the yeast protein interactome. *Proceedings of the National Academy of Sciences* of the United States of America, Vol. 98, No. 8, (April 2001), pp. 4569-4574, ISSN 1091-6490

Kaake, R.M., Wang, X. & Huang, L. (2010). Profiling of protein interaction networks of protein complexes using affinity purification and quantitative mass spectrometry. *Molecular & Cellular Proteomics*, Vol. 9, No. 8, (August 2010), pp. (1650-1665), ISSN 1535-9484

Kahle, J.J., Gulbahce, N., Shaw, C.A., Lim, J., Hill, D.E., Barabási, A.L. & Zoghbi HY. (2011). Comparison of an expanded ataxia interactome with patient medical records reveals a relationship between macular degeneration and ataxia. *Human Molecular Genetics*, Vo. 20, No. 3, (February 2011), pp. (510-527), ISSN 1460-2083

Kaltenbach, L.S., Romero, E., Becklin, R.R., Chettier, R., Bell, R., Phansalkar, A., Strand, A., Torcassi, C., Savage, J., Hurlburt, A., Cha, G.H., Ukani, L., Chepanoske, C.L., Zhen, Y., Sahasrabudhe, S., Olson, J., Kurschner, C., Ellerby, L.M., Peltier, J.M., Botas, J. & Hughes, R.E. (2007) Huntingtin interacting proteins are genetic modifiers of neurodegeneration. *PLoS Genetics*, Vol. 3, No. 5, (May 2007), pp. (e82), ISSN 1553-7390

Kettern, N., Dreiseidler, M., Tawo, R. & Höhfeld, J. (2010). Chaperone-assisted degradation: multiple paths to destruction. *Biological Chemistry*, Vol. 391, No. 5, (May 2010), pp. (481-489), ISSN 1437-4315

Koegl, M. & Uetz, P. (2007). Improving yeast two-hybrid screening systems. *Briefings in Functional Genomics and Proteomics*, Vol. 6, No.4, (December 2007), pp. (302-312), ISSN 2041-2647

Krauthammer, M., Kaufmann, C.A., Gilliam, T.C. & Rzhetsky, A. (2004). Molecular triangulation: bridging linkage and molecular-network information for identifying candidate genes in Alzheimer's disease. *Proceeding of the National Academy of Sciences of the United States of America*, Vol. 101, No. 42, (October 2004), pp. (15148-15153), ISSN 0027-8424

Krogan, N.J., Cagney, G., Yu, H., Zhong, G., Guo, X., Ignatchenko, A., Li, J., Pu, S., Datta, N., Tikuisis, A.P., Punna, T., Peregrin-Alvarez, J.M., Shales, M., Zhang, X., Davey, M., Robinson, M.D., Paccanaro, A., Bray, J.E., Sheung, A., Beattie, B., Richards, D.P., Canadien, V., Lalev, A., Mena, F., Wong, P., Starostine, A., Canete, M.M., Vlasblom, J., Wu, S., Orsi, C., Collins, S.R., Chandran, S., Haw, R., Rilstone, J.J., Gandi, K., et al. (2006). Global landscape of protein complexes in the yeast saccharomyces cerevisiae. *Nature*, Vol. 440, No. 7084 (March 2006), pp. (637-643), ISSN 0028-0836

Li, J., Uversky, V.N. & Fink, A.L. (2001). Effect of familial Parkinson's disease point mutations A30P and A53T on the structural properties, aggregation, and fibrillation of human alpha-synuclein. *Biochemistry*, Vol. 40, No. 38, (September 2001), pp.(11604-11613), ISSN 0006-2960

Lim, J., Hao, T., Shaw, C., Patel, A.J., Szabó, G., Rual, J.F., Fisk, C.J., Li, N., Smolyar, A., Hill, D.E., Barabási, A.L., Vidal, M. & Zoghbi, H.Y. (2006). A protein-protein interaction network for human inherited ataxias and disorders of Purkinje cell degeneration. *Cell*, Vol. 125, No. 4, (May 2006), pp. (801-814), ISSN 0092-8674

Liu, B., Jiang, T., Ma, S., Zhao, H., Li, J., Jiang, X. & Zhang, J. (2006). Exploring candidate genes for human brain diseases from a brain-specific gene network. *Biochemical* and *Biophysical* Research *Communications*, Vol. 349, No. 4, (November *2006*), pp. (1308-1314), ISSN 0006-291X

Maccioni, R.B., Farías, G., Morales, I. & Navarrete, L. (2010). The revitalized tau hypothesis on Alzheimer's disease. *Archives of Medical Research*, Vol. 41, No. 3, (April 2010), pp.(226-231), ISSN 0188-4409

Martin, I., Dawson, V.L. & Dawson, T.M. (2011). Recent advances in the genetics of Parkinson's disease. *Annual Review of Genomics and Human Genetics*, Vol. 22, No. 12, (September 2011), pp. (301-325), ISSN 1527-8204

Martin, L., Latypova, X. & Terro, F. (2011). Post-translational modifications of tau protein: implications for Alzheimer's disease. *Neurochemistry International*, Vol. 58, No. 4, (March 2011), pp. (458-471), ISSN 0197-0186

Martindale, D., Hackam, A., Wieczorek, A., Ellerby, L., Wellington, C., McCutcheon, K., Singaraja, R., Kazemi-Esfarjani, P., Devon, R., Kim, S.U., Bredesen, D.E., Tufarom F. & Hayden, M.R. (1998). Length of huntingtin and its polyglutamine tract influences localization and frequency of intracellular aggregates. *Nature Genetics*, Vol. 18, No.2, (February 1998), pp. (150-154), ISSN 1546-1718

Martinez-Vicente, M. & Cuervo, A.M. (2007). Autophagy and neurodegeneration: when the cleaning crew goes on strike. *Lancet Neurology*, Vol. 6, No. 4, (April 2007), pp. (352-361), ISSN 1474-4422

Meixner, A., Boldt, K., Van Troys, M., Askenazi, M., Gloeckner, C.J., Bauer, M., Marto, J.A., Ampe, C., Kinkl, N. & Ueffing, M. (2011). A QUICK screen for Lrrk2 interaction partners--leucine-rich repeat kinase 2 is involved in actin cytoskeleton dynamics. *Molecular & Cellular Proteomics*, Vol. 10, No. 1, (January 2011), pp. (M110.001172), ISSN 1535-9484

Morphy, R. & Rankovic, Z. (2007). Fragments, network biology and designing multiple ligands. *Drug Discovery Today*, Vol. 12, No. (3-4), (February 2007), pp. (156-160), ISSN 1359-6446

Mudher, A. & Lovestone, S. (2002). Alzheimer's disease-do tauist and Baptist finally shake hands? *Trends in Neurosciences*, Vol. 25, No. 1, (January 2002), pp. (22-26), ISSN 0166-2236

Nieznanski, K. (2010). Interactions of prion protein with intracellular proteins: so many partners and no consequences? *Cellular and Molecular Neurobiology*, Vol. 30, No. 5, (July 2010), pp. (653-666), ISSN 1573-6830

Norstrom, E.M., Zhang, C., Tanzi, R. & Sisodia, S.S. (2010). Identification of NEEP21 as a ß-amyloid precursor protein-interacting protein in vivo that modulates amyloidogenic processing in vitro. *Journal of Neuroscience*, Vol. 30, No. 46, (November 2010), pp. (15677-15685), ISSN 1529-2401

O'Brien, R.J. & Wong, P.C. (2011). Amyloid precursor protein processing and Alzheimer's disease. *Annual Review of Neuroscience*, Vol. 34, (July 2011), pp. (185-204), ISSN 0147-006X

Olzscha, H., Schermann, S.M., Woerner, A.C., Pinkert, S., Hecht, M.H., Tartaglia, G.G., Vendruscolo, M., Hayer-Hartl, M., Hartl, F.U. & Vabulas, R.M. (2011). Amyloid-like aggregates sequester numerous metastable proteins with essential cellular functions. *Cell*, Vol. 144, No. 7 (January 2011), pp. (67-78), ISSN 0092-8674

Pennuto, M., Palazzolo, I. & Poletti, A. (2009). Post-translational modifications of expanded polyglutamine proteins: impact on neurotoxicity. *Human Molecular Genetics*, Vol. 15, No. 18 (R1), (April 2009), pp. (R40-47), ISSN 1460-2083

Perreau, V.M., Orchard, S., Adlard, P.A., Bellingham, S.A., Cappai, R., Ciccotosto, G.D., Cowie, T.F., Crouch, P.J., Duce, J.A., Evin, G., Faux, N.G., Hill, A.F., Hung, Y.H., James, S.A., Li, Q.X., Mok, S.S., Tew, D.J., White, A.R., Bush, A.I., Hermjakob, H. & Masters, C.L. (2010). A domain level interaction network of amyloid precursor protein and Abeta of Alzheimer's disease. *Proteomics*, Vol. 10, No. 12, (June 2010), pp. (2377-2395), ISSN 1615-9861

Rajagopala, S.V. & Uetz, P. (2011). Analysis of protein-protein interactions using high-throughput yeast two-hybrid screens. *Methods in Molecular Biology*, Vol. 781, (2011), pp. (1-29), ISSN 1064-3745

Rigaut, G., Shevchenko, A., Rutz, B., Wilm, M., Mann, M. & Séraphin, B. (1999). A generic protein purification method for protein complex characterization and proteome exploration. *Nature Biotechnology*, Vol. 17, No. 10, (October 1999), pp. (1030-1032), ISSN 1087-0156

Rual, J.F., Venkatesan, K., Hao, T., Hirozane-Kishikawa, T., Dricot, A., Li, N., Berriz, G.F., Gibbons, F.D., Dreze, M., Ayivi-Guedehoussou, N., Klitgord, N., Simon, C., Boxem, M., Milstein, S., Rosenberg, J., Goldberg, D.S., Zhang, L.V., Wong, S.L., Franklin, G., Li, S., Albala, J.S., Lim, J., Fraughton, C., Llamosas, E., et al. (2005). Towards a proteome-scale map of the human protein-protein interaction network. *Nature*, Vol. 437, No. 7062, (October 2002), pp. (1173-1178), ISSN 0028-0836

Rumble, B., Retallack, R., Hilbich, C., Simms, G., Multhaup, G., Martins, R., Hockey, A., Montgomery, P., Beyreuther, K. & Masters, C.L. (1989). Amyloid A4 protein and its precursor in Down's syndrome and Alzheimer's disease. *The New England Journal of Medicine*, Vol 320, No. 22, (June 1989), pp. (1446-1452), ISSN 0028-4793

Schnack, C., Danzer, K.M., Hengerer, B. & Gillardon, F. (2008). Protein array analysis of oligomerization-induced changes in alpha-synuclein protein-protein interactions points to an interference with Cdc42 effector proteins. *Neuroscience*, Vol. 154, No. 4, (July 2008), pp. (1450-1457). ISSN 0306-4522

Seebacher, J., Gavin, A.C. (2011). SnapShot: Protein-protein interaction networks. *Cell*, Vol. 144, No. 6, (March 2011), pp. (1000), ISSN 0092-8674

Selbach, M. & Mann, M. (2006). Protein interaction screening by quantitative immunoprecipitation combined with knockdown (QUICK). *Nature Methods*. Vol. 3, No. 12, (December 2006), pp. (981-983), 1548-7091

Shao, J. & Diamond, M.I. (2007). Polyglutamine diseases: emerging concepts in pathogenesis and therapy. *Human Molecular Genetics*, Vol. 16, Spec No. 2, (October 2007), pp. (R115-123), ISSN 1460-2083

Singleton, A.B., Farrer, M., Johnson, J., Singleton, A., Hague, S., Kachergus, J., Hulihan, M., Peuralinna, T., Dutra, A., Nussbaum, R., Lincoln, S., Crawley, A., Hanson, M., Maraganore, D., Adler, C., Cookson, M.R., Muenter, M., Baptista, M., Miller, D., Blancato, J., Hardy, J. & Gwinn-Hardy, K. (2003). alpha-Synuclein locus triplication causes Parkinson's disease. *Science*, Vol. 302, No. 5646, (October 2003), pp. (841), ISSN 1095-9203

Soler-López, M., Zanzoni, A., Lluís, R., Stelzl, U. & Aloy, P. (2011). Interactome mapping suggests new mechanistic details underlying Alzheimer's disease. *Genome Research*, Vol. 21, No. 3, (March 2011), pp. (364-376), ISSN 1943-0264

Suzuki, H. (2006). Protein-protein interactions in the mammalian brain. *The Journal of Physiology*, Vol. 575, No. 2, (September 2006), pp. (373-377), ISSN 1469-7793

Tamayev, R., Zhou, D. & D'Adamio, L. (2009). The interactome of the amyloid beta precursor protein family members is shaped by phosphorylation of their intracellular domains. *Molecular Neurodegeneration*, Vol.14, No. 4 (July 2009), pp. (28), ISSN 1750-1326

Tyedmers, J., Mogk, A. & Bukau, B. (2010). Cellular strategies for controlling protein aggregation. *Nature Reviews Molecular Cell Biology*, Vol. 11, No. 11, (November 2010), pp. (777-788), ISSN 1471-0080

Uetz, P., Giot, L., Cagney, G., Mansfield, T.A., Judson, R.S., Knight, J.R., Lockshon, D., Narayan, V., Srinivasan, .M, Pochart, P., Qureshi-Emilim, A., Li, Y., Godwin, B., Conover, D., Kalbfleisch, T., Vijayadamodar, G., Yang, M., Johnston, M., Fields, S. & Rothberg, J.M. (2000). A comprehensive analysis of protein-protein interactions in Saccharomyces cerevisiae. *Nature*, Vol. 403, No. 6770, (February 2000), pp. (623-627), ISSN 0028-0836

Van Dam, D. & De Deyn, P. P. (2006). Drug discovery in dementia: the role of rodent models. *Nature Reviews Drug Discovery*, Vol. 5, No. 11, (November 2006), pp. (956-970), ISSN 147-41776

Venkatesan, K., Rual, J.F., Vazquez, A., Stelzl, U., Lemmens, I., Hirozane-Kishikawa, T., Hao, T., Zenkner, M., Xin, X., Goh, K.I., Yildirim, M.A., Simonis, N., Heinzmann, K., Gebreab, F., Sahalie, J.M., Cevik, S., Simon, C., de Smet, A.S., Dann, E., Smolyar, A., Vinayagam, A., Yu, H., Szeto, D., Borick, H., Dricot, A., Klitgord, N., Murray, R.R., Lin, C., Lalowski, M., Timm, J., Rau, K., Boone, C., Braun, P., Cusick, M.E., Roth, F.P., Hill, D.E., Tavernier, J., Wanker, E.E., Barabási, A.L. &Vidal, M. (2009). An empirical framework for binary interactome mapping. *Nature Methods*, Vol. 6, No. 1, (January 2009), pp. (83-90), ISSN 1548-7105

Vidal, M., Cusick, M.E. & Barabási, A.L. (2011). Interactome networks and human disease. *Cell*, Vol. 144, No. 6, (March 2011), pp. (986-998), ISSN 0092-8674

Wang, Y., Garg, S., Mandelkow, E.M. & Mandelkow, E. (2010). Proteolytic processing of tau. *Biochemical Society Transactions*, Vol. 38, No. 4, (August 2010), pp: (955-961), ISSN 1470-8752

West, M.W., Wang, W., Patterson, J., Mancias, J.D., Beasley, J.R. & Hecht, M.H. (1999). De novo amyloid proteins from designed combinatorial libraries. *Proceeding of the National Academy of Sciences of the United States of America*, Vol. 96, No. 20, (September 1999), pp. (11211-11216), ISSN 0027-8424

Wolfe, M.S. (2009). Tau mutations in neurodegenerative diseases. *The Journal of Biological Chemistry*, Vol. 284, No. 10, (March 2009), pp. (6021-6025), ISSN 1083-351X

Woods, W.S., Boettcher, J.M., Zhou, D.H., Kloepper, K.D., Hartman, K.L., Ladror, D.T., Qi, Z., Rienstra, C.M. & George, J.M. (2007). Conformation-specific binding of alpha-synuclein to novel protein partners detected by phage display and NMR spectroscopy. *The Journal of Biological Chemistry*, Vol. 282, No. 47, (November 2007), pp. (34555-34567), INSS 1083-351X

Yu, H., Braun, P., Yildirim, M.A., Lemmens, I., Venkatesan, K., Sahalie, J., Hirozane-Kishikawa, T., Gebreab, F., Li, N., Simonis, N., Hao, T., Rual, J.F., Dricot, A., Vazquez, A., Murray, R.R., Simon, C., Tardivo, L., Tam, S., Svrzikapa, N., Fan, C., de Smet, A.S., Motyl, A., Hudson, M.E., Park, J., Xin, X., Cusick, M.E., Moore, T., Boone, C., Snyder, M., Roth, F.P., Barabási, A.L., Tavernier, J., Hill, D.E. & Vidal, M. (2008). High-quality binary protein interaction map of the yeast interactome network. *Science*, Vol. 322, No. 5898, (October 2008) pp. (104-110), ISSN 1095-9203

Zheng, X.Y., Yang, M., Tan, J.Q., Pan, Q., Long, Z.G., Dai, H.P., Xia, K., Xia, J.H. & Zhang, Z.H. (2008). Screening of LRRK2 interactants by yeast 2-hybrid analysis. *Zhong Nan Da Xue Xue Bao Yi Xue Ban*, Vol. 33, No. 10, (October 2008), pp. (883-891), ISSN 1672-7347

10

Protein Interactions in S-RNase-Based Gametophytic Self-Incompatibility

Thomas L. Sims

Department of Biological Sciences, Northern Illinois University
USA

1. Introduction

With well over 200,000 documented species (Mora et al., 2011) flowering plants (angiosperms) are among the most successful taxa on the planet. A major reason for the success of the angiosperms is self-incompatibility, a genetic and biochemical barrier to inbreeding that promotes outcrossing and diversity in populations. Plants exhibiting self-incompatibility have the ability to recognize (species-specific) pollen as being "self" or "non-self", with self (incompatible) pollen being rejected and non-self (compatible) pollen being accepted. S-RNase-based Gametophytic Self-Incompatibility (GSI) has been characterized in the Solanaceae, Rosacaeae, and Plantaginaceae (McClure et al., 2011; Meng et al. 2011; Chen et al., 2010; Sims & Robbins 2009), with the genetic, physiological and molecular basis of this form of GSI described in detail. To date, over a dozen different proteins have been identified that function in different parts of the GSI response; most of these have been tested for protein interactions with other GSI response pathway proteins. The two key recognition proteins: S-RNase (the style-expressed recognition component) and SLF (the pollen-expressed recognition component) interact with each other, and with other components of a putative SCF^{SLF} E3 ubiquitin ligase complex. Recently Kubo et al., 2010 demonstrated the existence of multiple SLF variant classes. Multiple S-RNase and SLF alleles are present in breeding populations (Richman et al., 1995, 1996, 2000), and it now seems probably that collaborative interaction of different SLF alleles and classes with different S-RNase alleles governs self/non-self recognition in GSI. In this review, I summarize the genetic basis of GSI, describe the different proteins identified that are thought to function in the GSI pathway, and describe what is known with regard to protein interactions underlying the function of self-incompatibility. Most of the work discussed here comes from studies in the Solanaceae and Plantaginaceae. Gametophytic self-incompatibility has also been studied extensively in the Rosaceae (e.g. Sassa et al., 2010). Work that demonstrates possible differences in the mechanism of GSI in Solanaceae/Plantaginaceae versus Rosaceae will be discussed as appropriate.

2. Genetic studies of gametophytic self-incompatibility

The first description of gametophytic self-incompatibility was by none other than Charles Darwin. As Darwin (1891) observed:

"....protected flowers with their own pollen placed on the stigma never yielded nearly a full complement of seed; whilst those left uncovered produced fine capsules, showing that pollen from other plants must have been brought to them, probably by moths. Plants growing vigorously and flowering in pots in the green-house, never yielded a single capsule; and this may be attributed, at least in chief part, to the exclusion of moths."

Since Darwin's observation, self-incompatibility systems in general, and GSI in particular, have interested both molecular and evolutionary biologists. As an example of self/non-self recognition, GSI presents interesting challenges in terms of molecular interactions, how recognition specificity is determined, and what types of sequences determine allelism. In terms of population genetics, evolutionary biologists have investigated questions of how GSI haplotypes are established and maintained over evolutionary time (Kohn, 2008).

2.1 The genetic basis of S-RNase-based gametophytic self-incompatibility

Early studies (de Nettancourt, 1977; Linskens 1975, Mather, 1943) demonstrated that self/non-self recognition was encoded by a single genetic locus, the S-locus, with pistil and pollen recognition components (termed "pistil-S" and "pollen-S", respectively). Both pistil-S and pollen-S have multiple alleles, such that a given S-locus recognition phenotype is now termed a S-locus haplotype. The S-locus ribonuclease (S-RNase) is pistil-S, and the S-locus F-box protein (SLF; SFB in Rosaceae) has been demonstrated to be pollen-S (Sijacic et al., 2004). During pollination, a match between S-RNase and SLF haplotypes results in pollen rejection (incompatibility). Lack of a match results in pollen acceptance (compatibility) and fertilization (see Figure 1). Recognition specificity in GSI is a cell-autonomous response, in that rejection or acceptance is specific to individual pollen tubes, and is not an "all or none" phenomenon. This can be demonstrated by the existence of "half-compatible" pollinations (Figure 1). In this case, pollen tubes expressing a SLF-specificity matching the S-RNase in the style are rejected, while other pollen tubes in the same style, with no haplotype match, grow normally and function for fertilization and seed set.

2.2 Tetraploidy results in self-compatibility due to competitive interaction

An intriguing aspect of GSI is that tetraploidy, in heterozygous individuals, leads to self-compatibility (Figure 1). In this case, heteroallelic, diploid pollen (e.g. S1-SLF/S2-SLF) is compatible on either a tetraploid style ($S_1S_1S_2S_2$) or a diploid style (S_1S_2). Haploid pollen (e.g. S_1 or S_2) remains incompatible on tetraploid styles (Figure 1). This phenomenon has been termed "competitive interaction" (de Nettancourt 1977). Competitive interaction is only observed in situations where the parent plant was heterozygous for S-locus haplotype. Tetraploids that are homozygous at the S-locus (homozygous plants can be obtained by bud-pollination), do not show competitive interaction, but remain self-incompatible. Competitive interaction is most likely the cause of GSI breakdown (compatibility) in induced pollen-part mutants (Golz et al., 1999, 2001). In mutants induced by radiation, Golz et al. (1999, 2001) showed that GSI breakdown was associated with partial duplications of S-haplotypes, in which the compatible pollen presumably phenocopied the heteroallelic condition found in tetraploids. Competitive interaction has been used as a test for the identity of pollen-S (Kubo et al., 2010; Sijacic et al., 2004), since transgenic plants, having diploid, heteroallelic pollen (two different SLF haplotypes) are self-compatible (Figure 1 and sections below).

3. Pistil-S and pollen-S

Although the genetic "identities" of pistil-S and pollen-S have been known for many years, the identification of specific proteins corresponding to these entities has been a more recent phenomenon. The S-locus ribonuclease (S-RNase) was initially cloned in 1986 (Anderson et al., 1986) with its identity as pistil-S confirmed eight years later (Lee et al., 1994, Murfett et al., 1994). Identification of SLF as pollen-S is far more recent. SLF was first identified by chromosomal walking (Entani et al., 2003; Lai et al., 2002; Wang et al., 2004) and subsequently confirmed as pollen-S using a competitive-interaction assay (Sijacic et al., 2004). As will be explained, the molecular nature of pollen-S appears to be far more complex than originally envisioned.

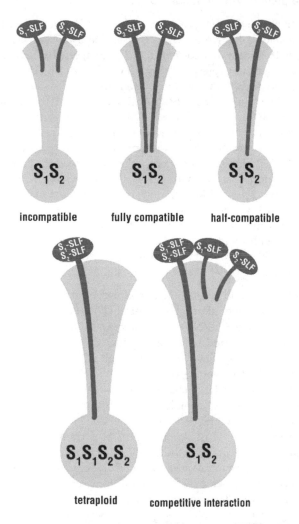

Fig. 1. **Genetic basis of gametophytic self-incompatibility.**

Figure 1 illustrates different types of pollinations with styles and pollen expressing different haplotypes at the S-locus. In an incompatible pollination (top left), a match of haplotypes between pollen and style results in incompatibility. No match of S-locus haplotypes (top, middle) results in full compatibility. A "half-compatible" cross results when half of the pollen carries a S-locus haplotype matching that of the style, but the other pollen is not matching. In this case, only the "matching" pollen tubes are rejected. The lower portion of the figure illustrates GSI breakdown in tetraploids (left figure) by competitive interaction. The figure at lower-right illustrates how competitive interaction can be used in transgenic assays to demonstrate that a particular gene (in this case, SLF) is pollen-S.

3.1 S-RNase is the pistil recognition component of GSI

The ability to selectively inhibit the growth of self pollen is determined in the style by a S-locus encoded ribonuclease known as the S-RNase. The S-RNase was first identified as a highly-expressed stylar protein that co-segregated with specific S-haplotypes (Anderson et al., 1986). The S-RNase gene is expressed at high levels late during development of the pistil (Clark et al., 1990), and encodes a secreted protein that accumulates to high levels in the transmitting tract of the style (Ai et al., 1990; Anderson et al., 1989). Comparative sequence analysis showed that S-RNase alleles have a high degree of sequence polymorphism, but that the polymorphism is not evenly distributed across the protein. Overall amino acid sequence identity can be less than 50% between allleles (Ioerger et al., 1991; Sims, 1993; Richman et al., 1995). Detailed sequence comparisons showed that S-RNase proteins have five highly conserved domains and two adjacent hypervariable domains, HVa and HVb (Ioerger et al., 1991; Sims, 1993). Although much of the sequence variability among S-RNase alleles is found in the two hypervariable regions, other portions of the protein are variable as well (Figure 2). The conserved domains C2 and C3 contain two histidine residues, His31 and His91, that along with Lys90 make up the catalytic site of the ribonuclease (Ida et al., 2001; Ishimizu et al., 1995). (Note that in different S-RNase alleles, the exact positions of these concerved amino acids vary by one or two positions).

Transgenic gain-of-function and loss-of-function experiments gave conclusive evidence that the S-RNase was the style-recognition component of GSI. Murfett et al. (1994) used a gain-of function approach, where the S_{A2}-RNase of *Nicotiana alata* (under control of a strong style-specific promoter) was transferred to a regenerable *N. lansgsdorfii* x *N. alata* hybrid. The transgenic plant remained compatible when pollinated with S_{C10} pollen from *Nicotiana alata*, but now showed the ability to reject S_{A2} pollen. Lee et al. (1994) used an antisense approach to down-regulate the *Petunia inflata* S_3-RNase in a S_2S_3 background. Plants with reduced levels of S_3-RNase were no longer capable of inhibiting S_3 pollen, although the transgenic plant showed otherwise normal GSI behavior. Lee et al. (1994) also used a gain-of-function approach, in which the S_3-RNase of *Petunia inflata* was transferred to a plant of the S_1S_2 genotype. Transgenic plants expressing the S_3-RNase at levels comparable to endogenous S-RNase had acquired the ability to reject S_3 pollen. These plants continued to reject S_1 and S_2 pollen, but set seed capsules when pollinated at an immature bud stage where the S-RNase is expressed at minimal levels (Clark et al., 1990; Lee et al., 1994). In these experiments, only the style recognition was altered. Pollen recognition specificity was not affected, confirming that a separate gene product from the S-RNase encoded the "pollen-S" component.

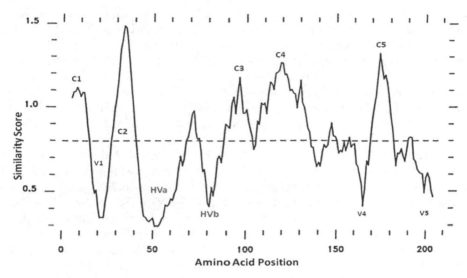

Fig. 2. **Graphical depiction of amino acid alignment among Solanaceae S-RNase alleles.**
Amino acid sequences for eighteen S-RNase alleles were aligned using PlotSimilarity. The
dotted line shows the average similarity score across the protein. Peaks above the line
represent conserved regions (labeled C1 through C5). Valleys below the line represent more
variable regions. Amino acids in hypervariable regions HVa and HVb were shown to be
sufficient for determining S-RNase recognition specificity (after Sims, 1993).

Further work either analyzing spontaneous mutants (Royo et al., 1994) or using transgenic
plants (Huang et al.,1994) demonstrated that ribonuclease activity of the S-RNase was
required for pollen rejection. Royo et al. (1994) cloned and sequenced the S-RNase allele
from a self-compatible S_cS_c accession of *Lycopersicon peruvianum* (now *'Solanum peruvianum'*,
http://solgenomics.net). The S_c allele sequence showed a change at amino acid 33 from the
conserved histidine to asparagine. No other sequence changes were observed, and the
authors concluded this change was correlated with both the loss of RNase activity in S_cS_c
styles and with self-compatibility. In related work, Huang et al. (1994) used *in vitro*
mutagenesis to construct a H93N variant of the *P. inflata* S_3-RNase, and introduced that
construct in the S_1S_2 background. Unlike the results obtained when the wild-type S_3-RNase
was transferred (Lee et al., 1994), the H93N S_3-RNase was unable to reject S_3 pollen.
Reinforcing the critical role of ribonuclease activity in S-RNase function were earlier
experiments indicting that degradation of pollen-tube RNA was associated with self-
incompatibility. In those experiments (McClure et al., 1990) pollen RNA was labeled *in vivo*
by watering plants with a solution containing ^{32}P-orthophosphate, then used for compatible
or incompatible pollinations. Pollen tube RNA was degraded following incompatible
pollination, but was not degraded following compatible pollination.

3.1.1 S-RNase recognition specificity is determined by hypervariable domains

Experiments investigating the basis for allele specificity in the S-RNase protein have focused
on the role of the hypervariable regions. In one approach, Zurek et al. (1997) constructed

chimeric S-RNase genes having different combinations of S_{A2} and S_{C10} conserved and variable domains, then expressed the chimeric proteins in transgenic plants. Although the transgenic styles had ribonuclease activity levels equivalent to self-incompatible controls, none of the chimeric S-RNase constructs could reject S_{A2} or S_{C10} pollen. By contrast, Matton et al. (1997) took advantage of two S-RNase allles in *Solanum chacoense* that were closely related in sequence, to make more limited alterations. The S_{11}- and S_{13}-RNase alleles of *S. chacoense* differ by only 10 amino acids, three of which are found in HVa and one in HVb. Matton et al. (1997) used *in vitro* mutagenesis to change the four S_{11} residues in the HVa and HVb regions to those found in S_{13}, then expressed the altered allele transgenically in the $S_{12}S_{14}$ background. Pollination with S_{11} and S_{13} pollen demonstrated that changing only these residues changed the recognition specificity of the transferred S-RNase from S_{11} to S_{13}. In an extension of this experiment (Matton et al., 1999), changing only two residues in HVa plus the HVb residue resulted in a "dual-specificity" S-RNase that retained the ability to reject S_{11} pollen while acquiring the ability to also reject S_{13} pollen. These experiments demonstrated that, at least for these two alleles, amino acid sequences in the hypervariable regions determine allelic specificity. The protein crystal structure has been determined for the S_{F11} S-RNase of *Nicotiana alata* (Ida et al., 2001). The two hypervariable regions are located on the surface of the S_{F11} S-RNase and readily accessible to solvent (Ida et al., 2001). These regions include all four of the equivalent residues to those targeted in the mutagenesis experiments of Matton et al. (1997, 1999). Another potential basis for allele specificity might be variability in carbohydrate modification of S-RNases, which are glycoproteins (Woodward et al., 1989). This does not appear to be the case, however as as elimination of the glycosylation site has no effect on the ability of S-RNase to reject pollen (Karunanandaa et al.,1994).

3.1.2 Both self and non-self S-RNases are imported into pollen tubes in vivo

Immunolocalization experiments, either using traditional TEM (Luu et al., 2000) or fluorescently-tagged antibodies hybridized to paraffin-embedded sections (Goldraij et al., 2006) demonstrate that both incompatible and compatible S-RNases are imported into pollen tubes. The authors of these two studies reached different conclusions about the location of S-RNase inside pollen tubes following compatible or incompatible pollinations. Luu et al., (2000) working with self-incompatible potato (*S. chacoense*), fixed pollinated styles, 18 hr post-pollination, with 0.5% glutaraldehyde, followed by embedding, hybridization with anti-S_{11} antibody and 20 nm gold-labeled secondary antibody, and visualization via TEM. S_{11}-RNase was taken up into both compatible and incompatible pollen tubes, and labeling was seen primarily in pollen-tube cytoplasm, with little labeling in the pollen-tube vacuole. Goldraij et al. (2006), working with *Nicotiana*, hybridized anti-S-RNase antibodies along with anti-callose, anti-aleurain (marker for vacuolar lumen) and anti-vPPase (marker for vacuolar membrane) to fixed, paraffin-embedded sections, then visualized fluorescence using confocal microscopy. These authors concluded that S-RNase was initially sequestered in a vacuolar compartment in pollen in both compatible and early-stage (16 hr) incompatible pollen tubes, but that this compartment broke down at later stages (36 hr) of incompatible pollinations, releasing S-RNase into the pollen-tube cytoplasm.

3.1.3 The cytotoxic model for pollen rejection

Current models for pollen rejection in GSI propose a cytotoxic mechanism, where RNA degradation in incompatible pollen tubes reduces protein synthesis resulting in a slowing or cessation of pollen tube growth and failure of incompatible pollen tubes to reach the ovary. This model is based on observations outlined in the previous sections, but is not without its caveats. S-RNases are imported into pollen tubes, and at least in self-incompatible pollinations (Goldraij et al., 2006) are freely distributed in the pollen-tube cytoplasm. The ribonuclease activity of S-RNases is required for pollen tube rejection, and generalized RNA degradation is associated with self-incompatibility, but not with cross-pollination. The ability to reject incompatible pollen tubes is also dependent on a threshold level of S-RNase expression and accumulation in the style. Both developmental (Clark et al., 1990; Shivanna, 1969) and transgenic assays (Lee et al., 1994; Murfett et al., 1994, Zurek et al., 1997) show that styles expressing the S-RNase at low-to-moderate levels are incapable of rejecting otherwise incompatible pollen. S-RNases, like other T2 ribonucleases, show no obvious substrate specificity, at least *in vitro* (Singh et al., 1991).

The cytotoxic model, while attractive and consistent with the majority of current evidence, cannot, however, fully explain some other observations. Grafting experiments (Lush & Clarke, 1996) where upper regions of incompatible styles were grafted onto compatible styles, and pollinated, showed that incompatible pollen tubes could recover, growing out of the incompatible style into the compatible style. Also, Walles & Han (1998) using conventional TEM, observed intact polysomes in incompatible pollen tubes after pollination. Last, there is little correlation between overall ribonuclease activity found in different styles with the level of self-incompatibility (Clark et al., 1990; Singh et al., 1991; Zurek et al., 1997), although it should be assumed that non S-RNase ribonucleases likely contribute to overall style RNase levels.

3.2 SLF is the pollen-recognition component of GSI

Although the S-RNase was identified and cloned early on, it took an additional 18 years before the S-locus F-box protein (SLF) was conclusively identified as "pollen-S". Even today, what, functionally constitutes "pollen-S" is not fully understood; recent work suggests that different SLF variants may act collaboratively to recognize S-RNases. Additionally, several other proteins are involved and/or required for recognition (see sections below).

Even before SLF was identified and cloned, the majority of experimental evidence pointed to pollen-S as an inhibitor of S-RNase activity in compatible pollen. Tetraploidy is associated with the breakdown of self-incompatibility in those cases where the parental diploid plant was heterozygous for two different S-locus haplotypes, but not when the parent plant was homozygous (Chawla et al., 1997; de Nettancourt, 1977; Entani et al., 1999). In early studies, Brewbaker & Natarajan (1960) induced pollen-part mutants of *Petunia* using irradiation (pollen part mutants are self-compatible, fertile as pollen parents, and show normal GSI behaviour in the style). Pollen-part mutants were obtained only when the irradiated parent was heterozygous, and were associated with centric chromosome fragments. Golz et al (1999, 2001) revisited this work, carrying out mapping and cytological analyses of pollen-part mutants of *Nicotiana alata* induced by gamma radiation. In all cases, the pollen-part mutants were associated with apparent duplications of part or all of the S-locus, either as centric chromosome fragments or as translocations. Luu et al. (2000), and later Goldraij et al. (2006) showed that both compatible and incompatible S-RNase proteins were imported into pollen tubes. Together, these results discredited a model where pollen-S was a "gatekeeper" preventing import of non-self S-RNases and suggested that compatible pollinations result from the specific inhibition of imported S-RNase proteins. According to models based on the results just described, pollen-S was an inhibitor of all S-RNases, except its own cognate S-RNase. Thus, compatible pollinations resulted from pollen-S inhibiting the action of any non-self S-RNase, while incompatible pollinations resulted from the inability of pollen-S to inhibit a co-evolved S-RNase.

3.2.1 Predictions for pollen-S

Prior to the actual isolation of pollen-S, there was a relative consensus with regard to the properties expected of this protein. Genetic studies had indicated that there was little or no recombination between pistil-S and pollen-S (de Nettancourt 1977), so both genes were expected to be tightly linked. That linkage, together with the observation that S-RNase alleles had diverged prior to speciation in the Solanaceae (some S-RNase alleles are more similar across species than within species) resulted in the assumption that S-RNase sequences and pollen-S sequences should be co-evolved. That is, most researchers fully expected that the degree of polymorphism among pollen-S alleles should be roughly equivalent to the polymorphism observed among S-RNase alleles (Kao & McCubbin, 1996). Pollen-S was also thought to interact directly with the S-RNase, with that interaction resulting in the inhibition of the action of the S-RNase in compatible pollinations. The sections below will illustrate that these assumptions were only partially correct.

3.2.2 Genetic and physical mapping of the S-Locus

The first attempt at mapping the S-locus was carried out by Tanksley and Loaiza-Figueroa (1985) who mapped the S-locus to chromosome I of *Lycopersicon peruvianum* using enzyme-linkage. RFLP mapping in potato (Gebhardt et al., 1991) demonstrated that chromosome I of tomato and potato were homeologous, and that the S-locus mapped to the same location in potato as in tomato. The S-locus was physically mapped in *Petunia hybrida* by fluorescence in-situ hybridization (FISH), using T-DNA inserts linked to the S-locus (ten Hoopen et al., 1998). Those experiments showed that in *Petunia hybrida*, the S-locus was located in a sub-centromeric region of chromosome III. Mapping of linked RFLP markers demonstrated synteny of the S-locus across four species in the Solanaceae (*Lycopersicon peruvianum, Nicotiana alata, Petunia hybrida* and *Solanum tuberosum*). Entani et al. (1999) carried out similar FISH experiments, but used cDNAs and genomic clones of the S-RNase instead of linked T-DNA inserts. Like ten Hoopen et al. (1998), Entani et al. (1999) found that the S-RNase gene was found in a subcentromeric region of chromosome III of *Petunia hybrida*. Both Li et al. (2000) and McCubbin et al. (2000) used RNA differential display to identify pollen-expressed cDNAs linked to the S-locus. Although not realized at the time, both of these differential display experiments identified cDNAs that would later turn out to true pollen-S genes. Part of the failure to recognize that these linked cDNAs did, in fact, encode pollen-S was the high degree of sequence identity between cDNAs isolated from different haplotypes as compared to the polymorphism previously observed for S-RNase alleles.

3.2.3 Gene walking identified pollen-S

The large amount of repetitive DNA sequences flanking S-RNase genes (Coleman & Kao, 1992; Matton et al., 1995), together with the subcentromeric location of the S-locus (ten Hoopen et al., 1998; Entani et al., 1999) were originally thought to preclude a map-based cloning approach for isolation of pollen-S (Kao & McCubbin 1996). Indeed, some of the early efforts to clone pollen-S involved T-DNA tagging (Harbord et al., 2000) yeast two-hybrid screens (Sims & Ordanic, 2001) or other protein-interaction methods (Dowd et al., 2000). Although these approaches provided important information regarding S-RNase-based GSI, none of them resulted in the identification of pollen-S. The first indication that pollen-S might be cloned using a map-based approach came from the work of Ushijima et al. (2001) who constructed an ~200 kb cosmid contig around the S-locus of *Prunus dulcis* (almond). When these authors carried out Southern blots with cosmid clones spanning the contig, with genomic DNA of different S-locus haplotypes, they observed that a ~70 kb region in the center of the contig was highly polymorphic across haplotypes, whereas either end of the contig showed a high degree of sequence similarity across haplotypes. This "island of polymorphism" presumably resulted from the known lack of recombination at the S-locus and was taken as defining the physical limits of the S-locus in *Prunus dulcis*.

Lai et al. (2002) were the first to report the isolation of the S-locus F-box gene, which would turn out to be pollen-S. Screening of a BAC library from *Antirrhinum hispanicum* with the S_2-RNase identified a 63 kb BAC clone, which was then fully sequenced. Of several putative ORFs identified, most were retrotransposons, however, the 'gene-11' ORF, when used to screeen a cDNA library, identified a pollen-expressed F-box protein, termed AhSLF-S_2. AhSLF-S_2 was located 9 kb downstream of the S_2-RNase gene, and appeared to be allele-specific, making it a good candidate for pollen-S. Similarly, Ushijima et al. (2003) sequenced the 70 kb region of *Prunus dulcis*, and identified a pollen-expressed, haplotype-specific, F-box gene, which they termed SFB. Using S-locus-specific cDNAs previously generated, and starting with BACs known to contain the S-RNase gene, Wang et al. (2004) identified an 881 kb contig surrounding the S-locus in *Petunia inflata*. Sequencing and analysis of a 328 kb region of this contig identified several genes, one of which, PiSLF, was highly similar to the F-box genes isolated from *Antirrhinum* and *Prunus*. Two previously identified S-linked F-box genes *A113* and *A134* (McCubbin et al., 2000) mapped outside of the 881 kb region.

3.2.4 Competitive interaction showed that SLF was pollen-S

The identity of SLF as pollen-S was confirmed by taking advantage of the phenomenon of competitive interaction in heteroallelic pollen (see section 2.2 and Figure 1). Sijajic et al., transferred the S_2-allele of SLF (PiSLF$_2$, but see nomenclature change in section 5.x, below) into a S_1S_1 line of *Petunia inflata*. First generation transgenic plants segregated two types of pollen, haploid S_1 pollen and heteroallelic S_1(PiSLF$_2$) pollen.. Self-pollination of the the S_1S_1(PiSLF$_2$) plant produced large fruits, indicating that the trangenic plant, formerly self-incompatible, was now self-compatible. Similarly, when S_1S_1(PiSLF$_2$) pollen was used to pollinate a non-transformed S_1S_1 plant, fruit set showed that the S_1S_1(PiSLF$_2$) pollen behaved as compatible pollen. Conversely, pollination of S_1S_1(PiSLF$_2$) styles with pollen from a non-transformed S_1S_1line produced no seed capsules, demonstrating that the loss of self-incompatibility in the transgenic plant was confined to the pollen. Analysis of progeny resulting from self-pollination demonstrated that all of the progeny inherited the transgene. Similar results were reported by Qiao et al. (2004b) who transferred the Ah-SLF$_2$ gene of *Antirrhinum hispanicum* into S$_{3L}$S$_{3L}$ *Petunia hybrida*. This experiment was conducted with two variations. In the first variation, a clone containing both Ah-S$_2$-RNase and Ah-SLF$_2$ was transferred to the host plant. Transgenic plants expressing both the *A. hispanicum* S-RNase and SLF were self-compatible, with the change in compatibility again confined to the pollen. Analysis of progeny confirmed that all inherited both the S$_2$-RNase and Ah-SLF$_2$. The change to the self-compatible phenotype was dependent on expression of the transgenes. In two individuals, neither the S$_2$-RNase transgene nor the endogenous S$_{3L}$-RNase was expressed at detectable levels, most likely due to co-suppression. Both of these progeny were completely self-incompatible. In the second variation, the Ah-SLF$_2$ cDNA alone, under control of the pollen-specific LAT52 promoter, was introduced into the S$_{3L}$S$_{3L}$ line. As above, the transgenic plants were self-compatible on the pollen side, but displayed normal self-incompatibility behavior when used as the style parent. These conversions of self-incompatibility to compatibility following pollen-specific expression of the SLF transgene is a direct phenocopy of the competitive interaction effect seen in heteroallelic pollen in tetraploid heterozygotes. One remarkable aspect of the work reported by Qiao et al (2004b) is that the *Antirrhinum* SLF protein can apparently cause competitive interaction in a completely different species.

3.2.5 SLF proteins appear to have different evolutionary history than S-RNases

Although many of the key predictions for the properties of pollen-S are indeed found for the SLF proteins (pollen-expression, interaction with S-RNases, competitive interaction), a surprising and confusing finding was the distinct lack of polymorphism among SLF proteins, together with the existence of multiple SLF-related proteins, originally termed SLFL (SLF-like) proteins. As will be discussed below (see section 8), many of these SLFL proteins may turn out to be true SLFs. Another confusing finding was apparent differences in the functional characteristics of SLF proteins in Solanaceae and Plantaginaceae compared with the equivalent SFB proteins in Rosaceae.

The first SLF proteins identified (Ah-SLF$_1$, Ah-SLF$_2$, Ah-SLF$_4$, Ah-SLF$_5$ in *Antirrhinum*) share 97% to 99% amino acid sequence identity. By contrast the related *Antirrhinum* S-RNase proteins share only 38% to 53% amino acid identity by pairwise BLASTp. Similarly, if not quite so dramatically, SLF proteins from *Petunia inflata* share ~ 90% amino acid sequence identify, while the corresponding S-RNase proteins share only about 70% amino acid sequence identity. Phylogenetic comparisons (e.g. Newbigin et al., 2008) present an even more striking picture. S-RNase sequences appear to be an ancient lineage; in the Solanaceae, S-RNases from one species are often more similar to a S-RNase from a different species than to other S-RNases within the same species. By contrast SLF sequences from an individual species cluster together. In addition, while the variability across S-RNases is clustered in variable and hypervariable regions (Figure 2), the variability across SLF alleles appears to be uniformly distributed across the protein. Because recombination between style and pollen recognition specificities is rarely if ever observed (de Nettancourt, 1977) the traditional assumption has been that pistil-S and pollen-S (S-RNase and SLF) have co-evolved and share the same evolutionary history. The actual observations, however appear to contradict that notion. One potential solution to this dilemma is that pollen-S may actually be comprised of multiple SLF protein variants, not a single SLF (see section 8).

4. Interaction assays identified other pollen proteins with roles in GSI

S-RNase and SLF are the pistil and pollen recognition components of GSI, however several other proteins with presumed or demonstrated roles in GSI have been identified and studied. Some of these proteins were first identified by protein-interaction screens with S-RNase or SLF, in other cases, a presumed role in GSI has been demonstrated using protein interaction assays. Figure 3 summarizes the interactions of pollen-expressed proteins with the S-RNase, specifics of these interactions are discussed below.

4.1 SBP1

Sims and Ordanic (2001) identified PhSBP1 (S-ribonuclease binding protein) in a yeast two-hybrid screen of a pollen cDNA library from S_1S_1 *Petunia hybrida*. The bait protein used for the screen was the N-terminal half of the S_1-RNase, containing domains C1 to C3 (see Figure 2). In subsequent pairwise interaction assays, PhSBP1 interacted with the same N-terminal construct of the S_3-RNase, and with subdomains (C2-HVa-HVb-C3, HVa-HVb) of both S_1- and S_3-RNases.

SBP1 did not show interaction with the C-terminal regions of either S-RNase (C4-V4-C5-V5 in Figure 2) nor with an unrelated bait protein, P53. ScSBP1 was isolated from *Solanum chacoense* (O'Brien et al., 2004), using a yeast two-hybrid screen with the HVa+HVb regions of the S_{11}-Rnase and the S_{13}-RNase as bait. Both the S_{11} and S_{13} baits interacted with ScSBP1, however a full-length S-RNase with a single amino acid change (H144L) at one of the active-site histidines failed to interact with SBP1. Similarly Hua & Kao (2006) used a partial bait (HVa-HVb-C3) of the *Petunia inflata* S2-RNase to screen a two-hybrid library and isolated PiSBP1. Similar to other reports (O'Brien et al., 2004; Sims & Ordanic, 2001) PiSBP1 did not interact with full-length S-RNase, with non-specific controls, or importantly, with a non-S-locus ribonuclease. Hua & Kao (2006) further showed that SBP1 interacted with $PiSLF_2$ and $PiSLF_1$ in both two-hybrid and pull-down assays, as well as with Cullin-1 and PhUBC1 (Sims, unpublished), an E2 conjugation enzyme protein from *Petunia hybrida*. Lee et al. (2008) used C-terminal domains of the style-transmitting-tract proteins TTS and 120K to screen a pollen two-hybrid library from *Nicotiana alata*, and also pulled out NaSBP1 from this screen. All of these reports (Hua & Kao, 2006; Lee et al., 2008; O'Brien et al., 2004; Sims & Ordanic 2001) showed that SBP1 was not pollen-specific, but was expressed in all tissues examined. SBP1 is non-allelic as well, as SBP1 isolated from S_1S_1 and S_3S_3 lines of *Petunia hybrida* are 100% identical. The SBP1 protein has two identifiable protein domains, a coiled-coil domain in the center of the sequence and a C-terminal RING-HC domain. RING-HC domains are characteristic of E3 ubiquitin ligases (Freemont 2000), and SBP1 has E3 ubiquitin ligase activity *in vitro* (Hua and Kao, 2008).

4.2 SSK1

Huang et al. (2006) used the *Antirrhinum hispanicum* SLF protein Ah-SLF$_2$ to screen a pollen yeast-two-hybrid library, and identified a SKP1-like protein that they named SSK1. AhSSK1 interacted with both Ah-SLF$_2$ and Ah-SLF$_5$ but not with proteins identified as SLF paralogs (Zhou et al., 2003). AhSSK1 futher interacted with a Cullin-1 protein. Sequence and phylogenetic analyses showed that AhSSK1 was related to, but distinct from canonical SKP1 proteins. In particular, AhSSK1 differed at several internal residues and also has a 7-residue C-terminal "tail" that extends beyond the "WAFE" sequence that terminates most plant SKP1 proteins (Huang et al., 2006; Zhao et al, 2010). Zhao et al. (2010) showed that SSK1 almost certainly plays a critical role in self-incompatibility. Using AhSSK1 as a guide, they isolated PhSSK1 from *Petunia hybrida*. PhSSK1, similar to AhSSK1 interacts with both SLF and Cullin-1 from *Petunia*. To directly test the role of PhSSK1, Zhao et al. (2010) used a RNAi construct of PhSSK1 to down-regulate this gene in $S_{3L}S_{3L}$ *Petunia hybrida*. When transgenic plants showing reduced levels of PhSSK1 in pollen were used as the pollen parent in crosses to S_1S_1 or S_vS_v *P. hybrida*, no seed capsules were produced. Conversely, when these same lines were used as pollen parent to a line defective for S-RNase expression (S_oS_o) normal seed capsules were produced, suggesting that down-regulation of SSK1 specifically affected cross-compatibility in the self-incompatibility response.

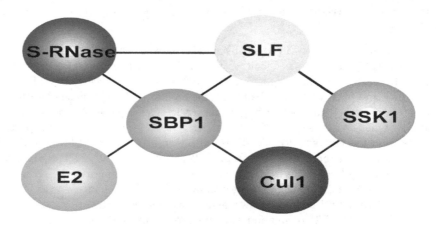

Fig. 3. **Protein interactions of pollen-expressed proteins in GSI.**
Lines between individual proteins indicate protein interactions that have been observed by yeast two-hybrid or pull-down assays. See the text for details.

5. Protein interactions between SLF and S-RNase proteins

S-locus F-box proteins were first cloned by chromosome-walking experiments to identify pollen-expressed proteins tightly linked to the S-RNase. Further studies examined the interaction between SLF proteins and S-RNases in detail. Qiao et al. (2004a) examined the interaction of the *Antirrhinum* Ah-SLF-S$_2$ protein with S-RNases using pull-down (His-tag), yeast two-hybrid, and co-immunoprecipitation assays. Pull-down assays demonstrated that the C-terminal portion of Ah-SLF-S$_2$ (lacking the F-box domain) interacted with S-RNase from style extracts of *Antirrhinum hispanicum*. The N-terminal F-box domain was incapable of interacting with S-RNase, and the full-length SLF protein could not be tested as it could not be expressed in *E. coli*. Similarly, the C-terminal portion of Ah-SLF-S_2 interacted with a full-length (lacking the signal peptide) S-RNase construct in yeast two hybrid assays; neither the F-box domain nor the full-length SLF protein showed interaction in the two-hybrid assays. Both of these assays also showed that Ah-SLF-S_2 interacted with different S-RNases, without any apparent allelic specificity. Co-immunoprecipitation assays where extracts from pollinated styles were immunoprecipitated using anti-Ah-SLF-S_2 antibody, then blotted with anti-S-RNase antibody showed that Ah-SLF-S_2 interacted with both S_2-and S_4-RNase *in vivo*. Qiao et al., (2004a) also tested interaction of Ah-SLF paralogs (Zhou et al., 2003) with S-RNase proteins. The SLF paralogs were identified as pollen-expressed SLF-like genes linked to the S-locus but distant from S-RNase or Ah-SLF-S_2. Similar SLF-like genes linked to the S-locus but outside of the core S-RNase-SLF contig had previously been identified in *Petunia inflata* (McCubbin et al., 2000; Wang et al., 2003). No interaction was observed between Ah-SLF-S_2 and any of the SLF paralogs. Recent data (see section 8) indicates the SLF-like genes (SLF paralogs) may, however be true SLF proteins, that recognize a subset of S-RNases rather than all S-RNases.

Hua and Kao (2006) also tested interactions between SLF and S-RNase allles in *Petunia inflata* using pull-down assays with His-tagged SLF and GST-tagged S-RNase constructs expressed in bacteria. These assays showed that both PiSLF$_1$ and PiSLF$_2$ interacted with the HVaHVbC3domain of the S_2-RNase, but that the non-self interactions (PiSLF$_1$:S_2-RNase) were far stronger than the self interactions (PiSLF$_2$:S_2-RNase) interactions. Similarly the reciprocal interactions of S_1- and S_2-RNase domains with His-tagged PiSLF$_1$ while showing some interaction in both cases, were far stronger for the non-self pairs than the self-pairs. Sims et al., (2010) used a fluorogenic substrate for β-galactosidase to quantify the strength of two-hybrid interactions between different domains of the S_1- and S_3-RNase of *Petunia hybrida* with SLF-S1 from *P. hybrida*. Similar to the results obtained by Hua & Kao (2006), both self and non-self S-RNases interacted with *P. hybrida* SLF-S1, but the interaction appeared stronger for the non-self interactions compared with the self-interactions.

One of the critical questions in GSI is that of how SLF and S-RNase alleles recognize each other as self versus non-self. This question has recently become even more complicated (see section 8) as it appears that proteins originally identified as SLF-like (SLFL), and not involved in GSI, may in fact be true SLF proteins. Chromosome-walking, differential display or degenerate PCR-cloning approaches (McCubbin et al., 2000; Wang et al., 2003,2004; Wheeler & Newbigin 2007; Zhou et al., 2003) in the Solanaceae and Plantaginaceae identified a number of SLF-like genes (SLFL) that were linked to the S-locus. These genes were thought not to be involved in self-incompatibility interactions specifically, since they did not show interaction with known S-RNases nor did they exhibit competitive interaction in transgenic assays (Hua et al., 2007; Meng et al., 2011). Hua et al., (2007) attempted to identify domains of SLF proteins that governed allelic specificity by iterative pairwise comparisons of SLF proteins with SLFL proteins. These comparisons identified three "SLF-specific" regions SR1, SR2 and SR3. Based on this identification, these authors then divided the SLF proteins into three domains: FD1, containing the F-box and SR1 (amino acids 1-110), FD2 (amino acids 111-259, including the SR2 region) and FD3 (amino acids 260-389, including SR3). Domain-swapping experiments, in which different chimeric proteins were tested for the ability to interact with the S_3-RNase in pull-down assays suggested that FD2 was the domain primarily responsible for SLF-S-RNase interactions. Testing chimeric constructs between PiSLF$_2$ and PiSLFLb-S$_2$ (a SLF-like protein in the same S$_2$ haplotype as PiSLF$_2$) failed to demonstrate functionality of the FD2 domain *in vivo* (Fields et al., 2010). That is, neither chimeric protein showed competitive interaction in transgenic assays. Given that most SLFL proteins now appear to be bona fide SLF variants (see section 8), the long-term significance of these assays is unclear. The different SLFL proteins used for sequence comparisons represent different classes of SLF variants (section 8) so that the "SLF-specific" domains identified may represent regions that are more similar within a particular SLF-variant class.

6. A role for ubiquitination in gametophytic self-incompatibility

The observed protein interactions described above, together with the properties of SLF, SBP1 and SSK1 all suggest that recognition of self versus non-self in gametophytic self-incompatibility involves ubiquitination pathways. Pollen-S (SLF), is an F-box protein, and F-box proteins are know the be the recognition component of SCF E3 ubiquitin ligases (Cardozo & Pagano, 2004; Hua et al., 2008; Sijacic et al., 2004). SBP1 (Hua & Kao 2006; Sims & Ordanic 2001; Sims et al., 2010) is a RING-HC protein. RING proteins are E3 ubiquitin ligases (Deshaies & Joazeiro 2009; Freemont 2000), and SBP1 has E3 ubiquitin ligase activity *in vitro* (Hua et al., 2007; Sims unpublished). AhSSK1 (Huang et al., 2006) and PhSSK1 (Zhao et al., 2010) are SKP1-like proteins (SKP1 is a scaffold component of SCF E3 ligases). Pollen extracts have been shown to ubiquitinate S-RNase proteins, albeit in an allele-independent manner (Hua & Kao 2006). Together, these results have lead to the proposal that a non-canonical SCF^{SLF}-like complex acts to recognize and ubiquitinate S-RNases, leading to the inhibition of S-RNase activity in compatible pollen tubes. This complex is proposed to differ from a canonical SCF complex, because neither SKP1 orthologues (Hua & Kao 2006; Huang et al., 2006; Zhao et al., 2010) or RBX1 (Hua & Kao 2006) interact with SLF or Cullin. Instead either (or both) SBP1 and SSK1 have been proposed to replace RBX1 and/or SKP1 in this complex (Sims 2007; Hua et al., 2008; Sims & Robbins 2009; Zhao et al., 2010). According to the simplest version of this model, recognition of non-self S-RNases by the SCF^{SLF} complex would lead to polyubiquitination and degradation of S-RNase by the 26S proteasome complex (Sims 2007; Hua et al., 2008). One prediction of the SCF^{SLF} ubiquitin ligase complex model is that down-regulation of SLF, SBP1 or SSK1 should render all pollen tubes incompatible, regardless of genotype. To date, down-regulation of SLF or SBP1 has not been reported. Down-regulation of PhSSK1 (Zhao et al., 2010) does, however, result in a switch from compatibility to incompatibility, in accordance with this model. Figure 4 summarizes the structure of the proposed SCF^{SLF} ubiquitin ligase complex.

7. Style-expressed proteins with roles in GSI

Several style-expressed proteins have been shown to either be required for pollen rejection, or to interact with S-RNase or SBP1 in different assays. Cruz-Garcia et al., (2005) immobilized the S_{c10}-RNase from *Nicotiana alata* on an Affi-Gel column, then tested the ability of different proteins from style extracts to bind to the immobilized S-RNase. NaTTS, Na120K and NaPELPIII stylar proteins all bound to the S_{c10}-RNase in a specific manner. All three of these are previously-characterized proteins that are secreted into the transmitting tract of the style and that interact with pollen tubes. Biochemical data suggests that these proteins and the S-RNase may form a complex that is taken up into pollen tubes. In an extension of these experiments, Lee et al., (2008) used the C-terminal domains of the TTS and the 120 K proteins in yeast two-hybrid screens of pollen cDNA libraries. In addition to interaction with SBP1 (see section 4.2) a putative cysteine protease, NaPCCP, interacted with both TTS and 120K. Figure 5 summarizes observed protein interactions of the style proteins.

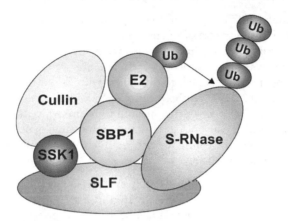

Fig. 4. **Proposed SCF^{SLF} ubiquitin ligase complex.**
Components of a proposed SCF^{SLF} complex are illustrated digrammatically. Specific contacts between components are based on protein-interaction assays summarized in Figure 3.

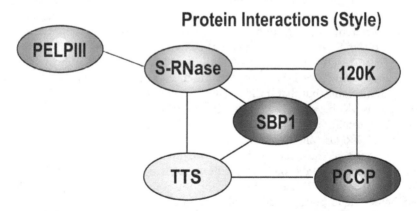

Fig. 5. **Interactions of the style proteins TTS, 120K and PELPIII with S-RNase and the pollen-expressed proteins SBP1 and PCCP.**

Two proteins, the 120k protein and a small asparagine-rich protein HT-B are required for the ability to reject pollen tubes in GSI (Hancock et al., 2005; McClure et al., 1999; O'Brien et al., 2002). Down-regulation of these genes by antisense (McClure et al., 1999) or RNAi (Hancock et al., 2005; O'Brien et al., 2002) resulted in an inability to block pollen tube growth, even though S-RNase proteins were expressed at normal levels. The 120K protein interacts with S-RNase and may be imported into pollen tubes in a complex with S-RNase and other style proteins (Cruz-Garcia et al., 2005). HT-B is also imported into pollen tubes (Goldraij et al., 2006), but whether in a complex with S-RNase and other style proteins, or separately, is not known.

8. Collaborative non-self recognition by SLF proteins in GSI

Although SLF proteins fulfill many of the expectations of pollen-S, different lines of evidence suggested that the nature of pollen-S may be more complicated than previously thought. First, as described, the evolutionary history of SLF proteins appeared far different than that of the linked S-RNase proteins, with the SLF proteins having evolved more recently and showing no evidence of co-evolution with S-RNase proteins as previously expected (Newbigin et al., 2008). Further, multiple SLF-like genes had been identified in *Petunia, Antirrhinum* and *Nicotiana* which appeared to be linked to the S-locus if not as close to the S-RNase as was SLF. It was unclear how the high degree of sequence identity among different SLF proteins could account for the ability to recognize and inhibit multiple different S-RNase alleles in populations.

Kubo et al. (2010) cloned additional SLF alleles from *Petunia hybrida*. [It should be noted here that *Petunia hybrida* is an "artificial" species created in the 19th century by crossing *Petunia axillaris* x *Petunia integrifolia*. *Petunia inflata* has been viewed either as synonymous with, a subspecies of, or very closely related to *Petunia integrifolia* (Stehmann et al., 2009). Thus S-locus haplotypes found in *Petunia hybrida* should be identical to those in either of the progenitor species.] When the SLF protein from *Petunia hybrida* S_7 haplotype was sequenced, it was found to be identical with the previously isolated SLF from *Petunia axillaris* S_{19}. What was striking, however was that the two S-RNases in these lines (S_7 versus S_{19}) were substantially different, having only 45% amino acid sequence identity. Reciprocal pollinations between S_7 and S_{19} confirmed that these two lines indeed had separate S-locus haplotypes. These results suggested that additional genes beyond SLF might constitute pollen-S. Further testing of SLF-S_7 showed that it could not cause competitive interaction in S_5S_7, S_7S_{11} or S_5S_{19} plants, but that it did show competitive interaction in S_7S_9 plants as well as in S_5S_{17} plants (Kubo et al., 2010 and supplemental material). Thus it appeared that individual SLF proteins could cause competitive interaction (i.e. act as pollen-S) with a limited subset of S-RNase alleles. Further analysis of proteins previously identified as SLF-like genes showed that these too, reacted with different subsets of S-locus haplotypes to cause competitive inhibition. Protein interaction assays demonstrated that there was direct correlation between SLF-S-RNase interaction and the ability to show competitive interaction in diploid heteroallelic pollen. Additional sequence comparisons demonstrated that SLF proteins could be grouped into at least six subclasses. Because Wheeler & Newbigin (2007) identified at least 10 different SLF-like genes in *Nicotiana*, and because not all of the previously-identified SLFL genes from *Petunia inflata* were included in the comparative sequence analysis of Kubo et al. (2010), it is possible that more than six SLF subclasses are present. As a result of this analysis, a new nomenclature for SLF proteins has been proposed. The original SLF isolates (e.g. PiSLF₁, PiSLF₂ etc.) have been renamed as the SLF1 variant class. Thus PiSLF₁ has been renamed S_1-SLF1, PiSLF₂ is now S_2-SLF1 and so on.

Genes previously identified as encoding SLFL proteins now comprise SLF2, SLF3, SLF4, SLF5, SLF6 and possibly additional SLF classes. The general nomenclature is thus $S_{haplotype}$ ID-SLF(class).

These results led Kubo et al. (2010) to propose a "Collaborative Recognition" model for the interaction of SLF variants with different S-RNases. According to this model, different SLF variants can recognize separate but partially overlapping subsets of S-RNases. Thus S_7-SLF1 reacts with S_{17}-RNase and S_9-RNase, but not with S_{11}- or S_{19}-RNase, S_7-SLF2 reacts with S_9-, S_{11}- and S_{19}-RNases, but not with S_{17}-RNase. One important area of future research (see below) will be to further dissect the complexities of protein interactions between different SLF variants and S-RNase proteins.

9. Models for pollen recognition and rejection in GSI

At present, two different, but non-exclusive models have been proposed to explain the mechanism of pollen acceptance versus pollen-rejection in gametophytic self-incompatibility. Both models presume that incompatible pollen tubes are rejected via the cytotoxic action of the S-RNase, and that self-compatibility (or cross-compatibility) results from inhibiting S-RNase action, consistent with the presumed role of pollen-S as an inhibitor. Where the models differ is in the primary mechanism for S-RNase inhibition, as well as the "default" condition of pollination. One model proposes that pollen-S (SLF) acts via the SCF^{SLF} E3 ubiquitin ligase complex to polyubiquitinate S-RNases, resulting in degradation by the 26 S proteasome complex. In this model, self-incompatibility (pollen rejection) would be the default pathway, unless the S-RNase is inhibited. The alternative model proposes that S-RNase imported into pollen tubes is sequestered in a vacuolar-like compartment. In this model, the default pathway is compatibility, unless SLF-S-RNase recognition leads to a breakdown of the compartment and release of the S-RNase.

9.1 The ubiquitination-degradation model

Much of the evidence for this model comes from protein-interaction assays, along with the known characteristics of the interacting proteins. Pollen-S (SLF), is an F-box protein, the recognition component of SCF E3 ubiquitin ligases (Cardozo & Pagano, 2004; Sijacic et al. 2004; Hua et al. 2008). SBP1 (Sims & Ordanic 2001; Hua & Kao 2006; Patel 2008, Sims et al., 2010) is a RING-HC protein. RING proteins are E3 ubiquitin ligases (Freemont 2000, Deshaies & Joazeiro 2009), and SBP1 has E3 ubiquitin ligase activity in vitro (Hua et al. 2007; Sims unpublished). AhSSK1 (Huang et al. 2006) and PhSSK1 (Zhao et al. 2010) are SKP1-like proteins (SKP1 is a scaffold component of SCF E3 ligases). Pollen extracts have been shown to ubiquitinate S-RNase proteins, albeit in an allele-independent manner (Hua & Kao 2006). Together, these results have lead to the proposal that a non-canonical SCF^{SLF}-like complex acts to recognize and ubiquitinate S-RNases, leading to the inhibition of S-RNase activity in compatible pollen tubes. This complex is proposed to differ from a canonical SCF complex, because neither SKP1 orthologues (Hua & Kao 2006; Huang et al. 2006; Zhao et al 2010) or RBX1 (Hua & Kao 2006) interact with SLF or Cullin. Instead either (or both) SBP1 and SSK1 have been proposed to replace RBX1 and/or SKP1 in this complex (Sims 2007; Hua et al. 2008; Sims & Robbins 2009; Zhao et al. 2010). According to the simplest version of this model, recognition of non-self S-RNases by the SCF^{SLF} complex would lead to polyubiquitination and degradation of S-RNase by the 26S proteasome complex (Sims 2007; Hua et al., 2008). One prediction of the SCF^SLF ubiquitin ligase complex model is that down-regulation of SLF, SBP1 or SSK1 should render all pollen tubes incompatible, regardless of genotype. To date, down-regulation of SLF or SBP1 has not been reported. Down-regulation of PhSSK1 (Zhao *et al.* 2010) does, however, result in a switch from compatibility to incompatibility, in accordance with this model.

Although the ubiquitination-degradation model is attractive, several predictions of this model remain untested, and other predictions may (depending on interpretation) be contradicted by current evidence. The pattern of ubiquitination of the S-RNase *in vivo* is not known. Because K48-linked or K63-linked polyubiquitination, or monoubiquitination leads to different cellular outcomes for the tagged proteins, it will be important to determine what ubiquitination patterns occur in reponse to SLF:S-RNase interaction. Also, it is not clear whether large-scale degradation of S-RNase proteins occurs in compatible pollinations. The high level of secreted extracellular S-RNase that accumulates in the transmitting tract make it challenging to monitor the level of S-RNase proteins in pollinated styles. As stated earlier, the degradation model predicts that inactivation or down-regulation of SCF^SLF E3 ubiquitin ligase components should result in pollen rejection. This prediction appears to be sustained in the case of SSK1. Different SFB mutants characterized in the Rosaceae, however (Marchese et al., 2007; Sonneveld et al., 2005; Ushijima et al., 2004; Vilanova et al., 2006), all of which either truncate or delete the SFB protein, are self-compatible. Although these data (along with some other differences between Solanaceae/Plantaginaceae versus Rosaceae) have been interpreted as suggesting that GSI has a different mechanistic basis in these taxa, there is also a large degree of similarity in how GSI functions in Solanaceae/Plantaginaceae versus Rosaceae (e.g., S-RNase, F-box proteins) such that it may be premature to make a definitive judgement on that point (McClure et al., 2011). Figure 6 summarizes the basic ubiquitination-degradation model.

9.2 The sequestration model

Evidence for this model comes primarily from the work of Goldraij et al. (2006), who reported that S-RNase was sequestered in a vacuolar compartment in compatible pollinations. These authors fixed and paraffin-embedded pollinated styles of *Nicotiana alata*, then hybridized sections to antibodies for callose (pollen-tube cell wall marker), S_{c10}-RNase, 120K protein, HT-B, aleurain (vacuolar lumen marker) or vPPase (vacuolar membrane marker). They concluded that in a compatible pollination, S-RNase inside pollen tubes remained in a ribbon-like vacuole bounded by the 120K protein. HT-B levels in compatible pollinations were low or undetectable. In later stages of incompatible pollinations, conversely, S-RNase appeared to be released into the cytoplasm of pollen tubes, and HT-B levels remained higher than in compatible pollinations (Goldraij et al., 2006; McClure et al., 2011). This model is consistent with the results of RNAi down-regulation of HT-B and 120K, which prevent rejection of incompatible pollen. What this model only incompletely explains is the required S-RNase::SLF interaction leading to compatibility or incompatibility. Both genetic evidence and the protein-interaction data summarized in previous sections show that S-RNase and SLF must interact to determine self/non-self recognition. If S-RNase is sequestered in a vacuolar compartment, however and SLF is cytoplasmic, it is not clear how

Ubiquitination Model for S-RNase-GSI

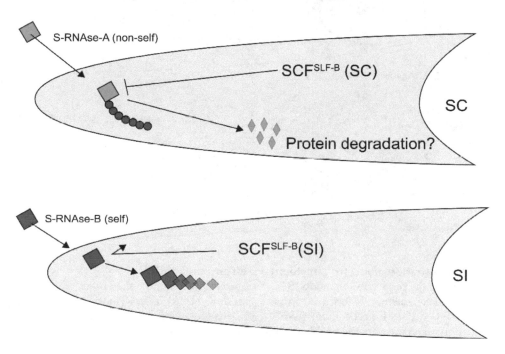

Fig. 6. **Ubiquitination-degradation model for gametophytic self-incompatibility.**
According to this model, both non-self (S-RNase-A) and self (S-RNase-B) proteins are imported into pollen tubes (the mechanism of import is not defined, but probably does not involve a specific receptor). In compatible (non-self) pollen tubes a SCF[SLF] E3 ubiquitin ligase complex targets ths S-RNase for polyubiquitination and degradation. In self-incompatible (self) pollen tubes, the SCF[SLF] complex is incapable of targeting the S-RNase, which acts to degrade pollen tube RNA and inhibit protein synthesis and growth.

this interaction can take place. McClure et al. (2011) suggest that a small amount of S-Rnase may be able to escape the vacuolar compartment, possibly by retrograde transport, to interact with the SCF[SLF] complex. In the case of an incompatible pollination, this interaction presumably leads to stabilization of HT-B, breakdown of the vacuolar compartment and release of the S-RNase. Figure 7 summarizes the essential aspects of the sequestration model.

Sequestration Model for S-RNase-GSI

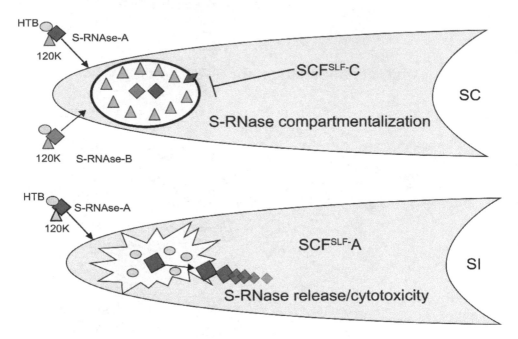

Fig. 7. Sequestration model for gametophytic self-incompatibility.
According to the sequestration model, S-RNases are imported into pollen tubes via an exocytotic mechanism, possibly in a complex with other style proteins (the complex shown with HT-B and 120K is speculative). S-RNase remains sequestered in a compatible pollination; in an incompatible pollination the vacuolar compartment breaks down releasing the S-RNase.

10. Current questions and future research

Although tremendous progress has been made in identifying genes and proteins involved in gametophytic self-incompatibility and in understanding much of the basic molecular biology of this phenomenon, many questions remain, and additional research is needed on nearly all aspects of GSI. In particular the collaborative recognition model raises the question of what is the exact nature of pollen-S? Do single SLF proteins interact one-on-one with individual S-RNases or can multiple SLFs interact simultaneously? Given the high degree of sequence identity with any specific class of SLF variants, what constitutes an allele? What protein interactions are required to make a determination of self versus non-self? If the sequestration model is correct, how do SLF and S-RNase even make contact? What is the specific role of ubiquitination in GSI interactions? Are S-RNases polyubiquitinated and degraded, or do different patterns of ubiquitination result in directing S-RNases to (or keeping them in) a membrane-bound compartment? What other proteins are needed for GSI interactions? Investigations not addressed in this review have suggested that proteins such as NaStEP (Busot et al., 2008) or Sli (Hosaka & Hanneman, 1998) may act as modifiers of the GSI response. What is the molecular basis for the quantitative, reversible, breakdown of GSI known as pseudo-self-compatibility, or PSC (Flaschenreim & Ascher 1979a, 1979b; Dana & Ascher 1985, 1986a, 1986b). What is the mechanism of uptake of S-RNases and other proteins into pollen tubes? Do Solanaceae/Plantaginaceae and Rosaceae really differ in fundamental mechanisms of GSI? More refined protein-interaction assays, suh as those using bi-molecular fluorescence complementation (Gehl et al., 2009), robust transgenic experiments, more complete information on genes and gene families involved in gametophytic self-incompatibility should all prove valuable in addressing these questions.

11. References

Ai Y, Singh A, Coleman CE, Ioerger TR, Kheyr-Pour A, Kao T-h. (1990) *Sexual Plant Reproduction*, 3, 130-138.

Anderson, M.A., Cornish, E.C., Mau, S.L., Williams, E.G., Hoggart, R., Atkinson, A., Bonig, I., Grego, B., Simpson, B., Roche, P.J., Haley, J.D., Penschow, J.D., Niall, H.D., Tregear, G.W., Coghlan, J.P., Crawford, R.J, & Clarke, A.E. (1986) *Nature*, 321, 38-44.

Anderson MA, McFadden GI, Bernatzky R, Atkinson A, Orpin T, Dedman H *et al.* (1989) The Plant Cell 1, 483-491.

Busot, G.Y., McClure, B., Ibarra-Sánchez, C.P., Jiménez-Durán, KI, Vázquez-Santana, S. & Cruz-Garcia, F. (2008) *Journal of Experimental Botany* 59, 3187-3201.

Cardozo T, Pagano M. (2004) *Nature Reviews Molecular Cell Biology*, 5, 739-751.

Chawla, B., Bernatzky, R., Liang, W. & Marcotrigiano, M. (1997) *Theoretical and Applied Genetics* 95, 992-996.

Chen, G., Zhang, B., Zhao, Z., Sui, Z., Zhang, H. & Xue, Y. (2010) *Journal of Experimental Botany* 61, 2027-2037.

Clark, KR, Okuley J, Collins PD, Sims TL. (1990) *The Plant Cell* 2, 815-826.

Coleman, C. & Kao, T.-H. (1992) *Plant Molecular Biology*, 18, 725-737.

Cruz-Garcia, F., Hancock, N., Kim, D. & McClure, B. (2005) *The Plant Journal*, 42, 295-304.

Dana MN, Ascher PD. (1985) *Journal of Heredity* 76, 468-470.

Dana MN, Ascher PD. (1986a) *Theoretical and Applied Genetics* 71, 573-577.

Dana MN, Ascher PD. (1986b) *Theoretical and Applied Genetics* 71, 578-584.

Darwin C (1891). *The Effects of Cross and Self Fertilisation in the Vegetable Kingdom*. (3rd ed). John Murray, London

de Nettancourt D. (1977) *Incompatibility in Angiosperms*, Springer-Verlag, ISBN 3-540-08112-7, Berlin

Deshaies RJ and Joazeiro CAP (2009) Annual Review of Biochemistry 78, 399-434.

Dodds PN, Ferguson C, Clarke AE and Newbigin E. (1999) *Sexual Plant Reproduction* 12, 76-87.

Dowd PE, McCubbin, AG., Wang, Verica, J.A., Tsukamoto, T., Ando, T. & Kao, T.-H., (2000) *Annals of Botany*, 85, 87-93.

Entani, T., Iwano, M., Shiba, H., Takayama, S., Fukui, K. & Isogai. A. (1999) *Theoretical and Applied Genetics* 99, 391-391.

Entani T, Iwano M, Shiba H, Che F-S, Isogai A Takayama S. (2003) *Genes to Cells* 8, 203-213.

Fields, A.M., Wang, N., Hua, Z. Meng, X. & Kao, T.-H., (2010) *Plant Molecular Biology* 74, 279-292.

Flaschenreim DR, Ascher PD. (1979a) *Theoretical and Applied Genetics* 54, 97-101

Flaschenreim DR, Ascher PD. (1979b) *Theoretical and Applied Genetics*, 55, 23-28

Freemont PS. (2000) *Current Biology* 10, 84-87.

Gebhardt, C., Ritter, E., Barone, A., Debener, T., Walkemeier, B., Schachtschabel, U., Kaufmann, H., Thompson, R.D., Bonierbale, M.W., Ganal, M.W., Tanksley, S.D. & Salamini, F. (1991) *Theoretical and Applied Genetics*, 83, 49-57.

Gehl C, Waadt R, Kudla J, Mendel RR and Hänsch (2009) *Molecular Plant* 2, 1051-1058.

Goldraij, A., Kondo, K., Lee, C.B., Hancock, C.N., Sivaguru, M., Vazquez-Santana, S., Kim, S., Phillips, T.E., Cruz-Garcia, F. & McClure, B. (2006) *Nature*, 439, 805-810.

Golz, J.F., Su, V., Clarke, A.E. & Newbigin, E. (1999) *Genetics* 152, 1123-1135.

Golz, J.F., Oh, H.Y., Su, V., Kusaba, M. & Newbigin, E. (2001) *PNAS* 98, 15372-15376.

Haglund K, Di Fiore PP, and Dikic I (2003) *Trends in Biochemical Sciences* 28, 598-604.

Hancock, N., Kent, L., & McClure, B. A. 92005) *The Plant Journal* 43, 716-723.

Harbord, R.M., Napoli, C.A. & Robbins, T.P. (2000) *Genetics* 154, 1323-1333.

Hosaka, K. & Hanneman, R.E. (1998) *Euphytica* 99, 191-197.

Hua Z, Kao T-H. (2006) *The Plant Cell* 18, 2531-2553.

Hua, Z., Meng, X. & Kao, T.-H. (2007) *The Plant Cell*, 19, 3593-3609.

Hua, Z. & Kao, T.-H. (2008) *Plant Journal* 54, 1094-1104.

Huang S, Lee H-S, Karunanandaa B, Kao T-H. (1994) *The Plant Cell* 6, 1021-1028.

Huang J, Zhao L, Yang Q, Xue Y. (2006) *The Plant Journal* 46, 780-793.

Ida K, Norioka S, Yamamoto M, Kumasaka T, Yamashita E, Newbigin E, Clarek AE,Sakiyama F, Sato M. (2001) *Journal of Molecular Biology* 314, 103-112.

Ioerger TR, Gohlke JR, Xu B, Kao T-h. (1991) *Sexual Plant Reproduction* 4, 81-87.

Ishimuzu, T., Miyagi, M., Norioka, S., Liui, Y.H., Clarke, A.e. & Sakiyama, F. (1995) *J. Biochem* 118, 1007-1013.

Kao, T.H., & McCubbin A.G. (1996) *PNAS* 93, 12059-12065.

Karunanandaa B, Huang S, Kao T-h. (1994) *The Plant Cell* 6, 1933-1940.

Kohn, J.R. (2008) What Geneologies of S-alleles Tell Us, In: *Self-Incompatibility in Flowering Plants, Evolution, Diversity, and Mechanisms*, Veronica E. Franklin-Tong, Editor, pp 103-121, Springer-Verlag, ISBN 978-3-540-68485-5, Berlin

Kubo, K.I., Entant, T., Takata, A., Wang, N., Fields, A.M., Hua, Z., Toyoda, M., Kawashima, S.I., Ando, T., Isogai, A., Kao, T.-H., & Takayama, S. (2010) *Science* 330, 796-799.

Lai Z, Ma W, Han B, Liang L, Zhang Y, Hong G, Xue Y. (2002) *Plant Molecular Biology* 50, 29-41.

Lee, C.B., Swatek, K.N. & McClure, B. (2008) *Journal of Biological Chemistry* 283, 26965-26973.

Lee, H.S., Huang, S. & Kao, T.-H., (1994) *Nature* 367, 560-563.

Li, J.H., Nass, N., Kusaba, M., Dodds, P.N., Treloar, N., Clarke, A.E. & Newbigin, E. (2000) *Theoretical and Applied Genetics*, 956-964.

Linskens, H.F. (1957) *Proceedings of the Royal Society of London, Series B.* 188,299-311.

Lush, W.M. & Clarke, A.E. (1996) *Sexual Plant Reproduction* 10, 27-35.

Luu DT, Qin KK Morse D, Cappadocia M. (2000) *Nature* 407, 649-651.

Marchese, A., Boskovic, R.I., Caruso, T., Raimondo, A., Cutuli, M. & Tobutt, K.R. (2007) *Journal of Experimental Botany* 58, 4347-4356.

Mather, K. (1943) *Journal of Genetics* 45, 215-235.

Matton DP Luu DT, Xike Q, Laublin G, O'Brien M, Maes O, et al. (1999) *The Plant Cell* 11, 2087-2098.

Matton DP, Maes O, Laublin G, Xike Q, Bertrand C, Morse D, Cappadocia M. 1997) *The Plant Cell* 9, 1757-1766.

McCubbin, A.G., Wang, X. & Kao, T.H. (2000) *Genome* 43, 619-627.

McClure, B.A., Gray, J.E., Anderson, M.A. & Clarke, A.e. (1990) *Nature* 347, 757-760.

McClure, B., Mou, B., Canevacsini, S. & Bernatzky, R. (1999) *PNAS* 96, 13548-13553.

McClure, B., Cruz-Garcia, F. & Romero, C. (2011) *Annals of Botany*, 108, 647-658

Meng, X., Sun, P. & Kao, T.-H. (2011) *Annals of Botany*, 108, 637-646.

Mora, C., Tittensor, D.P., Adl, S., Simpson, A.G.B. & Worm, B. (2011) *PLOS Biology*, 9, e1001127

Murfett, J., Atherton, T.L., Mou, B., Gasser, C.S. & McClure, B.A. (1994) *Nature* 367, 563-566.

Newbigin, E., Paape, T. & Kohn, J.R. (2008) *The Plant Cell.* 20, 2286-2292.

O'Brien, M., Kapfer, C., Major, G., Laurin, M., Bertrand, C., Kondo, K. Kowyama, Y. & Matton, D.P. (2002) *The Plant Journal* 32, 985-996.

O'Brien M, Major G, Chantha SC, Matton DP. (2004) *Sexual Plant Reproduction* 17, 81-87.

Patel, Avani (2008). Protein interactions between pistil and pollen components controlling gametophytic self-incompatibility. M.S. thesis, Northern Illinois University, United States -- Illinois. www.proquest.com

Qiao, H., Wang, H., Zhao, L., Zhou, J., Huang, J., Zhang, Y., & Xue, Y. (2004a) *The Plant Cell*, 16, 582-595.

Qiao H, Wang F, Zhao L, Zhou JL, Lai Z, Zhang YS. et al. (2004b) *The Plant Cell* 16, 2307-2322.

Richman, A.D., Kao, T.-H., Schaeffer, S.W. & Uyenoyama, M.K. (1995) *Heredity*, 75, 405-415.

Richman, A.D., Uyenoyama, M.K. & Kohn, J.R. (1996) *Heredity*, 76, 497-505.

Richman, A.R. & Kohn, J.R. (2000) *Plant Molecular Biology*, 42, 169-179.

Royo, J., Kunz, C., Kowyama, Y., Anderson, M., Clarke, A.e., & Newbigin, E. (1994) *PNAS* 91, 6511-6514.

Sassa, H., Kakui, H. & Minamkiawa, M. *Sexual Plant Reproduction* 23, 39-43.

Sijacic P, Wang X, Skirpan AL, Wang Y, Dowd PE, McCubbin AG, Huang S, Kao T-h. (2004) *Nature* 429, 302-305.

Sims TL (1993) Genetic Regulation of Self-Incompatibility. CRC Critical Reviews in Plant Sciences 12, 129-167.

Sims TL (2005) Pollen recognition and rejection in different self-incompatibility systems. Recent Research Developments in Plant Molecular Biology 2, 31-62.

Sims TL (2007) Mechanisms of S-RNase-based self-incompatibility. CAB Reviews, Perspectives in Agriculture, Veterinary Science, Nutrition and Natural Resources 2, No 058.

Sims TL and Ordanic M. (2001) *Plant Molecular Biology* 47, 771-783.

Sims, T.L. & Robbins, T.P. (2009) Gametophytic Self-Incompatibility in Petunia, In: *Petunia: Evolutionary, Developmental and Physiological Genetics*, 2nd Edition, Tom Gerats & Judith Strommer editors, pp 85-106, Springer, ISBN 978-0-387-84795-5, NY, USA.

Sims TL, Patel A and Shretsha P (2010) *Biochemical Society Transactions* vol 38, in press

Singh, A., Ai, Y. & Kao, T.-H. (1991) *Plant Physiology*, 96, 61-68.

Stehmann, J.R., Lorenz-Lemke, A.P., Freitas, L.B. & Semir J. (2009) The Genus Petunia, In: *Petunia: Evolutionary, Developmental and Physiological Genetics, 2nd Edition*, Tom Gerats & Judith Strommer editors, pp 1-28, Springer, ISBN 978-0-387-84795-5, NY

Sonneveld, T., Tobutt, K.R., Vaughan, S.P. & Robbins, T.P. (2005) *The Plant Cell* 17, 37-51.

Tanksley, S.D. & Loaiza-Figueroa, F. (1985) *PNAS* 82, 5093-5096

ten Hoopen, R., Harbord, R.M., Maes, T., Nanninga, N. & Robbins, ,T.P. (1998) *The Plant Journal* 16, 729-734.

Tsukamoto T, Ando T, Takahashi K, Omori T, Wataabe H, Kokubun H, Marchesi E and Kao T-H (2005) *Plant Molecular Biology* 57, 141-163.

Ushijima K, Sassa H, Tamura M, Kusaba M, Tao R, Gradziel TM et al. (2001) *Genetics* 158, 379-386.

Ushijima K, Yamane H, Watari A, Kakehi E, Ikeda K, Hauck NR et al. (2004) *The Plant Journal* 39, 573-586.

Ushijima K, Sassa H, Dandekar AM, Gradziel TM, Tao R, Hirano H. (2003) *The Plant Cell* 15, 771-781.

Vilanova, S., Badenes, M.L., Burgos, L., Martinez-Calvo, J., Llacer. G. and Romero, G. (2006) *Plant Physiology* 142, 629-641.

Walles, B. & Han S.P. (1998) *Physologia Plantarum* 103, 461-465.

Wang, Y., Wang, X., McCubbin, A.G. & Kao, T.-H., (2003) *Plant Molecular Biology* 53, 565-580.

Wang Y, Tsukamoto T, Yi K-W, Wang X, Huang A, McCubbin AG, Kao T-h. (2004) *Plant Molecular Biology* 54, 727-742.

Wheeler D. & Newbigin, E. (2007) *Genetics*, 177, 2171-2180.

Woodward, J.R., Bacic, A., Jahnen, W. & Clarke, A.E. (1989) *The Plant Cell*, 1, 511-514.

Zhao, L., Huang, J., Zhao, Z., Li, Q., Sims, T.L., & Xue, Y. (2010) *The Plant Journal*

Zhou J., Wang, F., MA, W., Zhang, Y., Han, B. & XUe, Y. (2003) *Sexual Plant Repoduction* 16, 165-177.

Zurek, D.M., Mou, B., Beecher, B. & McClure, B. (1997) *The Plant Journal* 11, 797-808

11

Biochemical, Structural and Pathophysiological Aspects of Prorenin and (Pro)renin Receptor

A.H.M. Nurun Nabi[1] and Fumiaki Suzuki[2]

[1]*Department of Biochemistry and Molecular Biology,*
University of Dhaka, Dhaka, Bangladesh;
[2]*Laboratory of Animal Biochemistry, Faculty of Applied Biological Sciences, Gifu*
University, 1-1 Yanagido, Gifu,
Japan

1. Introduction

Our knowledge of understanding the complex role of renin angiotensin system (RAS) or RA system in human physiology has been widened for more than 100 years and it is rapidly increasing day-by-day. Over the last two decades discoveries of angiotensin converting enzyme 2 (ACE 2), putative receptor for angiotensin (Allen et al., 2006; Aronsson et al. 1988; Bickerton & Buckley, 1961; Cooper et al., 1996; Deschepper et al., 1986; Dzau et al., 1986; Epstein et al., 1970), and (pro)renin receptor have laid the foundation of many new hypotheses in the context of their biochemical actions, physiological effects and activation of second messenger pathways. Thus, scientists have started to reconsider the complex biochemistry and physiology of RAS. The primary and main role of this system is to regulate homeostasis of body fluid that ultimately maintains the blood pressure (Kobori et al., 2006; Oparil & Haber, 1974a; Oparil & Haber; 1974b). This system catalyzes a liver product, angiotensinogen, to generate a small decapeptide, angiotensin-I (Ang-I). Angiotensin converting enzyme (ACE), thus, converts Ang-I into octapeptide, angiotensin-II (Ang-II). Ang-II acts directly within the central nervous system to increase blood pressure (Bickerton & Buckley, 1961). Injection of purified Ang-II peptide around the hypothalamus in rat brain stimulated thriving drinking response (Epstein et al., 1970). The physiological actions of the most potent hormone peptide are mediated via G-protein coupled angiotensin II type 1 (AT1) and angiotensin II type 2 (AT2) receptors. Ang-II facilitates vasoconstriction, cell proliferation, cell hypertrophy, anti-natriuresis, fibrosis, atherosclerosis using AT1 (Ito et al., 1995) while, via AT2 receptor, the peptide elicits vasodilation, anti-proliferation, anti-hypertrophy, anti-fibrosis, anti-thrombosis, anti-angiogenesis (Siragy & Carey, 1997; Goto et al., 1997; Gross et al., 2000). The classical renin angiotensin system with the generation of different peptides and their physiological effects has been presented in Figure 1.

The systemic or classical renin angiotensin system has usually been viewed as the blood-borne cascade whose ultimate product Ang-II plays the pivotal endocrine role. Plasma renin activity is the most accepted clinical marker of circulating RAS. However, circulating RAS remained unsuccessful to describe the autocrine and paracrine functions mediated by RAS within specific tissue sites particularly in heart, kidney, brain and vasculature. Transgenic

Classical Renin-Angiotensin System

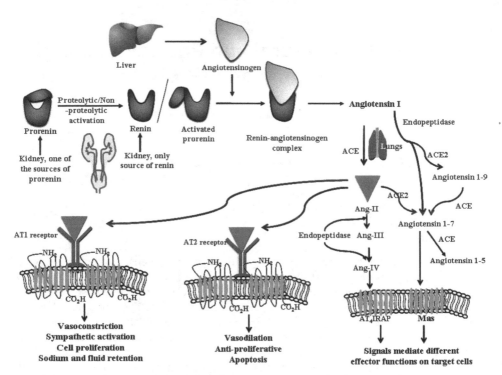

Fig. 1. Schematic diagram of the classical renin-angiotensin (RA) system shows angiotensin-II dependent pathway mediated different physiological effects via angiotensin type 1 (AT1) and type 2 (AT2) receptors. Renin, secreted from kidney, regulates the rate limiting step of this pathway by converting its liver originated macromolecule substrate angiotensinogen into a short peptide, angiotensin-I (Ang-I). Angiotensin-I, thus, is converted into angiotensin-II (Ang-II) by angiotensin converting enzyme. Other peptide products via stimulation of enzymes and receptor subtypes on target cells can also mediate physiological functions.

animals facilitate to demonstrate the existence of tissue RAS parallel to and independent of systemic RAS. Thus, local RA system has been ensured in intracellular compartments (de Mello, 1995; van Kesteren et al., 1997; Admiraal et al., 1999), interstitial fluids (de Lannoy et al., 1997, 1998), cardiac cells including fibroblasts, endothelial cells, myocytes, macrophages (de Lannoy et al., 1998; Hokimoto et al., 1996; Sun et al., 1994) as well as on the cell membrane (Danser et al., 1992; Neri Serneri et al., 1996). All the circulating components of renin angiotensin system i.e., renin, angiotensinogen, ACE, and Ang I and II though not produced but have been identified in cardiac tissue (Campbell et al., 1993; Danser et al., 1994). As a consequence, presence of local RAS in the heart could contribute to the pathogenesis of congestive heart failure, cardiac hypertrophy and remodeling, and reperfusion arrhythmias (Yusuf et al., 1991: Ruzicka et al., 1994; Schieffer et al., 1994; Van

Gilst et al., 1984). Direct action of Ang II within the central nervous system causes increased blood pressure. Also, presence of renin and endogenous production of angiotensin have established the existence of local RAS in the central nervous system (Bickerton and Buckley et al., 1961; Fischer-Ferraro et al., 1971; Ganten et al., 1971).

1.1 The main players associated with renin angiotensin system

The RA system is initiated by its rate-limiting enzyme renin (37 kDa with 340 amino acid residues) which catalyzes its only known substrate angiotensinogen. Renin is only secreted from kidney as preprorenin and levels of renin in the plasma of nephrectomized animals is not detectable. Professor Robert A. Tigerstedt and his student Per G. Bergman for the first time reported a "pressor" substance in the kidney extract more than 100 years ago, which caused increase in blood pressure in experimental animals and later, coined that substance as renin (Tigerstedt & Bergman, 1898). Renin, also known as aspartyl proteinase having an optimum pH of 5.5 to 7.5 instead of 2.0 to 3.4, has no known physiological effect other than the proteolysis of angiotensinogen (Murakami & Inagami, 1975; Inagami & Murakami, 1977; Matoba et al., 1978; Figueiredo et al., 1985: Dzau et al. 1979; Yokosawa et al., 1978; Yokosawa, 1980; Hirose, 1982; Pickens et al., 1965). The neutral pH is necessary to show its activity in plasma. The renin gene is also expressed in other tissues such as adrenal gland, gonads, placenta, pituitary, brain and hypothalamus (Hirose et al., 1978; Naruse et al., 1981, 1982; Pandey et al., 1984; Deschepper et al., 1986; Dzau et al., 1987; Paul et al., 1987; Suzuki et al., 1987; Tada et al., 1989). These extra renal renins have been thought to play a part in the tissue renin-angiotensin system proposed by several investigators (de Mello, 1995; van Kesteren et al., 1997; Admiraal et al., 1999; de Lannoy et al., 1997, 1998; Hokimoto et al., 1996; Sun et al., 1994; Danser et al., 1992, 1994; Neri Serneri et al., 1996; Campbell et al., 1993; Yusuf et al., 1991: Ruzicka et al., 1994; Schieffer et al., 1994; Van Gilst et al., 1984; Bickerton and Buckley et al., 1961; Fischer-Ferraro et al., 1971; Ganten et al., 1971).

Removal of the 23 amino acid residues from the C-terminus of preprorenin generates prorenin. Prorenin (45-47 kDa containing 406 amino acid residues), the pre-active form of renin, is predominantly synthesized by granular cells of the juxtaglomerular apparatus (JGA) in the terminal afferent arteriole (Schnermann & Briggs, 2008; Schweda et al., 2007) and principle cells of the collecting ducts (Prieto-Carrasquero et al., 2004; Rohrwasser et al., 1999; Kang et al., 2008). Prorenin is also synthesized in many other tissues like adrenal glands (Ganten et al., 1974, 1976; Ho and Vinson, 1998), zona glomerulosa (Doi et al., 1984; Deschepper, et al., 1986; Brecher et al., 1989), eye, Müller cells, mast cells (Krop et al., 2008), ovarian follicular fluid (Glorioso et al., 1986), and theca cells (Do et al., 1988), uterus (Derkx et al., 1987; Itskovitz et al., 1987), myometrium/decidual cells (Shaw et al., 1989), placenta (Lenz et al., 1991), chorionic cells, testis and leydig cells (Sealey et al., 1988). The submandibular gland in some mice strains produces a large amount of renin, which is a product of the Ren-2 renin gene distinct from the renal renin gene, Ren-1 (Cohen et al., 1972; Wilson et al., 1981; Holm et al., 1984) and this action is mediated by prorenin converting enzyme present in submandibular gland of the same mice strains (Kim et al., 1991). Prorenin, in the juxtaglomerular cells of the kidney, is converted to mature renin by the limited endoproteolysis after paired basic residues, Lys-Arg to remove the 43-amino acid residues containing prosegment sequence. The concentration of prorenin in human plasma is 10 times higher than that of mature renin though the physiological role of prorenin is still not clear and the relative concentration of prorenin to renin varies at different conditions. Thus, conversion of prorenin i.e., activation of prorenin to renin plays important role in the regulation of RA system. Certain proteases like trypsin or cathepsin were found to activate prorenin by cleaving the residue prosegment reversibly (Inagami et al., 1980; Shinagawa et al., 1990, 1994; Kikkawa et al., 1998; Jutras et al., 1999; Taugner et al., 1985; Wang et al., 1991; Jones et al., 1997). However, many tissues store prorenin but do not process it to active renin. Though extra-renal sources of prorenin are evident, kidney is the major source of plasma prorenin. Renin and prorenin have long been considered as the separate mediators of tissue and circulating systems (Sealy & Rubattu, 1989). *In vitro*, when prorenin is acidified at pH 3.3 or exposed to low temperature (< 4°C) or allowed to interact with antibodies designed from its prosegment sequences (protein-protein interaction), it mediates intrinsic catalytic activity without removal of the prosegment sequence from its N-terminus through a reversible change in conformation (Derkx et al., 1979, 1983, 1987a & b, 1992; Pitarresi et al., 1992, Suzuki et al., 2000, 2003).

Both renin and non-proteolytically activated prorenin catalyze angiotensinogen, a 6 kDa protein macromolecule found also in adipose tissues to generate a small decapeptide called angiotensin I. Both neonatal and adult rat cardiac cells express mRNA for angiotensinogen (Dostal et al., 1992; Malhotra et al., 1994; Zhang et al., 1995; Liang et al., 1998; Sadoshima et al., 1993), while van Kesteren and colleagues (1999) were unable to detect angiotensinogen in neonatal rat cardiac cells or in the conditioned medium of these cells using radioimmunoassay. Secreted angiotensinogen in the cultured medium of neuronal cells has been identified. Generated renin product, angiotensin I thus, further converted into angiotensin II by the action of ACE.

The (pro)renin receptor or (P)RR is now considering as another important regulatory component in renin-angiotensin system. However, ongoing research works have revealed its association both in angiotensin II-dependent and –independent pathways which also play pivotal role in the developmental processes.

2. (Pro)renin receptor, a new family member of renin angiotensin system

It's been almost a decade since the full length (pro)renin receptor or (P)RR was cloned (Nguyen et al., 2002). However, earlier the same group (Nguyen et al., 1996, 1998) reported high affinity binding of ^{125}I renin to primary and immortalized human mesangial cells (0.2 and 1.0 nM, respectively) in a time-dependent fashion that could attain saturable state. The (P)RR does not internalize the ligands inside the cells rather activates renin and prorenin after binding to generate angiotensin I and second messenger pathway by activating proteins involved in signaling. In the late nineties of the last century, the mannose 6-phosphate/insulin-like growth factor II (M6P/IGF2) receptor was found on rat cardiac myocytes (van Kesteren et al., 1997) and human endothelial cells (Admiraal et el., 1999; Saris et al., 2001) that could bind and internalize renin/prorenin (van Kesteren et al., 1997). However, such binding and internalization could not generate any angiotensin peptides intracellularly. Besides, existence of renin/prorenin receptor independent of mannose 6-phosphate such as renin binding protein (RnBP), renin/prorenin binding protein (ProBP) in rat tissues, vascular renin binding protein have also been confirmed (Takahashi et al., 1983;

Tada et al., 1992; Campbell et al., 1994; Sealy et al., 1996) that bind with different binding affinities to their ligands.

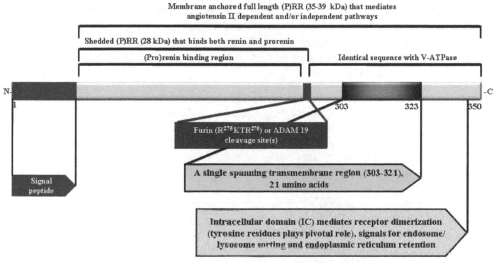

Fig. 2. Structure of (pro)renin receptor protein. The receptor protein is composed of three basic constituents with an N-terminal domain, which is the (pro)renin binding site, a single spanning transmembrane sequence that traverse through the plasma membrane and the intracellular cytoplasmic domain that recently has been identified as the important region required for the dimerization of (P)RR.

The (P)RR, expressed on the cell surface, is a 350 amino acid (39 kDa) containing protein with a single spanning transmembrane domain encoded from the X-chromosome. A short signal peptide is present at the N-terminus end of the unglycosylated large extracellular domain with ~310 residues and the transmembrane domain has putative 20-amino acid residues followed by a ~19-amino acid containing intracellular cytoplasmic (IC) domain (shown in Figure 2). Ubiquitous expression of (P)RR has been demonstrated with the highest amount of mRNA found in brain, heart and placenta while lower amount was expressed in liver, pancreas and kidney (Nguyen et al., 2002). It is reported that (P)RR expressed in VSMCs in human (P)RR transgenic rats can be recycled between intracellular compartment and cell membrane (Batenburg et al., 2007). The (P)RR is also localized on the membrane of stromal adipose cells (Achard et al., 2007), in the neurons of neonatal rats (Shan et al., 2008), on COS-7 cells (Nabi et al., 2007), in glomerular mesangial cells, the subendothelium of renal arteries, podocytes, and distal nephron cells (Nguyen et al., 2002, 1996, 1998) of human and rat kidneys; U937 monocytes (Feldt et al., 2008b) and also in intracellular compartments or on the surface of vascular smooth muscle cells (VSMCs) (Sakoda et al., 2007; Zhang et al, 2008), in endoplasmic reticulum (Schefe et al., 2006; Yoshikawa et al., 2011), golgi apparatus (Contrepas et al., 2009; Yoshikawa et al., 2011), cytosol (Contrepas et al., 2009; Cousin et al., 2009) and found in plasma (Cousin et al., 2009). Expression of (P)RR is also categorized in the subfornical organ (SFO), paraventricular nucleus, the supraoptic nucleus, the nucleus of the tractus solitarius (NTS), or the rostral ventrolateral medulla regions of brain that were believed to be involved in the central regulation of cardiovascular function and volume homeostasis (Contrepas et al., 2009), in the frontal lobe of human brain and pituitary (Takahashi et al., 2010). Retina is also a source of (P)RR (Satofuka et al., 2009). In particular, it is localized to pericytes in retinal vessels, endothelial cells, and, mostly in retinal ganglion cells and glia (Wilkison-Berka et al., 2010). Moreover, predominant expression of (P)RR using immunohistochemistry and *in situ* hybridization on the epical membrane of acid secreting cells in the collecting duct has been reported (Advani et al., 2009).

Full length rat and human recombinant (P)RR with transmembrane followed by cytoplasmic domains were expressed in baculovirus expression system, and identified in the cellular fraction (Nabi et al., 2006; Du et al., 2008; Kato et al., 2008). On the other hand, human (P)RR containing only the extracellular domain lacking transmembrane part was found secreted in the culture medium (Kato et al., 2009). Also, human (P)RR was successfully expressed in the Bombyx mori multiple nucleopolyhedrovirus (BmMNPV) and found in the silkworm larvae as well as in the fat body of silkworm larvae. ELISA and surface plasmon resonance technique in BIAcore assay system confirmed the renin/prorenin binding ability i.e., the functional bioactivity of (P)RR expressed and fractionated from silkworm and baculovirus expression system (Nabi et al., 2006; Du et al., 2008; Kato et al., 2008, 2009).

A protease called furin was found to be responsible for shedding of endogenous (P)RR in trans-golgi by cutting at positions $R^{275}KTR^{278}$ near the N-terminus of transmembrane sequence (Cousins et al., 2009). This soluble form of (P)RR [s(P)RR, 28 kDa] was detected in the conditioned cultured medium and also in human plasma using co-precipitation experiment with human renin. Another protease ADAM19 sheds intracellular (P)RR from golgi apparatus into the extracellular space (Yoshikawa et al., 2011). Moreover, constitutively secreted soluble form of (P)RR (~30 kDa) shedded from the cell surface was found in the cultured medium of human umbilical vein endothelial cells (HUVECs) (Biswas et al., 2011) that could also bind recombinant human prorenin with a nanomolar order, similar to what was reported for full length (P)RR on the cell surface (Nguyen et al., 2002; Batenburg et al., 2007; Nurun et al., 2007) or from the baculovirus expression system (Nabi et al., 2006). Non-proteolytic activation of prorenin occurred when it interacted with s(P)RR in the soluble phase and this was confirmed by Western blot analysis. Also, activated prorenin showed renin activity by generating Ang I from sheep angiotensinogen (Biswas et al., 2011). However, the enzymatic properties of renin after binding to (P)RR is yet to be determined. These phenomenons have been depicted in Figure 3 (vi).

C-terminal domain of (P)RR is identical to "M8-9," a truncated protein of 8.9 kDa that co-purified with a proton-ATPase of bovine chromaffin granule membranes (Ludwig et al., 1998). At the gene level (P)RR from human, rat, and mice showed 95% sequence homology, while at the amino acid level they showed 80% homology. Phylogenetic analyses also revealed that the sequences for (P)RR are not only conserved within the closely related species but also similar sequences are present in the remote species. The IC-domain of (P)RR mediates the signal transduction pathways and promyelocytic zinc finger (PLZF) protein has been identified as an associated protein that interacts with the IC-domain to down regulate expression of the receptor. (P)RR has also been reported to exist as a dimer *in vivo* (Schefe et al., 2006). Recent evidences suggest that short and relatively flexible loop of IC segment generates the driving force in the process of dimerization of (P)RR and tyrosine residues of IC contribute in dimerization dominantly (Zhang et al., 2011).

2.1 (Pro)renin receptor and its ligand: interaction of (pro)renin with (pro)renin receptor

Interaction of renin and prorenin with (pro)renin receptor instigates two pathways: one leads to generation of angiotensin II that ultimately contribute to the activation of local RA system via angiotensin II-dependent pathway as in case of classical circulating RA system and the other one leads to signal transduction mediated by angiotensin II-independent pathway outlined in Figure 3.

2.1.1 Binding mechanism and activation of renin angiotensin system

Binding of human renin to human (P)RR increases local angiotensin production as it is manifested by the increased (4/5-fold) substrate affinity of (P)RR-bound renin compare to free form of soluble renin (Nguyen et al., 2002). On the other hand, human renin bound to recombinant human/rat (P)RR and free form of soluble renin showed similar binding affinity for the substrate, sheep angiotensinogen at the micromolar order (Nabi et al, 2006; Nurun et al., 2007). However, kinetic data analyses revealed that prorenin preferentially binds to (P)RR and such binding initiates angiotensin I generation (Nabi et al., 2006; 2009b). Full length rat (P)RR expressed and isolated from the baculovirus expression system had almost 3 times higher binding affinity (K_D = 8.0 nM) for rat prorenin than that of mature rat renin (K_D = 20.0 nM) *in vitro* (Nabi et al., 2006). Receptor-bound rat prorenin also had similar affinity for the substrate (K_m = 3.3 μM) sheep angiotensinogen as it was for rat renin. On the other hand, receptor-bound renin showed higher molecular activity (10 nM·h) compared to free form of mature renin and receptor-bound activated prorenin (1.25 and 1.1 nM·h) (Nabi et al., 2006).

Ninety% of rat and fifty% of human prorenin (at 2.0 nM of initial concentration) bound to their respective (P)RR over expressed on the membrane of COS-7 cells and the K_Ds were estimated to be 0.89 and 1.8 nM, respectively. Receptor-bound rat and human prorenin showed 30% and 40% activity, respectively, in comparison with the activity of trypsinized prorenin molecules (Nurun et al., 2007). A similar binding and activation patterns of prorenin to human (pro)renin receptor expressed in VSMCs of transgenic rats (K_D = 6.0 nM) (Baternburg et al., 2007) and of rat prorenin by rat (P)RR expressed in cultured VSMCs were observed (Zhang et al., 2010). Differences in the K_D values of rat prorenin bound to the immobilized receptors on the synthetic surfaces and the membrane-anchored receptor could be due to the presence of some other associated proteins that might have stabilized the (P)RR on the membrane. Surface plasmon resonance technique in BIAcore assay system revealed almost four times higher binding affinity of human prorenin (1.2 nM) to the *in vitro* synthesized human recombinant (P)RR compared to that of human mature renin (4.4 nM) (Nabi et al., 2009b). The immobilized receptors bind recombinant human renin and prorenin with the dissociation constant (K_D) values of 1.2 and 4.4 nM, respectively. Also, the data obtained from the BIAcore kinetic study showed that association rate of prorenin to (P)RR is higher than that of mature renin (1.8 x 10^7 and 2.16 x 10^6 M^{-1}.s^{-1}, respectively) (Nabi et al., 2009b).

The binding mechanism of renin and prorenin to the (pro)renin receptor has also been proposed depending on the ground work led by Suzuki and colleagues who demonstrated the importance of "handle" ($I^{11P}FLKR^{15P}$) and "gate" ($T^{7P}FKR^{10P}$) region peptides designed from the prosegment sequence of prorenin in the non-proteolytic activation of prorenin via protein-protein interaction (Suzuki et al., 2003). Later, another peptide called "decoy" ($R^{10P}IFLKRMPSI^{19P}$ including the "handle" sequence) that mimics the N-terminus sequence of human prorenin prosegment showed its high binding affinity to the recombinant (P)RR and this affinity explains the probable reason for high binding affinity of prorenin for (P)RR. Decoy peptide has got binding affinity to (P)RR at the nanomolar order similar to that of prorenin (Nurun et al., 2007; Nabi et al., 2009a, 2009b). Even after 28 days of administration, fluorescent tagged handle region peptide (HRP) was recognized in the renal glomeruli and tubular lumen (Ichihara et al., 2006a; Kaneshiro et al., 2007). However, a signal of these fluorescent molecules is from the intact form of HRP or not is still arguable. This argument becomes even stronger from the findings of Leckie and Bottrill (2011). They synthesized part of prosegment sequence, RIFLKRMPSIR (it contains an additional arginine residue at the C-terminus of the decoy) and its scrambled sequence (SRRMIFPIKLR) to find out a novel binding sites in human umbilical vein endothelial cells using liquid chromatography coupled with tandem mass spectrometry (LC-MS/MS). Finally, they concluded that the binding of the human prorenin peptide $R^{10}IFLKRMPSIR^{20}$ to HUVEC proteins is not specific for amino acid sequence and probably involves a general peptide/protein uptake mechanism without detecting a specific prorenin prosegment binding sites (Leckie & Bottrill, 2011). Moreover, decoy peptide containing fluorescent component (carboxyfluorescein) either at N-terminus or C-terminus showed different binding affinity for (P)RR compared to that of wild type decoy *in vitro* (Nabi et al., 2010). Recombinant (P)RR coupled to CM5 sensor chips in BIAcore assay system (Nabi *et al*, 2009a, 2009b), immobilized on synthetic surfaces (Nabi *et al*, 2009b), (P)RR over expressed on COS-7 cells (Nurun *et al*, 2007) have revealed that decoy inhibits binding of renin/prorenin to (P)RR. The inhibitory constant (K_i) for the peptide was found at the nanomolar order. Also, subsequent *in vivo* studies have been carried out to show beneficial role of decoy peptide in ameliorating the end-stage organ damage related disorders by abolishing non-proteolytic activation of prorenin via inhibition of prorenin binding to (P)RR (Ichihara et al., 2004; 2006b & c; Kaneshiro et al., 2007; Satofuka et al., 2006, 2007, 2009; Wikinson-Berka et al., 2010).

Interestingly, decoy peptide has also been found to inhibit binding of renin to (P)RR and this action of decoy on renin is yet to be clarified. Based on these annotations and on the tertiary structure of renin as well as predicted tertiary structure of prorenin, the possibility of having a common site in both renin and prorenin through which these molecules can interact with the (P)RR other than the decoy peptide sequence was hypothesized. A new sequence ($S^{149}QGVLKEDVF^{158}$) that localizes in the flexible junctional region between the N- and C-domains of renin/prorenin termed as the "hinge" has recently been reported to have such pivotal role for renin/prorenin binding to (P)RR (Nabi et al., 2009b). The K_D for the binding of the "hinge" peptide to (P)RR was five times higher than that of the decoy and estimated to be 17 nmol/L. The "hinge" showed higher binding affinity to the (P)RR than that of another peptide ($A^{248}KKRLFDYVV^{257}$) from the C-domain of renin/prorenin molecule,. Like the decoy, "hinge" peptide also reduced the resonance signal of renin/prorenin binding to (P)RR as observed in BIAcore, and equilibrium state analysis revealed this paradigm as a competitive inhibition with the K_i of 37.1 and 30.7 nmol/L, respectively (Nabi et al., 2009b). Therefore, these data suggest that not only the decoy peptide but also the "hinge" region peptide together accounted for the higher binding affinity of prorenin and hence, prorenin molecule has at least two high affinity sites while renin has a single site for their binding to (P)RR. Considering the nanomolar binding affinities of renin/prorenin and handle region peptide, Duncan J Campbell in one of his review article suggested that the (pro)renin receptor may have at least two separate binding domains, one domain is for renin and the other one is for prorenin prosegment and/or HRP (Campbell, 2008). Though, prorenin has two regions to interact with (P)RR, but to confirm the existence of different binding sites within (P)RR for its ligands, three dimensional structure of (P)RR has to be elucidated.

Activation of renin angiotensin system or in other words, generation of Ang-I by (P)RR mediated non-proteolytically activated prorenin depends on the sources of prorenin. Human prorenin showed higher binding affinity to both human and rat (P)RR compared to that of rat prorenin (Biswas et al., 2010a). More interestingly, either bound to human or rat (P)RR, molecular activity of non-proteolytically activated human prorenin was 2-4 fold higher than that of rat prorenin (Biswas et al., 2010), which could be due to the slow activation rate through change in conformation of rat prorenin compared to that of human prorenin after protein-protein interaction. Contribution of prorenin prosegment in the non-proteolytic activation mechanism was reported earlier *in vitro* (Suzuki et al., 2000). Chimera of human renin and rat prosegment showed very slow activation like native rat prorenin compared to the chimera of rat renin and human prosegment. Thus, it could be proposed that the prosegment sequence of prorenin played a pivotal role for the activation of prorenin molecules. More concisely, species specific regions within the prorenin prosegment like "handle" (Nurun et al., 2007; Suzuki et al., 2003) and decoy peptides (Nurun et al., 2007; Nabi et al., 2009a, b) actually crucial for the interaction of prorenin with (P)RR and also, for the non-proteolytic activation mediated by protein-protein interaction. Activation of rat prorenin through change in conformation at acidic condition required long time, even days to month (Suzuki et al., 2000). However, (P)RR mediated activation of rat prorenin has been observed within hours using recombinant (P)RR on *in vitro* synthetic surface system (Nabi et al., 2006; Biswas et al., 2010a) or overexpressing (P)RR on COS-7 cells (Nurun et al., 2007) or on rat VSMCs (Batenburg et al., 2007; Zhang et al., 2008). This might be the result of quick conformational change of prosegment of rat prorenin exerted by the interaction of one protein (receptor) with the other (ligand).

Furthermore, while considering the binding mechanism of renin and prorenin to their receptor, (P)RR has not only been discussed from the ligand's point of view, rather primary structure of (P)RR has also got similar attention for explaining the possible mechanism of receptor's involvement in ligand binding. On the other hand, though three dimensional (3D) structure of renin (Dhanaraj et al., 1992) and a predicted 3D model of prorenin (Suzuki et al., 2003; Nabi et al., 2009b) are available but due to lack of 3D structure of the receptor, mechanism for interaction of (pro)renin can not be explained from the receptor's point of view. However, several anti-(P)RR antibodies designed from the middle part ([107]DSVANSIHSLFSEET[121] named as anti-107/121 antibody) and C-terminus [[221]EIGKRYGEDSEQFRD[235] and [237]SKILVDALQKFADD[250]; close to the N-terminus of transmembrane region of the receptor, named as anti-221/235 and 237/250 antibodies, respectively] regions of (pro)renin receptor have been used in many studies (Nabi et al, 2009a, 2009b; Nabi et al., 2012). Depending on the flexibility of the anti-(P)RR antibody associated (pro)renin receptor, it would show its binding affinity towards the ligands. The calculated binding affinities of prorenin were 2.9×10^{-9}, 1.2×10^{-9} and 1.74×10^{-9} nM, when (P)RR was immobilized or occupied by anti-107/121, anti-221/235 (Nabi et al, 2009a, 2009b) and 237/250 antibodies (Nabi et al., 2012), respectively. The recombinant (P)RR tagged with six histidine residues was synthesized in a cell free *in vitro* system using wheat germ lysate. It was hypothesized that the His tag sequence at the C-terminal end would retain the transmembrane characteristics of (P)RR *in vitro*. So, (P)RR occupied by the anti-His tag antibody would indicate its native binding pattern while interacting with the ligands. Study showed that the binding affinity of prorenin to anti-His tag antibody-bound (P)RR was 7.8 nM (Nabi et al., 2009a) and other studies using over expressed (P)RR on the cell surface showed comparable nanomolar order of binding affinity of prorenin to (P)RR (Nguyen et al., 2002; Nurun et al., 2007; Batenburg et al., 2007). Reports available so far indicate that binding region for prorenin within (P)RR resides possibly further upstream region of the amino acid residue at position 107, which could be more close to the N-terminal region(s) of the receptor.

2.1.2 Initiation of second messenger pathways

Binding of renin/prorenin to (P)RR initiates an intracellular signaling pathway that is independent of angiotensin II mediated pathway. Both renin and prorenin stimulated p42/p44 mitogen-activated protein kinase (MAPK) or ERK1/2 that leads to up-regulation of transforming growth factor-β1 release in mesangial cells, PAII, collagens, fibronectin (Huang *et al*, 2006; Huang *et al*, 2007; Sakoda *et al*, 2007) and cyclooxygenase 2 (Kaneshiro et al., 2006; Nguyen, 2006) as shown in Figure 3 (ii). Moreover, prorenin also activated p38 mitogen-activated protein kinase and simultaneously phosphorylate heat-shock protein-27 in cardiomyocytes (Sasris et al., 2006). Prorenin and renin induced activation of extracellular protein kinases (ERK) 1/2 in monocytes has also been reported (Feldt et al., 2008b). In the kidneys of diabetic mice, activation of all the three members of MAPK family including ERK, p38 and c-Jun NH_2-terminal kinase (JNK) was observed (Ichihara et al., 2006a), whereas another report (Sakoda et al., 2007) revealed activation of ERK not p38 and JNK upon activation of (P)RR via its ligand, prorenin. A protein called promyelocytic zinc finger (PLZF) has also been identified and this has been found to be associated with the cytoplasmic domain of the receptor (Schefe et al. 2006). Binding of prorenin to the receptor drives translocation of PLZF to the nucleus by stimulating P13K p85 pathway that ultimately generates short negative feedback loop to down regulate (P)RR expression [depicted in Figure 3 (iii)]. Furthermore, (P)RR is a component of the Wnt [wingless-type mouse mammary tumor virus (MMTV) integration site family] receptor complex. Wnt proteins are highly conserved secreted signaling molecules and regulators of multiple biological and pathological processes (Logan and Nusse, 2004). The signaling mechanism mediated by Wnt receptor in conjunction with (P)RR and H^+-VATPase has been explained in detail later in this chapter. The detail of the intracellular signaling pathway activated and mediated by the (pro)renin receptor has been depicted and categorically presented in Figure 3.

3. Pathophysiology of prorenin and (pro)renin receptor

Hepatocyte specific prorenin transgenic rat revealed direct pathophysiological action of prorenin. Prorenin is not activated in liver and less than 2% of the total circulating prorenin found to be active in plasma. Diabetic subjects with microalbuminuria had very high prorenin to renin ration. Before the onset of microalbuminuria levels of prorenin begins to increase, and in conjunction with the glycated haemoglobin, the prorenin levels in plasma could be used to predict the occurrence of later microalbuminuria (Deinum et al., 1999). The circulating prorenin is responsible for developing hypertrophy of cardiomyocytes, glomerulosclerosis and atherosclerosis of small to medium sized artery, indicating elevated prorenin itself causes cardiomyopathy, glomerulosclerosis and atherosclerosis. Use of angiotensin converting enzyme inhibitors and angiotensin-II type 1 receptor blockers play protective role in end-stage organ damage in patients with hypertension and diabetes by suppressing the circulating RA system. Yet, low amount of renin activity is still evident in the plasma of these under treatment diabetic and hypertensive subjects which could ultimately be attributed to the enhanced tissue RA system. Thus, reasons behind the direct involvement of prorenin in the pathology of hypertension, diabetes and heart failure remained unclear. Receptor- associated prorenin system (RAPS), a novel phenomenon, sheds light on this direct action of prorenin. (P)RR, the new member of the RA system, has set a new perception about the physiological functions, activation mechanism and pathophysiological roles of renin/prorenin by activating angiotensin II-dependent or - independent pathways [Figures 3 (i), (ii), (iii)]. It has its own intracellular signalling pathways. Non-proteolytic activation of prorenin after interacting with (P)RR hypothesized that this activation mechanism of prorenin plays a pivotal role in the regulation of tissue RA system and end-organ damage in diabetic animals. (P)RR mRNA and protein expression are up-regulated in the hearts and kidneys of rats with congestive heart failure (Hirose et al., 2009). Thus, (P)RR in the heart can act as a capturing molecule for renin/prorenin which ultimately explain the presence of local renin-angiotensin system in heart, which can't synthesize renin. In diabetes, enhanced activity of oxidative stress and AT1 receptor are associated with up-regulation of (P)RR and this could be suppressed using AT1 receptor blocker and NADPH-oxidase activity inhibitor (Siragy and Huang, 2008). (P)RR mediated stimulation of signal cascade (depicted in Figure 3) of transforming growth factor-β1 (TGF-β1) and connective tissue growth factor (CTGF) in renal glomeruli (Huang et al., 2011) and enhancement of renal production of the inflammatory cytokines- TNF-alpha and IL-1beta, independent of the effects of renal Ang-II (Matavelli et al., 2010), contributes to the development and progression of kidney disease in diabetes. Up-regulation of (P)RR expression by high glucose is mediated by both PKC-Raf-ERK and PKC-JNK-c-Jun signaling pathways. Also, nuclear factor-κB and activation protein-1 are involved in high-glucose-induced (P)RR up-regulation in rat mesangial cells (Huang and Siragy, 2010).

At 5–6 months of age, transgenic rats over expressing the human (P)RR gene nonspecifically developed glomerulosclerosis with proteinuria by three to seven times without elevating the blood pressure (Kaneshiro et al., 2007). Transgenic rats over expressing human (P)RR gene exclusively in smooth muscle cells developed hypertension at their 7 months of age (Burckle et al., 2006). (Pro)renin receptor mediated non-proteolytically activated prorenin contributes to the development and progression of nephropathy including proteinuria and glomerulosclerosis in diabetic animals with high plasma levels of prorenin by increasing angiotensin II tissue generation (Ichihara et al., 2004). An increased Ang-I content was observed in the heart of double transgenic mice over expressing human prorenin and angiotensinogen compared to the single-transgenic mice (Prescott et al., 2002). These results indicate how prorenin contribute to the generation of angiotensin peptides locally and tissue

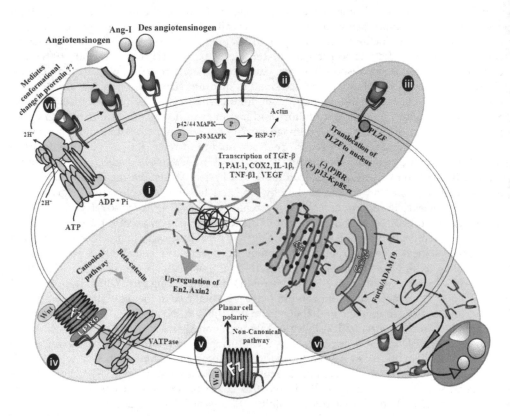

Fig. 3. Receptor associated prorenin system or RAPS mediated by prorenin and (pro)renin receptor [(P)RR] has set a new perception about the involvement of renin-angiotesin system in the pathophysiology of end-stage organ damage. Such nomenclature has been proposed due to the dual activation of tissue RAS (i) and RAS-independent signaling pathways (ii-vii). Augmentation of tissue RAS or RAPS via (P)RR initiates endocrine, paracrine or autocrine activities mediated by angiotensin II peptide. Binding of renin/prorenin to (P)RR initiates signal transduction via angiotensin II independent pathway by activating mitogen activated protein kinases (MAPKs) that induce expression of many regulatory proteins (ii). Translocation of promyelocytic zinc finger (PLZF) after prorenin binding to (P)RR leads to a short negative feed back loop that in turn suppresses (P)RR expression (iii). Also, (P)RR itself, independent of renin/prorenin, mediates Wnt-βcatenine (canonical, iv) and Ez/PCP (non-canonical) signaling pathways (v). (P)RR can be processed in the golgi apparatus by Furin or ADAM19 to its soluble form (vi). The shedded (P)RR through exocytosis can come outside of the cell and has been detected in human plasma and cultured cell medium. Shedded (P)RR binds (pro)renin. Prorenin bound to soluble (P)RR performs enzymatic activity (vi). V-ATPase participates in proton transport. C-terminal region of (P)RR is identical to the sequence of V-ATPase. This is not clear whether non-proteolytic activation of prorenin through conformational change after receptor binding is also mediated by the acidic environment created by membrane associated V-ATPase and (P)RR (vii).

damage after being taken up by tissues from circulation. (P)RR is up-regulated in kidneys of diabetic rats and renal mesangial cells exposed to high glucose concentration (Siragy and Huang, 2008; Huang and Syragi, 2010). Rapid phosphorylation at the serine residues of (P)RR in response to hyperglycemia up-regulates TGF-beta1-CTGF cascade (Huang et al., 2011), which initiates or augments kidney disease in diabetic rats. An increased Ang-I content was observed in the heart of double transgenic mice over expressing human prorenin in the liver and human angiotensinogen in the heart as compared to the single-transgenic mice (Prescott et al., 2002). These results indicate how prorenin contribute to the generation of angiotensin peptides locally and tissue damage after being taken up by tissues from circulation. Moreover, (P)RR by stimulating non-proteolytic activation of prorenin contribute to the development of renal and cardiac fibrosis in spontaneously hypertensive rats (SHRs) (Ichihara et al, 2006a,b). These data demonstrated the possible involvement of (P)RR in the pathogenesis of heart failure and kidney tissue damage. Association of (P)RR gene polymorphism with high blood pressure and left ventricular hypertrophy substantiate the important role of (P)RR in the pathogenesis of hypertension in humans (Hirose et al., 2011; Ott et al., 2011) In vitro and animal studies have shown that increased receptor expression could be linked to high blood pressure and to cardiac and glomerular fibrosis by activating mitogen-activated protein kinases and by upregulating gene expression of profibrotic molecules. Also, animal studies with angiotensin-II type 1a receptor deficiency showed that the (P)RR is involved in the development and progression of diabetic nephropathy through angiotensin-II independent pathway via activation of intracellular pathways (Ichihara et al., 2006c).

Association of prorenin and the (P)RR with the development of ocular pathology/diseases has been reported (Satofuka et al., 2006, 2007, 2008; Wikinson-Burka et al., 2011). Non-proteolytic activation of prorenin mediated by (P)RR is associated with retinal neovascularization in experimental retinopathy model of prematurity (Satofuka et al., 2007). Using the same model, the involvement of prorenin and (P)RR in the pathological angiogenesis, leukocyte accumulation and intracellular adhesion molecule-1 with vascular endothelial growth factor expression; retinal gene and protein expression of inflammatory mediators has also been demonstrated (Satofuka et al., 2006, 2008). RILLKKMPSV, a peptide sequence of rat prorenin prosegment, influences the vasculature, glia and neurons, and (pro)renin receptor expression in the retina (Wikinson-Burka et al., 2011).

4. Functions of (P)RR other than its involvement in renin-angiotensin system

The prototype of sequence homology between (P)RR from human and other species actually gave a clue regarding its plausible additional role in biological processes other than RAS. Two or more genes homologous to (P)RR have been found in C. elegans and Drosophila melanogaster that are phylogenetically distant from human. These species also express some components of RAS, which are not involved in homeostasis or electrolyte balance. Thus, (P)RR in these species may contribute to functions not related to RAS.

The C-terminal truncated fragment of (P)RR helps to assemble vacuolar H^+-proton adenosine triphosphatase (V-H^+-ATPase) (Ludwig et al., 1998). The (P)RR is also identical to endoplasmic reticulum–localized type 1 transmembrane adaptor precursor (CAPER) (Burckle & Bader, 2006; Campbell, 2006; Bader, 2007; Strausberg et al., 2002). Evolutionarily

V-H$^+$-ATPase is a highly conserved ancient enzyme in eukaryotic cells (Nelson et al., 2000) and this could be one of the most plausible reasons behind the sequence homology of C-terminus of human (P)RR with the evolutionarily close species like rat, mouse, chicken, drosophila, mosquito, zebra fish, frog and remote like *C. elegans* and bacteria *Ehrlichia chaffeensis*. (P)RR exists in truncated form composed of transmembarne region and the cytoplasmic tail that co-precipitates with V-ATPase may govern its function unrelated to RA system (Bader 2007). For this reason, (P)RR is also known as ATP6ap2 (adaptor protein type II vacuolar H$^+$-ATPase). The V-H$^+$-ATPase is expressed in the collecting ducts and distal tubules within the kidney, where it contributes to the urinary acidification as well as play pivotal role in endocytosis (Toei et al., 2010). Different subunits of V-ATPase perform different functions, notably, mutations in genes encoding C or D subunits in mice involved in embryonic lethality giving an evidence that V-ATPase plays an important role in development (Inoue et al, 1999; Miura et al, 2003), mutated B1 or A3 subunit involved in metabolic acidosis and osteoporosis in mice, respectively (Li et al., 1999; Finberg et al., 2005) and altered B1 or A4 subunit causes distal renal tubular acidosis in human (Karret, 1999; Smith, 2000). It is suggested that (P)RR and vacuolar H$^+$-ATPase are linked together in the kidney (Advani et al., 2009) while, for the assembly and function of vacuolar H$^+$-ATPase in the heart, (P)RR plays very pivotal roles (Kinouchi et al., 2010).

Recent evidences demonstrate that (P)RR is a component of the Wnt receptor complex (Cruciat et al., 2010). It is essential for *en2* expression because of its requirement in Wnt signaling. It also acts down stream of Wnts and upstream of β-catenin [Figures 3 (iv) and (v)]. Deletion of the cytoplasmic domain of (P)RR, which mediates renin signaling inside cell, showed no effect on Wnt receptor binding suggesting that (P)RR acts in a renin-independent manner as an adaptor between Wnt receptors and the V-ATPase complex. Moreover, malfunctioning (P)RR contributed to the abnormal tadpoles characterized by small heads, shortened tails, as well as defects in melanocyte and eye pigmentation at the early embryonic stage as (P)RR and V-ATPase are required to mediate Wnt signaling during antero-posterior patterning of *Xenopus's* early central nervous system development (Cruciat et al., 2010). A homologue of (pro)renin receptor in Drosophila [d(P)RR], localized mainly to the plasma membrane, has an evolutionarily conserved role at the receptor level for activation of canonical and noncanonical Wnt/Fz (frizzled) signaling pathways [Figure 3 (v)]. Attenuation of d(P)RR affects Wg target genes in cultured cells and *in vivo* (Buechling et al., 2010). Over expressed d(P)RR interacts with Fz and Fz2 receptors which is required for planar cell polarity in Drosophila epithelia and also for convergent extension movements in *Xenopus* gastrulae. Small interfering RNAs targeting human (pro)renin receptor significantly reduced Wnt-responsive TopFlash reporter activity in HEK293T cells. Thus, (P)RR has a conserved role in mediating Wnt signaling in human (Buechling et al., 2010). This data is also consistent with the findings of Cruciat et al (2010) who demonstrated the developmental role of (P)RR in *Xenopus*. Further, asymmetric subcellular localization of frizzled, a seven-pass transmembrane receptor that acts in both wingless (Wg) and planar cell polarity (PCP), is prerequisite for the proper functioning of PCP signaling pathway (Hermle et al., 2010). It has been demonstrated that the function of VhaPRR, an accessory subunit of the vacuolar (V)-ATPase proton pump in Drosophila and also known as the VhaM8-9 because of its sequence homology with V-ATPase, is tightly associated with Fz but not to other PCP core proteins. Fz fails to localize asymmetrically in absence of VhaPRR. It also acts as the modulators of canonical Wnt signaling pathway in larval and adult wing tissues. VhaPRR knock down caused multiple wing hair and hair mispolarization phenotypes (Hermle et al., 2010). These indicate the association of (P)RR in non-canonical (Fz/PCP) signaling pathways. Recent evidences regarding the association of (P)RR with H+-ATPase and Wnt signaling pathway shedding light on the reason behind the connection of non-proteolytic activation of prorenin by (P)RR with glomerulosclerosis, fibrosis, proteinuria. Though *in vitro* studies suggested non-proteolytic activation of prorenin mediated by (P)RR, but it is yet to determine whether the activation is mediated only by the protein-protein interaction, or by the co-operation of (P)RR and V-H+-ATPase, or only due to the acidic environment created as a result of proton transport *in vivo* [Figure 3 (vii)]. However, because Wnt signaling pathway promotes renal fibrosis, glomerulosclerosis and proteinuria, (He et al., 2009; Dai et al., 2009) it is possible that (P)RR might act in a combination of (P)RR-H+-ATPase-Wnt signaling pathway. Thus, (P)RR is involved in the Wnt/β-catenin canonical and Wnt/PCP non-canonical pathways in conjunction with V-H+-ATPase in a renin-independent fashion [Figures 3 (iv) and (v)].

Using zebra fish, the important association of (P)RR and V-ATPase in the development of brain and eye at the very early stage of embryonic development has been demonstrated (Amsterdam et al., 2004). A mutation in (P)RR is very lethal that causes death before the completion of embryonic stage by creating severe malformations of the central nervous system. In fact, while ACE is required to maintain fertility and ACE2 serves as a receptor for the SARS corona virus [causing factor for severe acute respiratory syndrome (SARS)], a single amino acid mutation in exon-4 of (P)RR mRNA associated with X-linked mental retardation and epilepsy (Ramser et al., 2005), and thus, (P)RR seems to be important for brain development and cognition. Also, another major finding (Contrepas et al., 2009) stated that (P)RR play essential role in neuronal cell differentiation. Other than the embryonic development, (P)RR gene polymorphism has been found to be associated with high blood pressure in Caucasian and Japanese male subjects (Hirose et al., 2011; Ott et al., 2011). Elevated blood pressure and increased heart rate in transgenic rats over expressing (P)RR in smooth muscles have been reported in their models (Burckle et al., 2006).

5. Inhibition of the activities of the components of renin angiotensin system: (pro)renin receptor as a new therapeutic target

Peptides mimicking the structural part of prorenin prosegment (pro-enzyme of renin) or N-terminal sequence of angiotensinogen containing the renin cleavage site were the first-generation of renin inhibitors (Boger et al., 1985; Hui et al., 1987; Bolis et al., 1987). Parenteral administration of these drugs efficiently reduced blood pressure by inhibiting renin activity in animals and in human being (Boger et al., 1985; Webb et al., 1985). However, due to their peptidic nature, these drugs had very poor oral bioavailability. Later, chemically modified CGP29287 achieved more attention as renin inhibitor due to its stability and longer duration of action when given orally (Wood et al., 1985). Further, development of other drugs like enalkiren (A 64662), CGP38560A, remikiren (Ro 425892) and zankiren (A 72517) with molecular weight of a tetra-peptide (Wood et al., 1994,1989; Maibaum et al., 2003) also failed to attract attention due to their low bioavailability (<2%), a short half-life and weak blood pressure lowering activity when administered orally (Wood et al., 1994; Nussberger et al., 2002; Rongen et al., 1995). On the other hand, an orally inactive peptide from snake venom established the important role of angiotensin converting enzyme (ACE) inhibitors in regulating blood pressure. This led to the development of Captopril, the first ACE inhibitor. Moreover, blood pressure lowering activity, to a great extent, depends on the inhibiting ability of plasma renin activity (PRA) and/or reducing plasma renin concentration (PRC). Thus, use of ACE inhibitors or angiotensin receptor blockers (ARBs) for inhibiting renin angiotensin system is not as effective as it should be because these inhibitors ultimately increase PRA or PRC (Mooser et al., 1990; Azizi et al., 2004). In addition, inhibition of ACE increases angiotensin I, which would be, via ACE-independent pathways by using cathepsins and tonins, converted into angiotensin II (Wolny et al., 1997; Hollenberg et al., 1998). Together these data indicate that direct renin inhibitors could be the superlative choice as an anti-hypertensive agent which would lower plasma renin activity.

Aliskiren, an octanamide, the first known representative of a new class of completely non-peptide, low-molecular weight, orally active transition-state renin inhibitor, that progressed to phase-III clinical trials (Wood et al., 2003). After oral dose of aliskiren (from 40 to 640 mg/day) in healthy volunteers, its plasma concentration increased dose dependently and the peak concentration reached after 3–6 hour with an average half life of 23.7 hour (Nussberger et al., 2002) making the compound suitable for once-daily administration. The oral bioavailability was 2.7%. Plasma steady-state concentrations were reached after 5–8 days of treatment. Aliskiren can inhibit enzymatic activities of receptor-bound renin and non-proteolytically activated prorenin, while it has no effect on the interaction of renin/prorenin with (P)RR. Also, aliskiren could not act as (P)RR blocker to inhibit renin/prorenin binding to (P)RR or failed to prevent (pro)renin signaling (Feldt et al., 2008b). Interestingly, when renin was incubated with aliskiren and then, allowed to bind to (P)RR, the binding affinity of renin to (P)RR decreased more than 1000 fold *in vitro* (Biswas et al., 2010b).

Also, an ideal blocker for (pro)renin receptor is indeed a necessity of time considering the direct association of (P)RR with increased blood pressure and its indirect involvement, via non-proteolytic activation of prorenin, in the pathogenesis of end-stage organ damage in hypertension, diabetes and ocular diseases. The efficacy of a peptidic blocker known as decoy peptide (R[10P]IFLKRMPSI[19P]) designed from the N-terminus of prorenin prosegment on the basis of the sequence of handle (I[11P]FLKR[15P]) region peptide was reported earlier for improving organ damage (Ichihara et al., 2004). Both human and rat decoy peptides inhibited the bindings of human and rat prorenins to their respective (P)RR expressed on the membranes of COS-7 cells with a similar K_i of 6.6 nM (Nurun et al., 2007). This peptide inhibited the bindings of not only prorenin but also renin to the preadsorbed receptors with the K_i values of 15.1 and 16.7 nM, respectively (Nabi et al., 2009b). Moreover, real-time bindings using surface plasmon resonance (SPR) technique in BIAcore assay system revealed evidence for the direct binding of native decoy peptide to the immobilized (P)RR with K_i of 3.5 nM (Nabi et al., 2009a, 2009b). The SPR technique displayed reduced resonance signal of prorenin binding to (P)RR while co-incubated with the decoy peptide.

The decoy proposition has also been tested *in vivo* using various transgenic models. Administration of HRP significantly inhibited increased levels of renal angiotensin II, the development of proteinuria and glomerulosclerosis in a model of diabetic nephropathy; rat

HRP completely prevented the development of diabetic nephropathy in heminephrectomized streptozotocin induced diabetic rats without affecting hyperglycemia (Takahashi et al., 2007). Urinary albumin excretion and the renal production of tumor necrosis factor-α and interleukine-β1 were decreased significantly when rat HRP was given directly into the renal cortical interstitium of diabetic rats (Matavelli et al., 2010). Prevention of the development of proteinuria, glomerulosclerosis, and complete inhibition of the activation of ERK1/2, p38, JnK in the kidney of diabetic angiotensin-II type-1a receptor-deficient mice was reported and thus, the role of (P)RR via angiotensin II independent pathway in association with prorenin was suggested (Ichihara et al., 2006c). Other investigators also confirmed the action of prorenin and (P)RR via angiotensin-II independent pathway (Huang et al., 2006; Muller et al., 2008; Feldt et al., 2008a, b). Moreover, HRP inhibits the development of retinal neovascularization by inhibiting non-proteolytic activation of prorenin caused by interaction with (P)RR in experimental retinopathy model of prematurity (Satofuka et al., 2007). Satofuka et al. using the same model, showed that the HRP suppressed the pathological angiogenesis, leukocyte adhesion and retinal expression of ICAM-1 and VEGF; also, reduced retinal gene and protein expression of inflammatory mediators (Satofuka et al., 2006, 2009). HRP also improved vascular disorder in a model of retinopathy of prematurity, but had detrimental effects on retinal neurons and glia. These effects occurred despite HRP not being detected in plasma. In young spontaneously hypertensive rats (SHR) under high salt-diet, HRP not completely but significantly attenuated glomerulosclerosis with proteinuria, cardiac hypertrophy with left ventricular fibrosis without affecting the development of hypertension (Ichihara et al, 2006 a, b). In addition, Susic et al made a further interesting observation by reporting reduced beneficial effects of decoy (PRAM-1) in SHR rat with normal diet (Susic et al., 2008).

On the contrary, many researchers are not satisfied about decoy's role as a fruitful (P)RR blocker (Batenburg et al., 2007; Muller et al., 2008; Feldt et al., 2008a, b; Mercure et al., 2009). Chronic HRP treatment did not improve target organ damage in renovascular Goldblatt hypertensive rats with high renin, prorenin and PRA that lead to Ang-II dependent target organ damage rather HRP counteracts the beneficial effects of aliskiren (van Esch et al., 2011). Also, HRP had no effects on the activation of signal transduction mediated by prorenin-(P)RR interaction (Feldt et al., 2008a). On the other hand, very recently, (P)RR siRNA technique and prolonged use of HRP or valsartan showed inhibition of rapid phosphorylation in the serine residues of (P)RR that ultimately suppressed inflammation in the kidneys (Huang et al., 2011). The concentration of HRP could not be measured in both blood and plasma of rats infused with either 0.1 or 1 mg/kg HRP per day, which suggested rapid metabolism of the peptide *in vivo* and this interpretation was supported by the finding that HRP was metabolized with a half-life of 5 minutes in EDTA-plasma at 37°C (Wikinson-Burka et al., 2011). Recycling of (P)RR between the cellular compartments and cell surface has been demonstrated earlier (Batenburge et al., 2007). Later, this annotation has been experimentally proved by the action of furin (Cousins et al., 2009) and ADAM19 (Yoshikawa et al., 2011), which till-to-date could be one of the most appropriate and acceptable explanation behind the useful execution of "decoy" as (P)RR blocker in some animal model or cell line while in other models, the "decoy" is not effective even at the same or sometimes higher concentration.

6. Conclusion and future direction

A sensitive enzyme-linked immunosorbent assay has been established to detect the level of soluble (P)RR in the medium of cultured cells and also in cell lysates (Kazal et al., 2011). It is now very important to set up such easily pursuable and sensitive method for the detection of (P)RR in human plasma. It may facilitate to diagnose specific disease or to measure degree of organ damage or to predict the end-stage organ damage. Three dimensional structure of (pro)renin receptor has to be resolved to clear the ambiguity of decoy hypothesis, to find out the binding site(s) of prorenin, renin and the decoy peptide within the molecule. Furthermore, a well accepted (P)RR blocker is now the demand of time to reduce the effects of (P)RR on end-stage organ damage. Thus, (P)RR, now-a-days, should be the novel target for developing new therapeutic approaches to ameliorate end-stage organ damage related disorders. However, considering the involvement of (P)RR in organ development specially in eye and brain development, more extensive studies should be performed before designing a (P)RR blocker.

7. References

Achard V., Boullu-Ciocca S., Desbriere R., Nguyen G., & Grino M. Renin receptor expression in human adipose tissue. *Am. J. Physiol – Regu. Physiol.*, vol. 292, no. 1, (January 2007), pp. (R274-R282).

Advani A., Kelly DJ., Cox AJ, White KE., Advani SL., Thai K., Connelly KA., Yuen D., Trogadis J., Herzenberg AM., Kuliszewski MA., Leong-Poi H., & Gilbert RE. The (Pro)renin receptor: site-specific and functional linkage to the vacuolar H+-ATPase in the kidney. *Hypertension*, vol. 54, no. 2, (August 2009), pp. (261-269).

Admiraal P.J., van Kesteren C.A., Danser A.H.J., Derkx F.H., Sluiter W., & Schalekamp M.A.D.H. Uptake and proteolytic activation of prorenin by cultured human endothelial cells. *J. Hypertens.*, vol. 17, no. 5, (May 1999), pp. (621-629).

Allen AM., Dosanjh JK., Erac M., Dassanayake S., Hannan RD., & Thomas WG. Expression of constitutively active angiotensin receptors in the rostral ventrolateral medulla increases blood pressure. *Hypertension*, vol. 47, no.6, (June 2006), pp. (1054-1061).

Amsterdam A., Nissen RM., Sun Z., Swindell EC., Farrington S., & Hopkins N. Identification of 315 genes essential for early zebra fish development. *Proc. Natl. Acad. Sci. USA*, vol. 10, no. 35, (August 2004), pp. (12792-12797).

Aronsson M., Almasan K., Fuxe K., Cintra A., Harfstrand A., Gustafsson JA., & Ganten D. Evidence for the existence of angiotensinogen mRNA in magnocellular paraventricular hypothalamic neurons. *Acta. Physiol. Scand.*, vol. 132, no. 4, (April 1988), pp. (585-586).

Azizi M., & Menard J. Combined blockade of the renin-angiotensin systemwith angiotensin-converting enzyme inhibitors and angiotensinII type1 receptor antagonists. *Circulation*, vol. 109, no. 21, (June 2004), pp. (2492-2499).

Bader M. The second life of the (pro)renin receptor. *Journal of Renin Angiotensin Aldosterone System*, vol. 8, no. 4, (December 2007), pp. (205-208).

Batenburg W.W., Krop M., Garrelds I.M., de Vries R., de Bruin RJ., Burcklé CA., Müller DN., Bader M., Nguyen G., Danser AH. Prorenin is the endogenous agonist of the (pro)renin receptor. Binding kinetics of renin and prorenin in rat vascular smooth muscle cells overexpressing the human (pro)renin receptor. *Journal of Hypertension,* vol. 25, no. 12, (December 2007), pp. (2441-2453).

Bickerton RK., & Buckley JP. Evidence for a central mechanism in angiotensin-induced hypertension. *Proc. Soc. Exp. Biol. Med.,* vol 106, no. , (1961), pp. (834–837).

Biswas K.B., Nabi A.N., Arai Y., Nakagawa T., Ebihara A., Ichihara A., Inagami T., & Suzuki F. Qualitative and quantitative analyses of (pro)renin receptor in the medium of cultured human umbilical vein endothelial cells. *Hypertens. Res.,* vol. 34, no. 6, (June 2011), pp. (735-739).

Biswas KB, Nabi AH, Arai Y, Nakagawa T, Ebihara A, Suzuki F. Species specificity of prorenin binding to the (pro)renin receptor in vitro. Front. Biosci., vol. 2, (June 2010a), pp. (1234-1240).

Biswas K.B., Nabi A.H., Arai Y., Nakagawa T., Ebihara A., Ichihara A., Watanabe T., Inagami T., & Suzuki F. Aliskiren binds to renin and prorenin bound to (pro)renin receptor in vitro. *Hypertens. Res.,* vol. 33, no. 10, (October 2010b), pp. (1053-1059).

Boger J., Payne L.S., Perlow D.S., Lohr N.S., Poe M., Blaine E.H., Ulm E.H., Schorn T.W., LaMont B.I., Lin, T.Y. *et al.* Renin inhibitors. Syntheses of subnanomolar, competitive, transition-state analogue inhibitors containing a novel analogue of statine. *J. Med. Chem.,* vol. 28, no. 12, (December 1985), pp. (1779–1790).

Bolis G., Fung AK., Greer J., Kleinert HD., Marcotte PA., Perun TJ., Plattner JJ. & Stein H. Renin inhibitors: dipeptide analogues of angiotensinogen incorporating transition-state, nonpeptidic replacements at the scissile bond. *J .Med. Chem.,* vol. 30, no. 10, (October 1987), pp. (1729–1237).

Buechling T., Bartscherer K., Ohkawara B., Chaudhary V., Spirohn K., Niehrs C., & Boutros M. Wnt/Frizzled signaling requires dPRR, the Drosophila homolog of the prorenin receptor. *Curr. Biol.,* vol. 20, no. 14, (July 2010), pp. (1263-1268).

Brecher A.S., Shier D.N., Dene H., Wang S.M., Rapp J.P., Franco S.R., & Mulrow P.J. Regulation of adrenal renin messenger ribonucleic acid by dietary sodium chloride. *Endocrinology,* vol. 124, no. 6, (June 1989), pp. (2907–2913).

Burckle C.A., Danser A.H.J., Muller D.N., Garrelds I.M., Gasc J.M., Popova E., Plehm R., Peters J., Bader M., & Nguyen G. Elevated blood pressure and heart rate in human renin receptor transgenic rats. *Hypertension,* vol. 47, no. 6, (March 2006), pp. (552-556).

Burckle, C., & Bader M. Prorenin and its ancient receptor. *Hypertension,* vol. 48, no. 4, (2006), pp. (549–551).

Campbell D.J. Critical review of prorenin and (pro)renin receptor research. *Hypertension,* vol. 51, no. 5, (March, 2008), pp. (1259–1264).

Campbell D.J., Kladis A., & Duncan A.M. Nephrectomy, converting enzyme inhibition and angiotensin peptides. *Hypertension.* Vol. 22, no. 4, (October, 1993), pp. (513-522).

Campbell D.J., & Valentijn A.J. Identification of vascular renin-binding proteins by chemical cross-linking: inhibition of binding of renin by renin inhibitors. *J. Hypertens.,* vol. 12, no. 8, (August 1994), pp. (879–890).

Cohen S., Taylor J.M., Murakami K., Michelakis A.M., & Inagami T. Isolation and characterization of renin-like enzymes from mouse submaxillary glands. *Biochemistry,* vol. 11, no. 23, (November 1972); pp. (4286-4293).

Cooper J.R., Bloom F.E., Roth R.H. (1996). *Biochemical Basis of Neuropharmacology*. Oxford Univ. Press, Oxford, UK.

Cousin C., Bracquart D., Contrepas A., Corvol P., Muller L., & Nguyen G. Soluble form of the (pro)renin receptor generated by intracellular cleavage by furin is secreted in plasma. *Hypertension*, vol. 53, no. 6, (June 2009), pp. (1077-1082).

Contrepas A., Walker J., Koulakoff A., Franek K.J., Qadri F., Giaume C., Corvol P., Schwartz C.E., & Nguyen G.A. Role of the (pro)renin receptor in neuronal cell differentiation. *Am. J. Physiol. Regul. Integr. Comp. Physiol.*, vol. 297, no. 2, (August 2009), pp. (R250-R257).

Dai C.., Stolz D.B., Kiss L.P., Monga S.P., Holzman L.B., & Liu Y. Wnt/beta-catenin signaling promotes podocyte dysfunction and albuminuria. *J. Am. Soc. Nephrol.*, vol. 20, no. 9, (September 2009), pp. (1997-2008).

Danser A.H.J., van Kats J.P., Admiraal P.J.J., Derkx F.H.M., Lamers J.M.J., Verdouw P.D., Saxena P.R., & Schalekamp M.A.D.H. Cardiac renin and angiotensins: uptake from plasma versus in situ synthesis. *Hypertension*. vol. 24, no. 1, (July 1994), pp. (37-48).

Danser A.H.J., Koning M.M.G., Admiraal P.J.J., Sassen L.M., Derkx F.H., Verdouw P.D., & Schalekamp M.A. Production of angiotensins I and II at tissue sites in the intact pig. *Am. J. Physiol.*, vol. 263, no. 2, (August 1992), pp. (H429-H437).

Derkx F.H.M., Tan-Tjiong H.L., Man in 't Veld A.J., Schalekamp M.P., & Schalekamp M.A.D.H. Activation of inactive plasma renin by plasma and tissue kallikreins. *Clin. Sci. (Lond).*, vol. 57, no. 4, (October 1979), pp. (351-357).

Derkx F.H.M., Tan-Tjiong L., Wenting G.J., Boomsma F., Man in 't Veld A.J and Schalekamp M.A.D.H. Asynchronous changes in prorenin and renin secretion after captopril in patients with renal artery stenosis. *Hypertension.*, vol. 5, no. 2, (March-April 1983), pp. (244-256).

Derkx F.H., Alberda A.T., de Jong F.H., Zeilmaker F.H., Makovitz J.W., Schalekamp & M.A. Source of plasma prorenin in early and late pregnancy: observations in a patient with primary ovarian failure. *J. Clin. Endocrinol. Metab.*, vol. 65, no. 2, (August 1987a), pp. (349-354).

Derkx F.H.M., Schalekamp M.P., & Schalekamp M.A.D.H. Two-step prorenin-renin conversion. Isolation of an intermediary form of activated prorenin. *J. Biol. Chem.*, vol. 262, no. 6, (February 1987b), pp. (2472-2477).

Derkx F.H.M., Deinum J., Lipovski M., Verhaar M., Fischli W., & Schalekamp M.A.D.H. Nonproteolytic "activation" of prorenin by active site-directed renin inhibitors as demonstrated by renin-specific monoclonal antibody. *J. Biol. Chem.*, vol. 267, no. 32, (November 1992), pp. (22837-22842).

Deschepper C.F., Mellon S.H., Cumin F., Baxter J.D., & Ganong W.F. Analysis by immunocytochemistry and *in situ* hybridization of renin and its mRNA in kidney, testis, adrenal, and pituitary of the rat. *Proc. Natl. Acad. Sci. U.S.A.*, vol. 83, no. 19, (October 1986), pp. (7552-7556).

Do Y.S., Sherrod A., Lobo R.A., Paulson R.J., Shinagawa T., Chen S.W., Kjos S., & Hsueh W.A. Human ovarian theca cells are a source of renin. *Proc. Natl. Acad. Sci. U. S. A.*, vol. 85, no. 6, (1March 988), pp. (1957-1961).

de Lannoy L.M., Danser A.H.J., van Kats J.P., Schoemaker R.G., Saxena P.R., & Schalekamp M.A.D.H. Renin-angiotensin system components in the interstitial fluid of the isolated perfused rat heart. Local production of angiotensin I. *Hypertension*, vol. 29, no. 6, (June 1997), pp. (1240–1251).

de Lannoy L.M., Danser A.H.J., Bouhuizen A.M.B., Saxena P.R., & Schalekamp M.A.D.H. Localization and production of angiotensin II in the isolated perfused rat heart. *Hypertension*, vol. 31, no. 5, (May 1998), pp. (1111–1117).

De Mello W.C. Influence of intracellular renin on heart cell communication. *Hypertension*, vol. 25, no. 6, (June 1995), pp. (1172–1177).

Deschepper C.F., Bouhnik J., & Ganong W.F. Colocalization of angiotensinogen and glial fibrillary acidic protein in astrocytes in rat brain. *Brain. Res.*, vol. 374, no. 1, (May 1986), pp. (195–198).

Dhanaraj V., Dealwis C.G., Frazao C., Badasso M., Sibanda B.L., Tickle I.J., Cooper J.B., Driessen H.P., Newman M., Aguilar C. X-ray analyses of peptide-inhibitor complexes define the structural basis of specificity for human and mouse renins. *Nature.*, vol. 357, no. 6378, (June, 1992), pp. (466-472).

Deinum J., Rønn B., Mathiesen E., Derkx F.H., Hop W.C., Schalekamp M.A. Increase in scrum prorenin precedes onset of microalbuminuria in patients with insulin-dependent diabetes mellitus.*Diabetologia*, vol. 42, no. 8, (August 1999), pp. (1006-1010).

Doi Y., Franco S.R. & Mulrow P.J. Evidence for an extra renal source of inactive renin in rats. *Hypertension*, vol. 6, no. 5, (September 1984) pp. (6627–6632).

Dostal D.E., Rothblum K.N., Chernin M.I., Cooper G.R., & Baker K.M. Intracardiac detection of angiotensinogen and renin: a localized renin — angiotensin system in neonatal rat heart. *Am. J. Physiol.*, vol. 263, no. 4 Pt 1, (October 1992), pp. (C838–C850).

Du D., Kato T., Nabi A.H.M.N., Suzuki F., & Park E.Y. Expression of functional human (pro)renin receptor in silkworm (*Bombyx mori*) larvae using BmMNPV bacmid. *Biotechnology and Applied Biochemistry*, vol. 49, no. 3, (March 2008), pp. (195-202).

Dzau V.J., Ellison K.E., Brody T., Ingelfinger J., & Pratt R.E. A comparative study of the distributions of renin and angiotensinogen messenger ribonucleic acids in rat and mouse tissues. *Endocrinology*, vol. 120, no. 6, (June 1987), pp. (2334-2338).

Dzau V.J., Ingelfinger J., Pratt R.E., & Ellison K.E. Identification of renin and angiotensinogen messenger RNA sequences in mouse and rat brains. *Hypertension*, vol. 8, no. 6, (June 1986), pp. (544–548).

Dzau V.J., Slater E.E., & Haber, E. Complete purification of human renin. *Biochemistry*, vol. 18, no. 23, (November 1979), pp. (524-528).

Epstein A.N., Fitzsimons J.T., & Rolls B.J. Drinking induced by injection of angiotensin into the rain of the rat. *J. Physiol.*, vol. 210, no. 2, (September 1970), pp. (457–474).

Feldt S. Maschke U. Dechend R., Luft F.C. & Muller D.N. The putative (pro)renin receptor blocker HRP fails to prevent (pro)renin signaling. *J. Am. Soc. Nephrol.*, vol. 19, no. 4, (April 2008a), pp. (743-748).

Feldt S. Batenburg W.W., Mazak I., Maschke U., Wellner M., Kvakan H., Dechend R., Fiebeler A., Burckle C., Contrepas A., Danser A.H.J., Bader M., Nguyen G., Luft F.C., & Muller D.N. Prorenin and renin-induced extracellular signal-regulated kinase 1/2 activation in monocytes is not blocked by aliskiren or the handle-region peptide. *Hypertension*, vol. 51, no. 3, (March 2008b), pp. (682-688).

Figueired, A.F.S., Takii Y., Tsuji, H., Kato K. & Inagami T. Rat kidney renin and chatepsin D: purification and comparison of properties. *Biochemistry*, vol. 22, no. 24, (November 1985), pp. (5476-5481).

K.E. Finberg, G.A. Wagner, M.A. Baileyetal. The B1-subunit of the H+ATPase is required for maximal urinary acidification. *Proc. Natl. Acad. Sci. USA*, vol.102, no.38, (September 2005), pp. (13616-13621).

Fischer-Ferraro C., Nahmod V.E., Goldstein D.J., & Finkielman S. Angiotensin and renin in rat and dog brain. *J. Exp. Med.*, vol. 133, no. 2, (February 1971), pp. (353-361).

Ganten D, Minnich JL, Granger P, Hayduk K, Brecht HM, Barbeau A, Boucher R, Genest J. Angiotensin-forming enzyme in brain tissue. *Science*, vol. 173, no. 991, (July 1971), pp. (64-65).

Ganten D., Ganten U., Kubo S., Granger P., Nowaczynski W., Boucher R., & Genest J. Influence of sodium, potassium and pituitary hormones on iso-renin in rat adrenal glands. *Am. J. Physiol.*, vol. 227, no. 1, (July 1974), pp. (224-229).

Ganten D., Hutchinson J.S., Schelling P., Ganten U., & Fischer H. The iso-renin angiotensin systems in extra-renal tissue. *Clin. Exp. Pharmacol. Physiol.*, vol. 3, no. 2, (March-April 1976), pp. (102-126).

Glorioso N., Atlas S.A., Laragh J.H., Jewelewicz R., & Sealey J.E. Prorenin in high concentrations in human ovarian follicular fluid. *Science*, vol. 233, no. 4771, (September 1986), pp. (1422-1424).

Goto M., Mukoyama M., Suga S.I., Matsumoto T., Nakagawa M., Ishibashi R., Kasahara M., Sugawara A., Tanaka I., & Nakao K. Growth-dependent induction of angiotensin II type 2 receptor in rat mesangial cells. *Hypertension*, vol. 30, no. 3, (September 1997), pp. (358-362).

Gross V., Schunck W.H., Honeck H., Milia A.F., Kärgel E., Walther T., Bader M., Inagami T., Schneider W., & Luft F.C. Inhibition of pressure natriuresis in mice lacking the AT receptor. *Kidney Int.*, vol. 57, no. 1, (January 2000), pp. (191-202).

He W., Dai C., Li Y., Zeng G., Monga S.P., Liu Y. Wnt/beta-catenin signaling promotes renal interstitial fibrosis. *J. Am. Soc. Nephrol.*, vol. 20, no. 4, (April 2009), pp. (765-776).

Hermle T., Saltukoglu D., Grünewald J., Walz G., & Simons M. Regulation of Frizzled-dependent planar polarity signaling by a V-ATPase subunit. *Curr. Biol.*, vol 20, no. 14, (July 2010), pp, (1269-1276).

Hirose T., Hashimoto M., Totsune K., Metoki H., Hara A., Satoh M., Kikuya M., Ohkubo T., Asayama K., Kondo T., Kamide K., Katsuya T., Ogihara T., Izumi S., Rakugi H., Takahashi K., & Imai Y. Association of (pro)renin receptor gene polymorphisms with lacunar infarction and left ventricular hypertrophy in Japanese women: the Ohasama study. *Hypertens. Res.*, vol. 34, no. 4, (April 2011), pp. (530-535).

Hirose T., Mori N., Totsune K., Morimoto R., Maejima T., Kawamura T., Metoki H., Asayama K., Kikuya M., Ohkubo T., Kohzuki M., Takahashi K., & Imai Y. Gene expression of (pro)renin receptor is upregulated in hearts and kidneys of rats with congestive heart failure. *Peptides*, vol. 30, no. 12, (December 2009), pp. (2316-2322).

Hirose S., Ohsawa T., Inagami T., & Murakami K., Brain renin from bovine anterior pituitary: isolation and properties. *J. Biol. Chem.*, vol. 257, no. 11, (June 1982), pp. (6316-6321).

Hirose S., Yokosawa H., & Inagami T. Immunochemical identification of renin in rat brain and distinction from acid protease. *Nature*, vol. 274, no. 5669, (July 1978), pp. (392-393).

Ho M.M., & Vinson G.P. Transcription of (pro)renin mRNA in the rat adrenal cortex, and the effects of ACTH treatment and a low sodium diet. *Journal of Endocrinology*, vol. 157, no. 2, (May 1998), pp. (217-223).

Hokimoto S., Yasue H., Fujimoto K., Yamamoto H., Nakao K., Kaikita K., Sakata R., & Miyamoto E. Expression of angiotensin-converting enzyme in remaining viable myocytes of human ventricles after myocardial infarction. *Circulation*, vol. 94, no. 7, (October 1996), pp. (1513-1518).

Hollenberg N.K., Fisher N.D., & Price, D.A. Pathways for angiotensin II generation in intact human tissue: evidence from comparative pharmacological interruption of the renin system. *Hypertension*, vol. 32, no. 3, (September 1998), pp. (387-392).

Holm I., Ollo R., Panthier J.J., & Rougeon F. Evolution of aspartyl proteases by gene duplication: the mouse renin gene is organized in two homologous clusters of four exons. *EMBO J.*, vol. 3, no. 3, (March 1984), pp. (557-562).

Huang Y., Wongamorntham S., Kasting J., McQuillan D., Owens R.T., Yu L., Noble N.A., & Border W. Renin increases mesangial cell transforming growth factor-beta1 and matrix proteins through receptormediated, angiotensin II-independent mechanisms. *Kidney Int.*, vol. 69, no. 1, (January 2006), pp. (105-113).

Huang J., Matavelli L.C., & Siragy M. Renal (pro)renin receptor contribute to the development of diabetic kidney disease through transforming growth factor-β1 and connective tissue growth factor signaling cascades. *Clinical and Experimental Pharmacology and Physiology*, vol. 38, no. 4, (April 2011), pp. (215-221).

Huang J., & Siragy H,M. Regulation of (pro)renin receptor expression by glucose-induced mitogen-activated protein kinase, nuclear factor-kappaB, and activator protein-1 signaling pathways. *Endocrinology*, vol. 151, no. 7, (July 2010), pp. (3317-3325).

Hui K.Y., Carlson W.D., Bernatowicz M.S., & Haber, E. Analysis of structure-activity relationships in renin substrate analogue inhibitory peptides. *J. Med. Chem.*, vol. 30, no. 8, (August 1987), pp. (1287-1295).

Ichihara A., Hayashi M., Kaneshiro Y., Suzuki F., Nakagawa T., Tada Y., Koura Y., Nishiyama A., Okada H., Uddin M.N., Nabi A.H.M.N., Ishida Y., Inagami T., & Saruta T. Inhibition of diabetic nephropathy by a decoy peptide corresponding to the "handle" region for non-proteolytic activation of prorenin. *J. Clin. Invest.*, vol. 114, no. 8, (October 2004), pp. (1128-1135).

Ichihara A., Kaneshiro Y., & Takemitsu T. Contribution of nonproteolytically activated prorenin in glomeruli to hypertensive renal damage. *J. Am. Soc. Nephrol.*, vol. 17, no. 9, (September 2006a), pp. (2495-2503).

Ichihara A., Kaneshiro Y., Takemitsu T., Sakoda M., Suzuki F., Nakagawa T., Nishiyama A., Inagami T., & Hayashi M. Non proteolytic activation of prorenin contributes to development of cardiac fibrosis in genetic hypertension. *Hypertension*, vol. 47, no. 5, (May 2006b), pp. (894-900).

Ichihara A., Suzuki F., Nakagawa T., Kaneshiro Y., Takemitsu T., Sakoda M., Nabi A.H.M.N., Nishiyama A., Sugaya T., Hayashi M., & Inagami T. Prorenin receptor blockade inhibits development of glomerulosclerosis in diabetic angiotensin II type

1a receptor deficient mice. *J. Am. Soc. Nephrol.*, vol. 17, no. 7, (July 2006c), pp. (1950–1961).

Imagawa M., Chiu R., & Karin M. Transcription factor AP-2 mediates induction by two different signal-transduction pathways: protein kinase C and cAMP. *Cell*, vol. 51, no. 2, (October 1987), pp. (251-260).

Inagami T. & Murakami K. Prorenin. *Biomed. Res.*, vol. 1, no. , (1980), pp. (456-475).

Inagami T. & Murakami K. Pure renin: isolation from hog kidney and characterization. *J. Biol. Chem.*, vol. 252, no. 9, (May 1977), pp. (2978-2983).

Inoue H., Noumi, T. Nagata M., Murakami H., & Kanazawa H. Targeted disruption of the gene encoding the proteolipid subunit of mouse vacuolar H-ATPase leads to early embryonic lethality. *Biochimica et Biophysica Acta.*,vol. 1413, no. 3, (November 1999), pp. (130–138).

Ito M., Oliverio M.I., Mannon P.J., Best C.F., Maeda N., Smithies O., & Coffman T.M. Regulation of blood pressure by the type 1A angiotensin II receptor gene. *Proc. Natl. Acad. Sci. USA*, vol. 92, no. 8, (April 1995), pp. (3521–3525).

Itskovitz J., & Sealey J.E. Ovarian prorenin-renin-angiotensin system. *Obstet. Gynecol. Surv.*, vol. 42, no. 9, (September 1987), pp. (545-551).

Jones C.A., Petrovic N., Novak E.K., Swank R.T., Sigmund C.D. & Gross K.W. Biosynthesis of renin in mouse kidney tumor As4.1 cells. *Eur. J. Biochem.*, vol. 243, no. 1-2, (January 1997), pp. (181-190).

Jutras I., & Reudelhuber T.L. Prorenin processing by cathepsin B in vitro and in transfected cells. *FEBS Lett.*, vol. 443, no. 1, (January 1999), pp. (48-52).

Kaneshiro Y., Ichihara A., Sakoda M., Takemitsu T., Nabi A.H.M.N., Uddin M.N., Nakagawa T., Nishiyama A., Suzuki F., Inagami T., & Itoh H. Slowly progressive, angiotensin II-independent glomerulosclerosis in human-renin/prorenin-receptor-transgenic rats. *J. Am. Soc. Nephrol.*, vol. 18, no. 6, (June 2007), pp. (1789–1795).

Julie K.J., Toma I., Sipos A., Elliott J., Sarah M., Vargas L., & Peti-Peterdi J. The collecting duct is the major source of p rorenin in diabetes. *Hypertension*, vol 51, no. 6, (June 2008), pp. (1597–1604).

Kato T., Du D., Suzuki F., & Park E.Y. Localization of human (pro)renin receptor lacking the transmembrane domain on budded baculovirus of Autographa californica multiple nucleopolyhedrovirus. *Appl. Microbiol. Biotechnol.*, vol. 82, no. 3, (March, 2009), pp. (431-437).

Kato T., Kageshima A., Suzuki F., & Park E.Y. Expression and purification of human (pro)renin receptor in insect cells using baculovirus expression system. Protein Expr Purif., vol. 58, no. 2, (April, 2008), pp. (242-248).

Karet F.E., Finberg K.E., & Nelson R.D., Nayir A., Mocan H., Sanjad S.A., Rodriguez-Soriano J., Santos F., Cremers C.W., Di Pietro A., Hoffbrand B.I., Winiarski J., Bakkaloglu A., Ozen S., Dusunsel R., Goodyer P., Hulton S.A., Wu D.K., Skvorak A.B., Morton C.C., Cunningham M.J., Jha V., Lifton R.P.. Mutations in the gene encoding B1 subunit of H+-ATPase cause renal tubular acidosis with sensor in neural deafness. *Nature Genetics*, vol. 21, no. 1, (January 1999), pp. (84–90).

Kim W.S., Nakayama K., Nakagawa T., Kawamura Y., Haraguchi K., & Murakami K. Mouse submandibular gland prorenin-converting enzyme is a member of glandular kallikrein family. *J. Biol. Chem.*, vol. 266, no. 29, (October 1991), pp. (19283-19287).

Kobori H., Ozawa Y., Suzakietal Y., Prieto-Carrasquero M.C., Nishiyama A., Shoji, T., Cohen E.P., & Navar L.G. Young scholars award lecture. Intratubular angiotensinogen in hypertension and kidney diseases. *Am. J. Hypertens.*, vol. 19, no. 5, (May 2006), pp. (541–550).

Kikkawa Y., Yamanaka N., Tada J., Kanamori N., Tsumura K., & Hosoi K. Prorenin processing and restricted endoproteolysis by mouse tissue kallikrein family enzymes (MK1, MK9, MK13, and MK22). *Biochim. Biophys. Acta Protein Struc. Mol. Enzymol.*, vol. 1382, no. 1, (January 1998), pp. (55-64).

Kinouchi K., Ichihara A., Sano M., Sun-Wada G.H., Wada Y., Kurauchi-Mito A., Bokuda K., Narita T., Oshima Y., Sakoda M., Tamai Y., Sato H., Fukuda K., & Itoh H. The (pro)renin receptor/ATP6AP2 is essential for vacuolar H^+-ATPase assembly in murine cardiomyocytes. *Circ. Res.*, vol. 107, no. 1, (July 2010), pp. (30–34).

Leckie B.J., Bottrill A.R. A specific binding site for the prorenin propart peptide Arg[10]-Arg[20] does not occur on human endothelial cells. J Renin Angiotensin Aldosterone Syst., vol. 12, no. 1, (May 2011), pp. (36-41).

Lenz T., James G.D., Laragh J.H., & Sealey J.E. Prorenin secretion from human placenta perfused in vitro. American Journal of Physiology, Endocrinology and Metabolism. *Am. J. Physiol.*, vol. 260 no. 6, (June 1991), pp. (E876-E882).

Liang F., & Gardner D.G. Autocrine/paracrine determinants of strain-activated brain natriuretic peptide gene expression in cultured cardiac myocytes. *J. Biol. Chem.*, vol. 273, no. 23, (1998), pp. (14612–14619).

Li Y.P., Chen W., Liang Y., Li E., & Stashenko P. ATP6i-deficient mice exhibit severe osteopetrosis due to loss of osteoclast-mediated extracellular acidification. *Nature Genetics.*, vol. 23, no. 4, (December 1999), pp. (447–451).

Logan C.Y., & Nusse R. The Wnt signaling pathway in development and disease. *Annu. Rev. Cell Dev. Biol.*, vol. 20, (2004), pp. (781-810).

Ludwig J., Kerscher S., Brandt U., Pfeiffer K., Getlawi F., Apps D.K., & Schägger H. Identification and characterization of a novel 9.2-kDa membrane sector-associated protein of vacuolar proton-ATPase from chromaffin granules. *J. Biol. Chem.*, vol. 278, no. 18, (May 1998), pp. (10939–10947).

Maibaum J., & Feldman D.L. Renin inhibitors as novel treatments for cardiovascular disease. *Expert. Opin. Ther. Patents*, vol. 13, (2003), pp. (589–603).

Manne K., & Danser A.H.J. Circulating versus tissue renin-angiotensin system: On the origin of (pro)renin. *Curr. Hypertense. Rep.*, vol. 10, no., 2, (April 2008), pp. (112-118).

Matavelli L.C., Huang J., & Siragy H.M. (Pro)renin receptor contributes to diabetic nephropathy by enhancing renal inflammation. *Clin. Exp. Pharmacol. Physiol.*, vol. 37, no., 3, (March 2010), pp. (277-282).

Miura G.I., Froelick G.J., Marsh D.J., Stark K.L., & Palmiter R.D. The d subunit of the vacuolar ATPase (Atp6d) is essential for embryonic development. *Transgenic Research.*, vol. 12, no. 1, (February 2003), pp. (131–133).

Matoba T., Murakami K. & Inagami, T. Rat renin: purification and characterization. *Biochim. Biophys. Acta.*, vol. 526, no. 2, (October 1978), pp. (560-571).

Ménard J, Boger R.S., Moyse D.M., Guyene T.T., Glassman H.N., & Kleinert H.D. Dose-dependent effects of the renin inhibitor zankiren HCl after a single oral dose in mildly sodium-depleted normotensive subjects. *Circulation*, vol. 91, no. 2, (January 1995), pp. (330–338).

Mercure C., Prescott G., Lacombe M.J., Silversides D.W., & Reudelhuber T.L. Chronic increases in circulating prorenin are not associated with renal or cardiac pathologies. *Hypertension.*, vol. 53, no. 6, (June 2009), pp. (1062-1069).

Mooser V., Nussberger J., Jullierat L. Reactive hyperreninemia is a major determinant of plasma angiotensin II during ACE inhibition. *J. Cardiovasc. Pharmacol.*, vol. 15, no. 2, (February, 1990), pp. (276–282).

Murakami K., & Inagami T. Isolation of pure and stable renin from hog kidney. *Biochem. Biophys. Res. Commun.*, vol. 62, no. 3, (February, 1975), pp. (757-763).

Muller D.N., Klanke B., Feldt S., Cordasic N., Hartner A., Schmieder R.E., Luft F.C., & Hilgers K.F. (Pro)renin receptor peptide inhibitor "handle-region" peptide does not affect hypertensive nephrosclerosis in goldblatt rats. *Hypertension*, vol. 51, no. 3, (March 2008), pp. (676-681).

Nabi A.H.M.N., Biswas K.B., Nakagawa T., Ichihara A., Inagami T., & Suzuki F. 'Decoy peptide' region (RIFLKRMPSI) of prorenin prosegment plays a crucial role in prorenin binding to the (pro)renin receptor. *Int. J. Mol. Med.* Vol. 24, no. 1, (July 2009a), pp. (83-89).

Nabi A.H.M.N., Biswas K.B., Nakagawa T., Ichihara A., Inagami T., & Suzuki F. Prorenin has high affinity multiple binding sites for (pro)renin receptor. *Biochim. Biophys. Acta.*, vol. 1794, no. 12, (December 2009b), pp. 1838-1847.

Nabi A.N., Biswas K.B., Arai Y., Nakagawa T., Ebihara ., Islam L.N., Suzuki F. (Pro)renin receptor and prorenin: their plausible sites of interaction. *Front Biosci.*, January 2012, In press.

Nabi A.H.M.N., Kageshima A., Uddin M.N., Nakagawa T., Park E.Y., & Suzuki F. Binding properties of rat prorenin and renin to the recombinant rat renin/prorenin receptor prepared by a baculovirus expression system. *Int. J. Mol. Med.*, vol. 18, no. 3, (2006), pp. (483-488).

Nabi A.H.M.N., Biswas K.B., Arai Y., Uddin M.N., Nakagawa T., Ebihara A., Ichihara A., Inagami T., & Suzuki F. Functional characterization of the decoy peptide, $R^{10P}IFLKRMPSI^{19P}$. *Front. Biosci.*, vol. 2, (June, 2010), pp. (1211-1217).

Naruse, K.; Takii, Y.; Inagam, T., Immunohistochemical localization of luteinizing hormone producing cells of rat pituitary. Proc. Nat. Acad. Sci., U.S.A. vol. 78, no. , (1981), pp. (7579-7583).

Naruse M., & Inagami T., Markedly elevated specific renin level in the adrenal in genetically hypertensive rats. Proc. Natl. Acad. Sci. U.S.A.78, (1982) 3295-3299.

Nelson N., Perzov N., Cohen A., Hagai K., Padler V., Nelson H. The cellular biology of proton-motive force generation by V-ATPases. *J. Exp. Biol.*, vol. 203, no. 1, (2000), pp. (89–95).

Neri Serneri G.G., Boddi M., Coppo M., Chechi T., Zarone N., Moira M., Poggesi L., Margheri M., Simonetti I. Evidence for the existence of a functional cardiac renin-angiotensin system in humans. *Circulation*, vol. 94, no. 8, (October 1996), pp. (1886–1893).

Nguyen G. Increased cyclooxygenase-2, hyperfiltration, glomerulosclerosis, and diabetic nephropathy: put the blame on the (pro)renin receptor? *Kidney International*, vol. 70, no. 4, (August 2006), pp. (618–620).

Nguyen G., Delarue F., Berrou J., Rondeau E., Sraer J.D. Specific receptor binding of renin on human mesangial cells in culture increases plasminogen activator inhibitor-1 antigen. *Kidney Int.*, vol. 50, no. 6, (1996), pp. (1897-1903).

Nguyen G., Bouzhir L., Delarue F., Rondeau E., & Sraer J.D. Evidence of a renin receptor on human mesangial cells: effects on PAI1 and cGMP. *Nephrologie*, vol. 19, no. 7, (1998), pp. (411-416).

Nguyen G., Delarue F., Burckle C., Bouzhir L., Giller T. & Sraer J.D. Pivotal role of the renin/prorenin receptor in angiotensin II production and cellular responses to renin. *J. Clin. Invest.*, vol. 109, no. 11, (June 2002), pp. (1417-1427).

Nurun NA., Uddin MN., Nakagawa T., Iwata H., Ichihara A., Inagami T., & Suzuki F. Role of "handle" region of prorenin prosegment in the non-proteolytic activation of prorenin by binding to membrane anchored (pro)renin receptor. *Front. Biosci.*, vol. 12, (September 2007), pp. (4810-4817).

Nussberger J., Wuerzner G., Jensen C., & Brunner, H.R. Angiotensin II suppression in humans by the orally active renin inhibitor aliskiren (SPP100). Comparison with enalapril. *Hypertension*, vol. 39, no. 1, (January 2002), pp. (e1-e8).

Oparil S., & Haber E. The renin–angiotensin system (first of two parts). *N. Engl. J. Med.*, vol. 291, no. 8, (August 1974a), pp. (389-401).

Oparil S., & Haber E. The renin–angiotensin system (second of two parts). *N. Engl. J. Med.*, vol. 291, no. 9, (August 1974b), pp. (446-457).

Ott C., Schneider M., Delles C., Schlaich M.P., Hilgers K.F., & Schmieder R.E. Association of (pro)renin receptor gene polymorphism with blood pressure in Caucasian men. *Pharmacogenetics and Genomics*, vol. 21, no. 6, (June 2011), pp. (347-349).

Pandey K.N., Melner N. M., Paramentier M., & Inagami T. Demonstration of renin activity in purified rat Leydig cells: evidence for the existence of endogenous inactive (latent) form of enzyme. *Endocrinol.*, vol. 115, no. 5, (November 1984), pp. (1753-17590).

Paul P., Wagner D., Metzger R., Ganten D., Lang R.E., Suzuki F., Murakami K., Burbach J.H.P., & Ludwig G. Quantification of renin mRNA in various mouse tissues by a novel solution hybridization assay. *J. Hypertens.*, vol. 6, no. 3, (March 1988), pp. (247-252).

Pickens P.T., Bumpus F.M., Lloyd AmM., Smeby R.R., & Page I.H. Measurements of renin activity in human plasma. *Circ. Res.*, vol. 17, no. 5, (November 1965), pp. (438-448).

Pitarresi T.M., Rubattu S., Heinrikson R. & Sealey J.E. Reversible cryoactivation of recombinant human prorenin. *J. Biol. Chem.*, vol. 267, no. 17, (June 1992), pp. (11753–11759).

Prescott G., Silversides D.W., & Reudelhuber T.L. Tissue activity of circulating prorenin. *Am. J. Hypertens.*, vol. 15, no. 3, (March 2002), pp. (280–285).

Prieto-Carrasquero M.C., Harrison-Bernard L.M., Kobori H., Ozawa Y., Hering-Smith K.S., Hamm L.L., Navar L.G. Enhancement of collecting duct renin in angiotensin II-dependent hypertensive rats. *Hypertension.*, vol. 44, no. 4, (August 2004), pp. (223–229).

Rohrwasser A., Morgan T., Dillon H.F., Zhao L., Callaway C.W., Hillas E., Zhang S., Cheng T., Inagami T., Ward K., Terreros D.A., & Lalouel J.M. Elements of a paracrine tubular renin angiotensin system along the entire nephron. *Hypertension.*, vol. 34, no. 6, (December 1999), pp. (1265-1274).

Rongen G.A., Lenders J.W.M., Smits P. & Thien T. Clinical pharmacy okinetics and efficacy of renin inhibitors. *Clin. Pharmacokinet.*, vol. 29, no. 1, (July 1995), pp. (6-14).

Ruzicka M., Yuan B., & Leenen F.H.M. Effects of enalapril versus losartan on regression of volume overload–induced cardiac hypertrophy in rats. *Circulation.*, vol. 90, no. 1, (July 1994), pp. (484-491).

Sadoshima J., Xu Y., Slayter H.S., & Izumo S. Autocrine release of angiotensin II mediates stretch-induced hypertrophy of cardiac myocytes in vitro. *Cell*, vol. 95, no. 5, (December 1993), pp. (977-984).

Saris J.J., Derkx F.H.M., de Bruin R.J.A., Dekkers D.H., Lamers J.M., Saxena P.M., Schalekamp M.A., & Danser A.H.J. High-affinity prorenin binding to cardiac man-6-P/IGF-II receptors precedes proteolytic activation to renin. *Am. J. Physiol.-Heart Circ. Physiol.*, vol. 280, no. 4, (April 2001), pp. (H1706–H1715).

Saris J.J., 't Hoen P.A., Garrelds I.M., Dekkers D.H., den Dunnen J.T., Lamers J.M., Danser A.H.J. Prorenin induces intracellular signalling in cardiomyocytes independently of angiotensin II. *Hypertension*, vol. 48, no. 4, (October 2006), pp. (564-571).

Satofuka S., Ichihara A., Nagai N., Yamashiro K., Koto T., Shinoda H., Noda K., Ozawa Y., Inoue M., Tsubota K., Suzuki F., Oike Y., & Ishida S. Suppression of ocular inflammation in endotoxin-induced uveitis by inhibiting nonproteolytic activation of prorenin. *Invest. Ophth. Vis. Sci.*, vol. 47, no. 6, (June 2006), pp. (2686-2692).

Satofuka S., Ichihara A., Nagai N., Koto T., Shinoda H., Noda K., Ozawa Y., Inoue M., Tsubota K., Itoh H., Oike Y., & Ishida S. Role of nonproteolytically activated prorenin in pathologic, but not physiologic, retinal neovascularization. *Invest. Ophth. Vis. Sci.*, vol. 48, no. 1, (January 2007), pp. (422-429).

Satofuka S., Ichihara A., Nagai N., Noda K., Ozawa Y., Fukamizu A., Tsubota K., Itoh H., Oike Y., & Ishida S. (Pro)renin receptor promotes choroidal neovascularization by activating its signal transduction and tissue renin-angiotensin system. *Am. J. Pathol.*, vol. 173, no. 6, (December 2008), pp. (1911–1918).

Satofuka S., Ichihara A., Nagai N., Noda K., Ozawa Y., Fukamizu A., Tsubota K., Itoh H., Oike, Y., & Ishida S. (Pro)renin receptor-mediated signal transduction and tissue renin-angiotensin system contribute to diabetes-induced retinal inflammation. *Diabetes*, vol. 58, no. 7, (July 2009), pp. (1625-1633).

Schefe J.H., Menk, M., Reinemund J., Effertz K., Hobbs R.M., Pandolfi P.P., Ruiz P., Unger T., Funke-Kaiser H. A novel signal transduction cascade involving direct physical interaction of the renin/prorenin receptor with the transcription factor promyelocytic zinc finger protein. *Circ. Res.*, vol. 99, no. 12, (December 2006), pp. (1355-1366).

Schieffer B., Wirger A., Meybrunn M., Seitz S., Holtz J., Riede U.N., & Drexler H. Comparative effects of chronic angiotensin-converting enzyme inhibition and angiotensin II type 1 receptor blockade on cardiac remodeling after myocardial infarction in the rat. *Circulation*, vol. 89, no. 5, (May 1994), pp. (2273-2282).

Schnermann J., & Briggs J.P. Function of the juxtaglomerular apparatus: control of glomerular hemodynamics and renin secretion. In: Alpern RJ, Hebert SC, editors. *The Kidney Physiology and Pathophysiology.* Burlington-San Diego-London: Elsevier Academic Press; 2008. pp. 589–626.

Schweda F., Friis U., Wagner C., Skott O., & Kurtz A. Renin release. *Physiology (Bethesda),* vol 22, (October 2007), pp. (310–319).

Sealey J.E., Catanzaro D.F., Lavin T.N., Gahnem, F., Pitarresi T., Hu L.F., & Laragh J.H. Specific prorenin/renin binding (ProBP). Identification and characterization of a novel membrane site. *Am. J. Hypertens.,* vol. 9, no. 5, (May 1996), pp. (491-502).

Sealey J.E., & Rubattu S. Prorenin and renin as separate mediators of tissue and circulating systems. *Am. J. Hypertens.,* vol. 2, no. 5, (May 1989), pp. (358-366).

Sealey J.E., Goldstein M., Pitarresi T., Kudlak T.T., Glorioso N., Fiamengo S.A., & Laragh J.H. Prorenin Secretion From Human Testis: No Evidence for Secretion of Active Renin or Angiotensinogen. J. Clin. Endocrinol. Metab., vol. 66, no. 5, (May 1988), pp. (974-978).

Shaw K.J., Do Y.S., Kjos S., Anderson P.W., Shinagawa T., Dubeau L., Hsueh W.A. Human decidua is a major source of renin. *J. Clin. Invest.,* vol. 83, no. 6, (June 1989), pp. (2085–2092).

Shinagawa, T., Do., Y.S., Daxter, J.D., Carili, C , Schilling, J. & Hseuh, W. Identification of an enzyme in human kidney that correctly processes prorenin. *Proc. Natl. Acad. Sci.,* U.S.A., vol. 87, no. 5, (March 1990), pp. (1927-1931).

Shinagawa T., Nakayama K., Uchiyama Y., Kominami E., Doi Y., Hashiba K., Yano K., Hseu W.A., & Murakami K. Role of cathepsin B as prorenin processing enzyme in human kidney. *Hypertens. Res.,* vo. 18, no. 2, (June 1995), pp. (131-136).

Strausberg R.L., Feingold E.A., Grouse L.H., Derge, J.G.; Klausner, R.D.; Collins, F.S.; Wagner, L.; Shenmen, C.M.; Schuler, G.D.; Altschul, S.F.; Zeeberg, B.; Buetow, K.H.; Schaefer, C.F.; Bhat, N.K.; Hopkins, R.F.; Jordan, H.; Moore, T.; Max, S.I.; Wang, J.; Hsieh, F.; Diatchenko, L.; Marusina, K.; Farmer, A.A.; Rubin, G.M.; Hong, L.; Stapleton, M.; Soares, M.B.; Bonaldo, M.F.; Casavant, T.L.; Scheetz, T.E.; Brownstein, M.J.; Usdin, T.B.; Toshiyuki, S.; Carninci, P.; Prange, C.; Raha, S.S.; Loquellano, N.A.; Peters, G.J.; Abramson, R.D.; Mullahy, S.J.; Bosak, S.A.; McEwan, P.J.; McKernan, K.J.; Malek, J.A.; Gunaratne, P.H.; Richards, S.; Worley, K.C.; Hale, S.; Garcia, A.M.; Gay, L.J.; Hulyk, S.W.; Villalon, D.K.; Muzny, D.M.; Sodergren, E.J.; Lu, X.; Gibbs, R.A.; Fahey, J.; Helton, E.; Ketteman, M.; Madan, A.; Rodrigues, S.; Sanchez, A.; Whiting, M.; Madan, A.; Young, A.C.; Shevchenko, Y.; Bouffard, G.G.; Blakesley, R.W.; Touchman, J.W.; Green, E.D.; Dickson, M.C.; Rodriguez, A.C.; Grimwood, J.; Schmutz, J.; Myers, R.M.; Butterfield, Y.S.; Krzywinski, M.I.; Skalska, U.; Smailus, D.E.; Schnerch, A.; Schein, J.E.; Jones, S.J.; Marra, M.A. Generation and initial analysis of more than 15,000 full-length human and mouse cDNA sequences", *Proc. Natl. Acad. Sci. USA,* vol. 99, no. 26, (December 2002), pp. (16899–16903).

Smith A.N., Skaug J., & Choateetal K.A. Mutations in ATP6N1B, encoding a new kidney vacuolar proton pump 116-kD subunit, cause recessive distal renal tubular acidosis with preserved hearing. Nature Genetics., vol. 26, no. 1, (September 2000), pp. (71–75).

Siragy H.M., & Carey R.M. The subtype-2 (AT) angiotensin receptor mediates renal production of nitric oxide in conscious rats. *J. Clin. Invest.*, vol. 100, no. 2, (1997), pp. (264–269).

Siragy H.M., & Huang J. Renal (pro)renin receptor upregulation in diabetic rats through enhanced angiotensin AT_1 receptor and NADPH oxidase activity. *Exp. Physiol.*, vol. 93, no. 5, (May 2008), pp. (709–714).

Susic D., Zhou X., Frohlich E.D., Lippton H., & Knight M. Cardiovascular effects of prorenin blockade in genetically hypertensive rats (SHR) on normal and high salt diet. *Am. J. Physiol. Heart Circ. Physiol.*, vol. 295, no. 3, (September 2008), pp. (H1117-H1121).

Sun Y., Cleutjens J.P.M., Diaz-Arias A.A., & Weber K.T. Cardiac angiotensin converting enzyme and myocardial fibrosis in the rat. *Cardiovasc. Res.*, vol 28, no. 9, (September 1994), pp. (1423–1432).

Suzuki F., Ludwig G., Hellmann W., Paul M., Lindpaintner K., Murakami K., & Ganten D. Renin gene expression in rat tissues: a new quantitative assay method for rat renin mRNA using synthetic cRNA. *Clin. Exp. Hyper. A.*, vol. 10, no. 2, (1987), pp. (345-359).

Suzuki F., Nakagawa T., Kakidachi H., Murakami K., Inagami T., & Nakamura Y. The dominant role of the prosegment of prorenin in determining the rate of activation by acid or trypsin: studies with molecular chimeras. Biochem. Biophys. Res. Commun., vol. 267, no. 2, (January 2000), pp. (:577-580).

Suzuki F., Hayakawa M., Nakagawa T., Uddin M.N., Ebihara A., Iwasawa A., Ishida Y., Nakamura Y., & Kazuo M. Human prorenin has 'gate and handle' regions for its non-proteolytic activation. *J. Biol. Chem.*, vol. 278, no. 25, (June 2003), pp. (22217–22222).

Tada, M.; Fukamizu, A.; Seo, M.S.; Takahashi, S.; Murakami, K. Renin expression in the kidney and brain is reciprocally controlled by captopril. *Biochem. Biophys. Res. Commun.*, vol. 159, no. 3, (March 1989), pp. (1065-1071).

Tada M., Takahashi S., Miyano M., & Miyake Y. Tissue-specific regulation of renin-binding protein gene expression in rats. *The J. Biochem.*, vol. 112, no. 2, (August 1992), pp. (175–182).

Takahashi H., Ichihara A., Kaneshiro Y., Inomata K., Sakoda M., Takemitsu T., Nishiyama A., & Itoh H. Regression of nephropathy developed in diabetes by (Pro)renin receptor blockade. *J. Am. Soc. Nephrol.*, vol 18, no. 7, (July 2007), pp. (2054-2061).

Takahashi K., Hiraishi K., Hirose T., Kato I., Yamamoto H., Shoji I., Shibasaki A., Kaneko K., Satoh, F., & Totsune K. Expression of (Pro)renin Receptor in the Human Brain and Pituitary, and Co-localisation with Arginine Vasopressin and Oxytocin in the Hypothalamus. *J. Neuroendocrinol.*, vol. 22, no. 5, (May 2010), pp. (453–459).

Takahashi S., Ohsawa T., Miura R., & Miyake Y. Purification and characterization of renin binding protein (RnBP) from porcine kidney. *The J. Biochem.*, vol. 93, no. 6, (June 1983), pp. (1583-1594).

Tigerstedt R., & Bergman P.G. Niere and Kreislauf. *Scand Arc Physiol.*, vol. **8**, no. , (1898), pp. (223-271).

Toei M., Saum R., & Forgac M. Regulation and isoform function of the V-ATPases. *Bochemistry*, vol. 49, no. 23, (June 2010), pp. (4715–4723).

van Esch J.H.M., van Veghel R, Garrelds I.M., Leijten F, Bouhuizen A. M., & Danser A.H.J. Handle Region Peptide Counteracts the Beneficial Effects of the Renin Inhibitor Aliskiren in Spontaneously Hypertensive Rats. *Hypertension*, vol. 57, no. 4, (February 2011), pp. (852-858).

Van Gilst W.H., de Graeff P.A., Kingma J.H., Wesseling H., & de Langen C.D.J. Captopril reduces purine loss and reperfusion arrhythmias in the rat heart after coronary artery ligation. *Eur. J. Pharmacol.*, vol. 100, no. 1, (April 1984), pp. 113-117.

van Kesteren C.A., Saris J.J., Dekkers D.H., Lamers J.M., Saxena P.R., Schalekamp M.A., Danser A.H.J. Cultured neonatal rat cardiac myocytes and fibroblasts do not synthesize renin or angiotensinogen: evidence for stretch-induced cardiomyocyte hypertrophy independent of angiotensin II. *Cardiovasc. Res.*, vol. 43, no. 1, (July 1999), pp. 148-156.

van Kesteren C.A.M., Danser A.H.J., Derkx F.H., Dekkers D.H., Lamers J.M., Saxena P.M., & Schalekamp M.A. Mannose 6-phosphate receptor-mediated internalization and activation of prorenin by cardiac cells. *Hypertension*, vol. 30, no. 6, (December 1997), pp. (1389–1396).

Wang P.H., Do Y.S., Macaulay L., Shinagawa T., Anderson P.W., Baxter J.D., & Hseuh W.A. (1991) Identification of renal cathepsin B as a human prorenin-processing enzyme. *J. Biol. Chem.*, vol. 266, no. 19, (July 1991), pp. (12633-12638).

Webb, D.J., Manhem, P.J., Ball, S.G., Inglis, G., Leckie, B.J., Lever, A.F., Morton, J.J., Robertson, J.I., Murray, G.D., Ménard, J., et al. A study of the renin inhibitor H142 in man. *J. Hypertens.*, vol. 3, no. 6, (December 1985), pp. (653–658).

Wilkinson-Berka, J.L., Heine R., Tan G., Cooper M.E., Hatzopoulos K.M., Fletcher E.L., Binger K.J., Campbell D.J., & Miller A.G. RILLKKMPSV influences the vasculature, neurons and glia, and (pro)renin receptor expression in the retina. *Hypertension*, vol. 55, no. 5, (June 2010), pp. (1454-1460).

Wilson C.M., Cherry M., Taylor B.A., & Wilson J.D. Genetic and endocrine control of renin activity in the submaxillary gland of the mouse. *Biochem. Genet.*, vol. 19, no. (5-6), (June 1981), pp. (509-523)

Wolny A., Clozel J.P., Rein J., Mory P., Vogt P., Turino M., Kiowski W. F., Ischli W. Functional and biochemical analysis of angiotensin II-forming pathways in the human heart. *Circ.Res.*, vol. 80, no. 2, (February 1997), pp. (219-227).

Wood J.M., Gulati N., Forgiarini P., Fuhrer W. & Hofbauer, K.G. Effects of a specific and long-acting renin inhibitor in the marmoset. *Hypertension*, vol. 7, no. 5, (October 1985), pp. (797–803).

Wood J.M., Criscione L., de Gasparo M., Bühlmayer P., Rüeger H., Stanton J.L., Jupp R.A. & Kay J. CGP 38 560: orally active, low-molecular-weight renin inhibitor with high potency and specificity. *J. Cardiovasc. Pharmacol.*, vol. 14, no. 2, (August 1989), pp. (221–226).

Wood J.M., Cumin F. & Maibaum J. Pharmacology of renin inhibitors and their application to the treatment of hypertension. *Pharmacol. Ther.*, vol. 61, no. 3, (1994) pp. (325-344).

Wood J.M., Maibaum J., Rahuel J., Grütter MG., Cohen NC., Rasetti V., Rüger H., Göschke R., Stutz S., Fuhrer W., Schilling W., Rigollier P., Yamaguchi Y., Cumin F., Baum HP., Schnell CR., Herold P., Mah R., Jensen C., O'Brien E., Stanton A., & Bedigian

MP. Structure-based design of aliskiren, a novel orally effective renin inhibitor. Biochem. Biophys. Res .Commun., vol. 308, no. 4, (September 2003), pp. (698-705).

Yokosawa H., Inagami T., & Haas E. Purification of human renin. *Biochem. Biophys. Res. Commun.*, vol. 83, no. 1, (July 1978), pp. (306-312).

Yokosawa H., Holladay L.A., Inagami T., Haas E. & Murakami K. Human renal renin: complete purification and characterization. *J. Biol. Chem.*, vol. 255, no. 8, (April 1980), pp. (3498-3502).

Yoshikawa A., Kusano Y.K., Kishi F., Kishi F,; Susumu T., Iida S., Ishiura S., Nishimura S., Shichiri M., & Senbonmatsu T. The (pro)renin receptor is cleaved by ADAM19 in the Golgi leading to its secretion into extracellular space. *Hypertens. Res.*, vol. 34, no. 5, (May 2011), pp. (599-605).

Yusuf S., on behalf of the SOLVD investigators. Effect of enalapril on survival in patients with reduced left ventricular ejection fractions and congestive heart failure. *N. Engl. J. Med.*, vol. 325, no. 5, (August 1991), pp. (293-302).

Zhang X., Dostal D.E., Reiss K., et al. Identification and activation of autocrine renin—angiotensin system in adult ventricular myocytes. *Am. J. Physiol.*, vol. 269, no. 5, (November 1995), pp. (H1791–H1802).

Zhang J., Noble N.A., Border W.A., Owens R.T., & Huang Y. Receptor-dependent prorenin activation and induction of PAI-1 expression in vascular smooth muscle cells. *Am. J. Physiol. Endocrinol. Metab.*, vol. 295, no. 4, (October 2008), pp. (E810–E819).

12

Cholesterol-Binding Peptides and Phagocytosis

Antonina Dunina-Barkovskaya
*Belozersky Institute of Physico-Chemical Biology
at Moscow Lomonosov State University
Russia*

1. Introduction

Phagocytosis is an important cellular process that in multicellular organisms ensures a defence against microbial invasion and removal of effete/apoptotic cells. Phenomenologically, phagocytosis is a process of internalization or engulfment by a cell of particles of a certain size (more than 0.5 µm) (Ofek et al., 1995; Pratten & Lloyd, 1986; Koval et al., 1998; Aderem & Underhill, 1999; Morrissette et al., 1999; Tjelle et al., 2000; May & Machesky, 2001; Djaldetti et al., 2002). After the contact of a particle with a phagocytozing cell, named "phagocyte" in the 19th century (*see* Heifets, 1982; Gordon, 2008), plasma membrane underneath the particle forms either invagination or extensions (pseudopods) surrounding the particle and eventually forms a vesicle (phagosome) that delivers the particle inside the cell.

The ability to phagocytoze is an integral feature of eucariotic cells, starting from single-celled animals to the higher vertebrates. In mammals, most of differentiated cells are able to phagocytoze to a certain extent; specialized cells named "professional phagocytes" (Rabinovitch, 1995) (monocytes, macrophages, neutrophils) do this most efficiently, but the activity of "non-professional" phagocytes is also very important both for anti-microbial defence and for tissue development, remodeling, and repair. For instance, macrophage-mediated phagocytosis plays a significant role in muscle tissue regeneration (Tidball & Wehling-Henricks, 2007). Senescent erythrocytes, mostly removed from circulation by macrophages, are also phagocytozed by epithelial cells of thyroid gland and urinary bladder (Aderem & Underhill, 1999). Fibroblasts incorporate solid particles – fragments of bone or prosthetic materials (Grinnell, 1984; Knowles et al., 1991). Lung epithelium cells can take up foreign particles inhaled with air (Kato et al., 2003; Saxena et al., 2008). Cells of retinal pigmented epithelium (RPE) phagocytoze and digest the shed outer segment membranes of rods (Rabinovitch, 1995; Aderem & Underhill, 1999; Krigel et al, 2010), and so on.

The vast diversity of the tasks and performances of the phagocytes may account for the fact that impairments in the phagocytic machinery accompany a number of serious illnesses, such as immunodeficiency (review of Lekstrom-Himes & Gallin, 2000), rheumatoid arthritis (Turner et al., 1973), retinal dystrophies (Gal et al., 2000), paroxysmal arrhythmia (James, 1994), cystic fibrosis and bronchiectasis (Vandivier et al., 2002). Macrophages are shown to be involved in promoting tumor angiogenesis, an essential step in the tumor progression to malignancy (Lin et al., 2006). Defective phagocytic clearance of apoptotic cells and macrophages as such are involved in the development of the aterosclerotic lesions that initiate acute thrombotic and vascular diseases, including myocardial infarction and stroke (Lucas & Greaves, 2001; Takahashi et al., 2002). Therefore, understanding the molecular mechanisms of phagocytosis is very important and should help to solve a number of medicinal problems and elaborate new approaches for regulation and control/correction of the phagocytic process.

By now, it is accepted that mechanism of phagocytosis implicates such processes as exocytosis, endocytosis, and adhesion (Aderem & Underhill, 1999; Botelho et al., 2000; Booth et al., 2001; Dunina-Barkovskaya, 2004; Lee at al., 2007; Fairn et al., 2010). A detailed list of the molecular participants that accomplish the initial membrane reorganization after the contact with the particle, subsequent formation and pinching-off of the phagosomal vesicle, and the components involved in the phagolysosome maturation and recycling has been created (Araki et al., 1996; Hackam et al., 1998; Morrissette et al., 1999; Garin et al., 2001; May & Machesky, 2001; Booth et al., 2001; Grinstein, 2010). This list includes receptors, membrane lipids, enzymes, cytoskeletal elements, ion-transporting systems (channels, exchanges, and pumps), and accessory cytoplasmic proteins required for membrane fusion, vesicle fission, and oxidative burst. However, molecular mechanism-based tuning of the phagocytic process in vivo and in vitro remains a challenge for the contemporary life sciences. This mini-review will briefly outline the role of cholesterol in the phagocytic process and consider cholesterol-binding peptides as potential tools for modulations and studies of the phagocytic process.

2. Cholestrol-dependence of the early stages of phagocytosis: What is cholesterol-dependent?

It has long been shown that the phagocytic process is cholesterol-dependent (Werb & Cohn, 1972) and very sensitive to sterols (Schreiber et al., 1975). Werb & Cohn, 1972, in their studies of the membrane composition changes following phagocytosis of latex particles showed that the ability to phagocytoze is regained several hours after the particle engulfment provided that the recovery medium contains cholesterol. Depleting plasma membrane cholesterol considerably inhibits phagocytosis (Peyron et al., 2000; Gatfield & Pieters, 2000). These observations raise a question: what molecular components accomplishing the phagocytic process are cholesterol-dependent?

2.1 Examples of cholesterol-dependence of phagocytic receptors

It is generally agreed that phagocytosis is triggered as a result of binding of cell membrane "phagocytic" receptors with their ligands on the particle surface. Ligand-receptor binding is followed by lateral clustering of the ligand–receptor complexes and by an unexplained way initiates a biochemical cascade leading to the actin polymerization at the sites of the vesicle formation. It is also assumed that the type of ligand (and the receptor involved) determines the "scenario" of the phagocytic process (Aderem & Underhill, 1999; Tjelle et al., 2000; Greenberg, 2001).

There are many reviews cataloging various types of phagocytic receptors and corresponding signaling cascades leading to the phagosome formation and detachment (Mosser, 1994;

Greenberg, 1995; Ofek et al., 1995; Aderem & Underhill, 1999; Astarie-Dequeker et al., 1999; Greenberg, 1999; Peyron, 2000; Tjelle et al., 2000; Greenberg, 2001; Djaldetti et al., 2002). In human phagocytes, the best studied are the receptors recognizing host serum immunoglobulin G (IgG) and complement C3 components (C3b and iC3b). These immune humoral factors are termed opsonins, and particles covered with opsonins are termed opsonized. Phagocytic receptor recognizing Fc-domain of IgG is termed FcγR, and that recognizing C3b and iC3b, complement 3 receptor, or CR3. In real life, however, phagocytes have to deal with non-opsonized particles, for example, in open wounds or in organs directly contacting with the environment (respiratory tract, gastro-intestinal tract) (Mosser, 1994; Ofek, 1995; Peyron, 2000; Djaldetti et al., 2002). In these cases an important role belongs to various receptors, such as mannose or beta-glucan receptors that bind integral components of the microorganism surface. To this group belong several receptors of macrophages, and in particular CD14, a receptor recognizing bacterial surface components, including lipopolysacharide (LPS); scavenger receptor A, as well as receptors CD36 and CD68 (macrosialin) that participate in phagocytosis of apoptotic cells. There is also phosphatidylserine receptor (PSR) recognizing phosphadidylserine that relocates from the inner to the outer monolayer of the plasma membrane of apoptotic cells (Fadok et al., 1992; Pradhan, 1997; Devitt et al., 1998; Giles et al., 2000; Li et al., 2003).

Although phagocytic receptors (as receptors in general, by definition) are considered "specialized", they are multispecific and multifunctional (Aderem & Underhill, 1999). This means that they recognize different ligands or certain molecular configurations ("pattern receptors") and can mediate other processes, such as endocytosis or adhesion. A striking example is CD36, multiligand scavenger receptor of class B. These ligands include thrombospondin-1, long-chain fatty acids, modified LDL, retinal photoreceptor outer segments, *Plasmodium falciparum* malaria-parasitized erythrocytes, sickle erythrocytes, anionic phospholipids, apoptotic cells, and collagens I and IV (Febbraio et al., 2001 and refs. therein). Another example is a complement receptor CR3 – also known as CD11b/CD18 and $\alpha_M\beta_2$, β_2-integrin – that functions not only as a membrane receptor recognizing iC3b but also as an adhesion molecule and binds diverse ligands, for example, intercellular adhesion molecule-1 (ICAM-1) (Ross & Větvicka, 1993; Ofek et al, 1995). Even Fcγ receptors may trigger either endocytosis or phagocytosis, depending on the size of the ligand-receptor cluster (Koval et al., 1998; Huang, 2006).

There are a number of works indicating that phagocytic processes involving certain phagocytic receptors are cholesterol-dependent. For example, Han et al., 1997 and Han et al., 1999 showed that lipoprotein lipids and cholesterol can upregulate the expression of the CD36 gene and protein. Moreover, according to Febbraio et al., 2001, CD36 colocalizes with caveolin-1 in specialized cholesterol- and sphingolipid-enriched microdomains of plasma membrane termed rafts. There are plenty of comprehensive reviews considering molecular structure, biophysics, and the roles of rafts in cell physiology (Simons & Ikonen, 1997; Brown & London, 2000; Pike, 2003; Lingwood & Simons, 2010, and references therein). In brief, rafts are defined as protein-lipid domains enriched with cholesterol and sphingomyelin. Rafts feature resistance to non-polar detergents (like Triton X-100), which points to strong interactions between molecules in a raft. At the same time, rafts are dynamic structures: they move laterally in the plane of the membrane, and big rafts can split into smaller ones, which in turn can fuse with each other. In artificial systems, in the absence of protein, rafts can be of micron size, while in cell membrane, raft size was reported to be several tens on nanometers. Cholesterol-sequestering agents (e.g., nystatin, filipin, β-cyclodextrin, etc.) or those interfering with its synthesis and metabolism (e.g., progesterone) prevent raft formation and inhibit cell processes in which rafts are involved, such as caveolar endocytosis. It is proposed that rafts may serve to concentrate signaling molecules and facilitate the integration of signaling cascades.

Another example of cholesterol- (and raft-) dependence concerns CR3-mediated phagocytosis. According to Peyron et al., 2000, nystatin and other cholesterol-sequestering agents (filipin, methyl-β-cyclodextrin) notably inhibit CR3-mediated phagocytosis of non-opsonized bacteria *Mycobacterium kansasii* by neutrophils. Moreover, phagocytosis is blocked if glycosylphosphatidylinositol- (GPI-) anchored proteins are removed with phosphatidyl inositol phospholipase C. The authors suggested that CR3-mediated phagocytosis of *Mycobacterium kansasii* requires binding of CR3 with GPI-ancored proteins localized in rafts. Once CR3 is not associated with rafts, it can mediate phagocytosis of zymozan or opsonized zymozan but not *Mycobacterium kansasii*. The observations suggest that the phagocytic "scenario" involving a given receptor, CR3, depends on the presence of cholesterol and interaction of the receptor with the lipid.

Fcγ-receptor-mediated phagocytosis is also cholesterol-dependent. Clustering of Fcγ receptors induced by binding to multiple opsonic ligands on a particle leads to phosphorylation of the Fcγ-immunoreceptor tyrosine-based activation motif (ITAM) by members of the Src family of kinases and followed by recruitment of the kinase Syk. Syk activation in turn initiates a signaling cascade, including activation of phosphatidylinositol 3-kinase (PI 3-kinase) and of the small GTPases Rac and Cdc42, which coordinate actin remodeling (Henry et al., 2004). As was reported (Kwiatkowska & Sobota, 2001; Katsumata et al., 2001; Kono et al., 2002; Kwiatkowska et al., 2003), one of the earliest signal events after the cross-binding of Fcγ receptors is lateral raft assembly that occurs before the activation of kinases of the Src family and independent of their activity. To trigger the lateral assembly of rafts, sufficient was the expression of ligand-binding monomer FcγR, without signal subunits carrying the activating fragment with tyrosine. Moreover, expression of the ligand-binding fragment of the receptor triggered fast mobilization of calcium. The authors (Kono et al., 2002; Kwiatkowska et al., 2003) suggested that lateral assembly of rafts is caused by ligand-binding subunits of the Fcγ-receptor and that it is the raft coalescence that triggers the signaling cascade of the biochemical reactions leading to rearrangements of membrane and cytoskeleton and eventually, to the formation of a membrane vesicle containing the particle. The importance of the integrity of the plasma membrane detergent-resistant microdomains for IgG-dependent phagocytosis was also shown by Marois et al., 2011. The authors also reported that phagocytosis of IgG-opsonized zymosan by human neutrophils required an extracellular influx of calcium that was blocked only by antibodies against FcγRIIIb. These data revive the question wheather FcγR may function as ligand-dependent channels (Young et al., 1983; Young et al., 1985). A quick local change of the ion channel activity at the site of the contact of the phagocyte with a particle remains a possible (but yet unexplored) step in the phagocytic signal cascade. In our hands, phagocytosis of non-opsonized beads by IC-21 macrophages is sensitive to methyl-β-cyclodextrin and carbenoxolon – glucocorticoid and a connexin channel blocker (Golovkina et al., 2009; Vishniakova et al., 2011).

Thus, even a brief overview of the very early stages of the phagocytic process shows that at least some of phagocytic receptors depend on cholesterol.

2.2 Phosphoinositides and lipid rafts in phagocytosis

It is recognized that successful phagosome formation requires local actin polymerization/depolymerization (Aderem & Underhill, 1999; Tjelle et al., 2000; May & Machesky, 2001; Greenberg, 2001; Lee at al., 2007) that dependes on the phosphatidylinsiotol metabolism. A detailed quantitative assessment of membrane remodeling during FcγR-mediated phagocytosis revealed marked changes in membrane composition that concerned the localization and metabolism of phosphoinositides (Botelho et al., 2000; Lee at al., 2007; Fairn et al., 2010). It was found in particular that at the onset of phagocytosis phosphatidylinositol (PI) 4,5-bisphosphate (PI4,5P$_2$) accumulates at sites where pseudopods are formed. Following the closure and fission of the phagosome, the phosphoinositide concentration drops, while phospholipase Cγ (PLCγ) is mobilized and local concentration of DAG increases. The authors suggested that this localized increase in PI4,5P$_2$ serves as a platform for the robust actin polymerization required for pseudopod extension. Recent works (Fairn et al., 2010; Grinstein, 2010) demonstrated in detail a highly localized sequence of changes in the level of several phosphoinositides as well as phosphatidylserine. The net changes in the content of these anionic phospholipids notably altered the surface charge of the membrane and caused the relocation of membrane-associated proteins due to electrostatic interactions. The authors hypothesize that modification of the membrane surface charge may play a role of an "electrostatic switch" that attracts or repulses proteins carrying polycationic or polyanionic motifs. Perhaps this hypothesis can be further extended: the role of such electrostatic switch may play a charged particle that touches the cell and triggers the phagocytic process.

Are there any correlations beween changes in the phosphoinisitide content and lipid raft assembly during phagocytosis? Lee at al., 2007 reported that there was an obvious clearance of the raft marker YFP-GPI from the base of forming phagosomes within minutes of particle contact and that this clearance resulted from the focal insertion of unlabeled endomembranes that are delivered focally by directed exocytosis. The authors interpreted these results as evidence against the raft involvement in the phagocytic process. However, these findings do not exclude the possibility that the raft formation preceded the observed changes in the phosphatidylinositide contents and possibly triggered these changes. The insertion of the new membranes should displace laterally and/or dilute the molecules and molecular ensembles (and rafts in particular) that initiated the phagocytic signaling cascade.

A number of works suggest close relations between rafts and phosphoinositides. For example, Hope & Pike, 1996 showed that polyphosphoinositide phosphatase, but not several other phosphoinositide-utilizing enzymes, is highly enriched in a low-density Triton-insoluble membrane fraction that contains caveolin; this fraction is also enriched in polyphosphoinositides, containing approximately one-fifth of the total cellular phosphatidylinositol (4,5)P$_2$. Defacque et al, 2002 suggested that PI(4,5)P$_2$ may exist in raft-like microdomains on latex bead phagosomes after isolation. On activation of cells with agonists or addition of ATP to the in vitro actin assay, PIPs are rapidly synthesized and may aggregate laterally into larger raft domains. The authors speculate that rafts may provide a platform for the proteins and lipids necessary for actin assembly to occur locally on the membrane of latex bead phagosomes.

Thus, the importance of phosphoinositides does not exclude the involvement of rafts in the phagocytosis mechanisms; in contrast, the dynamic functions of these lipid components of the plasma membrane appear to be highly coordinated in time and space throughout the course of the phagocytic process.

3. Cholesterol-binding sites in integral proteins involved in phagocytosis

Once phagocytosis is cholesterol-dependent and at least some of the phagocytic receptors aggregate in lipid rafts, it is important to know what structure(s) in these integral proteins makes them cholesterol-sensitive. Epand, 2006 reviewed the structural features of a protein that favour its association with cholesterol-rich domains. One of the best documented of these is certain types of lipidations; relatively new are the sterol-sensing domain (SSD) and the cholesterol recognition/interaction amino acid consensus (CRAC) domain. The latter was first described by Li & Papadopolous, 1998 for the peripheral-type benzodiazepine receptor (PBR). The CRAC sequence was formulated as $-L/V-(X)_{1-5}-Y-(X)_{1-5}-R/K-$; the presence of this site in C-terminus of PBR was necessary and sufficient for the cholesterol transport (Li & Papadopolous, 1998; Li et al., 2001). Moreover, the authors found a similar sequence in some other proteins known to interact with cholesterol (apolipoprotein A-I, some enzymes of steroid metabolism (e.g., P450scc (side-chain cleavage P450)), annexin, and some other proteins. Among them was caveolin – an integral protein accompanying caveolae, i.e., domains of plasma membrane containing lipid rafts and participating in caveolin-dependent endocytosis (Pelkmans et al., 2002).

A search for CRAC-like domains in the transmembrane areas of some integral proteins related to phagocytosis was performed by Cheshev et al., 2006. Apart from phagocytic receptor FcγR, the list of the molecules tested included proteins presumably involved in the regulation of the ionic composition in perimembrane cytoplasm during phagocytosis. Activation of phagocytes is accompanied by changes in the activity of the potassium channels of the IRK family (Eder, 1998; Arkett & Dixon, 1992; DeCoursey & Cherny, 1996; Colden-Stanfield, 2002). Ionotropic purinoreceptors P2X7 are found in monocytes/macrophages (North, 2002; Gudipatyet al., 2001); they cluster (Connon et al., 2003) and segregate in rafts (Torres et al., 1999). Gap-junction proteins are also known to accumulate in cholesterol-rich rafts (Schubert et al., 2002; Lin et al., 2003; Lin et al., 2004; Dunina-Barkovskaya, 2005). Connexin Cx43 is found in macrophages (Beyer & Steinberg, 1991; Beyer & Steinberg, 1993; Anand et al., 2008); colocalization of purinoreceptors P2X7 with connexin Cx43 in macrophages was reported (Beyer & Steinberg, 1991; Beyer & Steinberg, 1993; Fortes et al., 2004).

Figure 1 shows the fragments of the proteins aligned versus the CRAC sequence. Most of the proteins studied contained a conservative sequence of hydrophobic amino acids Val, Leu, Tyr, and Trp. This sequence can be described as follows: $(L/V)_{1-2}-(X)_{1-4}-Y/F-(X)_{1-3}-W$, which is close to the CRAC consensus $(L/V-(X)_{1-5}-Y-(X)_{1-5}-R/K)$. In contrast to PBR that has cholesterol-binding consensus in a cytoplasmic C-terminus, in the integral proteins (Fig. 1) the consensus sequence was always localized in the transmembrane domain. It may account for a slight difference between the PBR cholesterol-binding consensus and the sequences found.

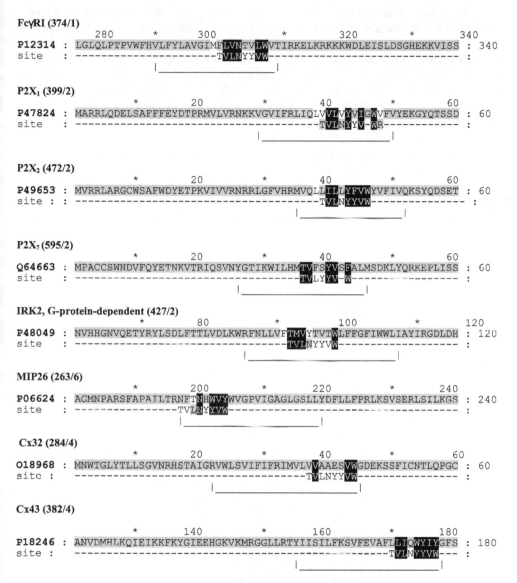

Fig. 1. Proposed cholesterol-binding sites in transmembrane domains of some proteins. Fragments of the proteins studied are highlighted with *grey colour*; a number at the left is the protein Uniport code (AC); numbers (and asterisks) above the fragments show the amino-acid residue numbers. Cholesterol-binding consensus (*TVLNYYVW*) is shown below the protein fragment (in the line marked "site"). Shown are the protein fragments (~60 amino-acid residues), in which the amino-acid sequence was comparable with the cholesterol-binding consensus. Identical or similar amino-acid residues are marked *black*. A square bracket below shows the position of transmembrane domains.

The presence in a transmembrane domain of the cholesterol-binding site that is potentially able to intract with cholesterol indicates that at least such interaction is possible. Binding of cholesterol at the level of the transmembrane domain of the receptor may not only account for segregation of the receptors in the lipid rafts at the early stages of phagocytosis (Kono et al., 2002; Kwiatkowska et al., 2003) but also favour an optimal FcγR configuration required for further interactions with kinases. As regards the channel proteins, binding of cholesterol may regulate the state of the ion-conducting pore. The channel molecules studied may be involved in the phagocytic process and their cholesterol-dependence may contribute to the cholesterol-dependence of the phagocytic process.

The presence of cholesterol-binding sequence in a number of integral proteins not only explains cholesterol-dependence of their functions. It seems very likely that experimental expression (and overexpression in particular) of integral proteins possessing cholesterol-binding sites may produce a cholesterol-sequestering effect in the transformed cells, similar to the effect of such cholesterol-sequestering agent as nystatin, methyl-β-cyclodextrin, etc. (Cheshev et al., 2006).

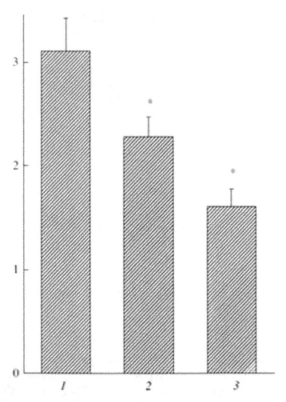

Fig. 2. Effect of cholesterol-binding peptide VLNYYVW on phagocytic activity of IC-21 macrophages (Dunina-Barkovskaya et al., 2007). Results of typical experiment are shown. 1, Control; 2, 1% DMSO; 3, VLNYYVW (100 μg/ml) in DMSO (1%). *Ordinate*, mean number of beads per cell. *, $p < 0.05$ vs. control.

The effect of a cholesterol-binding peptide VLNYYVW corresponding to the consensus cholesterol-binding sequence (Fig.1) on phagocytic activity was tested on IC-21 macrophages (Dunina-Barkovskaya et al., 2007). Phagocytosis was assessed by fluorescent microscopy, using 2-µm non-opsonized fluorescent latex beads. The peptide wase dissolved in DMSO, which turned to affect phagocytic activity by itself: in the presence of 0.5–1.3% DMSO the number of beads per cell was lower by 20–30% than in the absence of DMSO. Peptide VLNYYVW (5–100 µg/ml) augmented the inhibitory action of DMSO (Fig. 2). This result suggests that cholesterol-binding peptide may indeed affect the phagocytic process. What is the mechanism of this effect remains to be determined. Peptides may interfere with the interactions between integral proteins and membrane cholesterol and thus hinder the segregation of the proteins in cholesterol-rich domains. Another possibility is a cholesterol-sequestering effect similar to that exerted by nystatin, methyl-β-cyclodextrin and analogous substances. The formation of non-functional rafts in the membranes, like what was observed in artificial systems by Epand et al., 2003 is also possible. The authors showed that peptide LWYIK, a fragment of cholesterol-binding sequence, induced raft formation in phosphatidylcholine–cholesterol bilayer membranes.

4. Conclusion

Phagocytosis is an important cellular process underlying the innate and acquired immunity and involved in tissue remodelling throughout development or repair. Phagocytosis is a multi-stage process that engages endocytosis, exocytosis, and adhesion mechanisms. Highly coordinated local and dynamic rearrangements of the membrane underneath the target particle result in its engulfment and intracellular processing. Phagocytosis is a cholesterol-dependent process. One of the reason of this cholesterol dependency is the formation of cholesterol-enriched domains in plasma membrane, where phagocytic receptors (FcγR, in particular) may cluster and form supramolecular complexes required to set off and perform a further cascade of biochemical reactions leading to the rearrangements of cytoskeleton and formation of a membrane vesicle containing the particle. A molecular basis for direct interaction between integral proteins and membrane cholesterol can be provided by a cholesterol recognition/interaction amino acid consensus (CRAC) -L/V-$(X)_{1-5}$-Y-$(X)_{1-5}$-R/K-, described for a cholesterol-binding site of the peripheral-type benzodiazepine receptor (Li & Papadopolous, 1998; Li et al., 2001). Alignment of this site with amino acid sequences of a phagocytic receptor FcγRI and some ionic channels that may be involved in the phagocytic process and/or are capable of clustering in rafts (e.g., purinoreceptors and connexins) revealed that most of the proteins studied possessed a relatively conservative hydrophobic amino-acid sequence (Val-Leu---Tyr---Trp) analogous to that in the PBR cholesterol-binding site. This sequence was always localized in a transmembrane domain of a protein (Cheshev et al., 2006). Functional activity of a cholesterol-binding peptide VLNYYVW was tested and confirmed on cultured macrophages IC-21. Cholesterol-binding peptides can thus be a useful tool for further investigations and possibly serve for correction of phagocytosis and other cholesterol-dependent processes.

5. Acknowledgments

I appreciate enthusiasm and support of my colleaugues Kh.S. Vishniakova and I.I.Kireev.

6. References

Aderem & Underhill, 1999 Aderem, A. & Underhill, D.M. (1999). Mechanisms of Phagocytosis in Macrophages, *Ann. Rev. Immunol.*, vol. 17, pp. 593–623.

Anand R.J., Dai S., Gribar S.C., Richardson W., Kohler J.W., Hoffman R.A., Branca M.F., Li J., Shi X.H., Sodhi C.P., Hackam D.J. (2008). A Role for Connexin43 in Macrophage Phagocytosis and Host Survival after Bacterial Peritoneal Infection, *J. Immunol.*, vol. 181 (12), pp. 8534–8543.

Araki, N., Johnson, M.T. & Swanson, J.A. (1996). A Role for Phosphoinositide 3-Kinase in the Completion of Macropinocytosis and Phagocytosis by Macrophages, *J. Cell Biol.*, vol. 135, pp. 1249–1260.

Arkett, S.A., Dixon, S.J. & Sims, S.M. (1992). Substrate Influences Rat Osteoclast Morphology and Expression of Potassium Conductances, *J. Physiol.*, vol. 458, pp. 633–653.

Astarie-Dequeker, C., N'Diaye, E.-N., Le Cabec, V., Rittig, M.G., Prandi, J., Maridonneau-Parini, I. (1999). *Infection and Immunity*, vol. 67, pp. 469–477.

Beyer, E.C. & Steinberg, T.H. (1991). Evidence That the Gap Junction Protein Connexin–43 Is the ATP-Induced Pore of Mouse Macrophages, *J. Biol. Chem.*, vol. 266 (13), pp. 7971–7974.

Beyer, E.C. & Steinberg, T.H. (1993). Connexins, Gap-Junction Proteins, and ATP-Induced Pores in Macrophages, *Progress in Cell Research*, vol. 3. Gap Junctions. Ed. Hall J.E., Zampighi G.A., Davis R.M. Amsterdam: Elsevier Sci. Publ., pp. 71–74.

Booth, J.W., Trimble, W.S. & Grinstein, S. (2001). Membrane Dynamics in Phagocytosis, *Seminars Immunol.*, vol. 13, pp. 357–364.

Botelho, R.J., M. Teruel, R. Dierckman, R. Anderson, A. Wells, J.D. York, T. Meyer, and S. Grinstein. (2000). Localized Biphasic Changes in Phosphatidylinositol-4,5-bisphosphate at sites of Phagocytosis, *J. Cell Biol.*, vol. 151, pp. 1353–1368.

Brown, D.A. & London, E. (2000). Structure and Function of Sphingolipid- and Cholesterol-Rich Membrane Rafts, *J. Biol. Chem.*, vol. 275, pp. 17221–17224.

Cheshev, D.A., Chekanov, N.N. & Dunina-Barkovskaya, A.Ya. (2006). Cholesterol1Binding Sites in Transmembrane Domains of Integral Membrane Proteins Involved in Phagocytosis and/or Capable of Clustering in Lipid Rafts, *Biologicheskie membrany* (Rus.), vol. 23 (1), pp. 69–73.

Colden-Stanfield, M. (2002). Clustering of Very Late Antigen-4 Integrins Modulates K^+ Currents to Alter Ca^{2+}-mediated Monocyte Function, *Am. J. Physiol. Cell Physiol.*, vol. 283 (3), pp. C990–C1000.

Connon, C.J., Young, R.D. & Kidd, E.J. (2003). P2X7 Receptors Are Redistributed on Human Monocytes after Pore Formation in Response to Prolonged Agonist Exposure, *Pharmacology*, vol. 67 (3), pp. 163–168.

DeCoursey, T.E. & Cherny, V.V. (1996). Voltage-Activated Proton Currents in Human THP-1 Monocytes, *J. Membr. Biol.*, vol. 152, pp. 131–140.

Defacque, H., Bos, E., Garvalov, B., Barret, C., Roy, Ch., Mangeat, P., Shin, Hye-Won, Rybin, V. & Griffiths, G. (2002). Phosphoinositides Regulate Membrane-dependent Actin Assembly by Latex Bead Phagosomes. *Molecular Biology of the Cell*, vol. 13, 1190–1202.

Devitt, A., Moffatt, O.D., Raykundalia, C., Capra, J.D., Simmons, D.L. & Gregory, C.D. (1998). Human CD14 Mediates Recognition and Phagocytosis of Apoptotic Cells, *Nature*, vol. 392, pp. 505–509.

Djaldetti, M., Salman, H., Bergman, M., Djaldetti, R. & Bessler, H. (2002). Phagocytosis--the Mighty Weapon of the Silent Warriors. *Microscopy Research and Technique*. vol. 57, pp. 421–431.

Dunina-Barkovskaya, A.Y. (2004). Phagocytosis – Three in One: Endocytosis, Exocytosis, and Adhesion, *Biologicheskie membrany* (Rus.), vol. 21 (4), pp. 245–272.

Dunina-Barkovskaya, A.Ya. (2005). Are Gap Junctions Protein–Lipid Rafts? *Biologicheskie membrany* (Rus.), vol. 22 (1), pp. 27–33.

Dunina-Barkovskaya, A.Ya., Vishniakova, Kh.S., Cheshev, D.A., Chekanov, N.N., Bujurina, I.M. (2007). Effects of DMSO and Cholesterol-binding Peptides on Phagocytic Activity of Cultured Macrophages IC-21, *Biologicheskie membrany* (Rus.), vol. 24 (6), pp. 451–456.

Eder, C. (1998). Ion channels in microglia, *J. Cell Biol.*, vol. 275, pp. C327–C342.

Epand, R.M. (2006). Cholesterol and the Interaction of Proteins with Membrane Domains, *Prog Lipid Res.*, vol. 45 (4), pp. 279–294.

Epand, R.M., Sayer, B.G. & Epand R.F. (2003). Peptide-induced Formation of Cholesterol-rich Domains, *Biochemistry*, vol. 42 (49), pp. 14677–14689.

Fadok, V.A., Voelker, D.R., Campbell, P.A., Cohen, J.J., Bratton, D.L. & Henson, P.M. (1992). Exposure of Phosphatidylserine on the Surface of Apoptotic Lymphocytes Triggers Specific Recognition and Removal by Macrophages, *J. Immunol.*, vol. 148, pp. 2207–2216.

Fairn, G.D., Ogata, K., Botelho, R.J., Stahl, Ph.D., Anderson, R.A., de Camilli, P., Meyer, T., Wodak, Sh. & Grinstein, S. (2010). An Electrostatic Switch Displaces Phosphatidylinositol Phosphate Kinases from the Membrane during Phagocytosis, *J. Cell Biol.*, vol. 187 (5), pp. 701–714.

Febbraio, M., Hajjar, D.P. & Silverstein, R.L. (2001). CD36: A Class B Scavenger Receptor Involved in Angiogenesis, Aterosclerosis, Inflammation and Lipid Metabolism, *J. Clin. Invest.*, vol. 108, pp. 785–791.

Fortes, F.S.A., Pecora, I.L., Persechini1, P.M., Hurtado, S., Costa, V., Coutinho,1S.R., Braga, M.B.M., Silva-Filho, F.C., Bisaggio, R.C., de Farias, F.P., Scemes, E., Campos de Carvalho, A.C. & Goldenberg, R.C.S. (2004). Modulation of Intercellular Communication in Macrophages: Possible Interactions between Gap Junctions and P2 Receptors, *J. Cell Sci.*, vol. 117, pp. 4717–4726.

Fuki, I.V., Meyer, M.E. & Williams K.J. (2000). Transmembrane and Cytoplasmic Domains of Syndecan Mediate a Multi-step Endocytic Pathway Involving Detergent-insoluble Membrane Rafts, *Biochem. J.*, vol. 351 (3), pp. 607–612.

Gal, A., Li, Y., Thompson, D.A., Weir, J., Orth, U., Jacobson, S.G., Apfelstedt-Sylla, E. & Vollrath, D. (2000). Mutations in MERTK, the Human Orthologue of the RCS Rat Retinal Dystrophy Gene, Cause Retinitis Pigmentosa, *Nat. Genet.*, vol. 26 (3), pp. 270–271.

Garin, J., Diez, R., Kieffer, S., Dermine, J.F., Duclos, S., Gagnon, E., Sadoul, R., Rondeau, C. & Desjardins, M. (2001). The Phagosome Proteome: Insight into Phagosome Functions, *J. Cell Biol.*, vol. 152, pp. 165–180.

Gatfield, J. & Pieters, J. (2000). Essential Role for Cholesterol in Entry of Mycobacteria into Macrophages, *Science*, vol. 288 (5471), pp. 1647–1650.

Giles, K.M., Hart, S.P., Haslett, C., Rossi, A.G. & Dransfield, I. (2000). An Appetite for Apoptotic Cells? Controversies and Challenges, *Br. J. Haematol.*, vol. 109, pp. 1–12.

Golovkina, M.S., Skachkov, I.V., Metelev, M.V., Kuzevanov, A.V., Vishniakova, Kh.S., Kireev, I.I. & Dunina-Barkovskaya, A.Ya. (2009). Serum-Induced Inhibition of the Phagocytic Activity of Cultured Macrophages IC-21. *Biologicheskie membrany* (Rus.), vol. 26 (5), pp. 379–386 [Translated version in: *Biochemistry (Moscow) Suppl. Series A: Membrane and Cell Biology* (2009), vol. 4 (3), pp. 412–419].

Gordon, S. (2008). Elie Metchnikoff: Father of Natural Immunity, *Eur. J. Immunol.*, vol. 38 (12), pp. 3257–3264.

Greenberg, S. (2001). Diversity in Phagocytic Signalling, *J. Cell Science*, vol. 114, pp. 1039–1040.

Greenberg, S. (1995). Signal Transduction of Phagocytosis, *Trend Cell Biol.*, vol. 5, pp. 93–99.

Greenberg, S. (1999). Modular Components of Phagocytosis, *J. Leukoc. Biol.*, vol. 66, pp. 712–717.

Grinnell, F. (1984). Fibroblast Spreading and Phagocytosis: Similar Cell Responses to Different-sized Substrata, *J. Cell Physiol.*, vol. 119, pp. 58–64.

Grinstein, S. (2010) Imaging Signal Transduction during Phagocytosis: Phospholipids, Surface Charge, and Electrostatic Interactions, *Am. J. Physiol. Cell Physiol.*, vol. 299 (5), C876–C881.

Gudipaty, L., Humphreys, B.D., Buel, G. & Dubyak, G.R. (2001). Regulation of P2X 7 nucleotide receptor function in human monocytes by extracellular ions and receptor density, *Am. J. Physiol. Cell Physiol.*, vol. 280, pp. C943–C953.

Hackam, D.J., Rotstein, O.D., Sjolin, C., Schreiber, A.D., Trible, W.S. & Grinstein, S. (1998). v-SNARE-Dependent Secretion is Required for Phagocytosis, *Proc. Natl. Acad. Sci. USA.*, vol. 95, pp. 11691–11696.

Han, J., Hajjar, D.P., Febbraio, M. & Nicholson, A.C. (1997). Native and Modified Low Density Lipoproteins Increase the Functional Expression of the Macrophage Class B Scavenger Receptor, CD36, *J. Biol. Chem.*, vol. 272, pp. 21654–21659.

Han, J., Hajjar, D.P., Tauras, J.M. & Nicholson, A.C. (1999). Cellular Cholesterol Regulates Expression of the Macrophage Type B Scavenger Receptor, CD36, *J. Lipid. Res.*, vol. 40, pp. 830–838.

Heifets, L. (1982). Centennial of Metchnikoff's Discovery, *J.Reticuloendothel. Soc.*, vol. 31 (5), pp. 381–391.

Henry, R.M., Hoppe, A.D., Joshi, N. & Swanson, J.A. (2004). The Uniformity of Phagosome Maturation in Macrophages, J. Cell Biol., vol. 164 (2), pp. 185–194.

Hope, H.R. & Pike, L.J. (1996). Phosphoinositides and Phosphoinositide-utilizing Enzymes in Detergent-insoluble Lipid Domains, *Mol. Biol. Cell*, vol. 7 (6), pp. 843–851.

Huang, Z.Y., Barreda, D.R., Worth, R.G., Indik, Z.K., Kim, M.K., Chien, P. & Schreiber, A.D. (2006). Differential Kinase Requirements in Human and Mouse Fcγ Receptor Phagocytosis and Endocytosis, *J. Leukoc. Biol.*, vol. 80 (6), pp. 1553–1562.

James, T.N. (1994). Normal and Abnormal Consequences of Apoptosis in the Human Heart from Postnatal Morphogenesis to Paroxysmal Arrhythmias, *Circulation*, vol. 90, pp. 556–573.

Kato, T., Yashiro, T., Murata, Y., Herbert, D.C., Oshikawa, K., Bando, M., Ohno, S. & Sugiyama, Y. (2003). Evidence that Exogenous Substances Can Be Phagocytized by Alveolar Epithelial Cells and Transported into Blood Capillaries, *Cell Tissue Res.*, vol. 311, pp. 47–51.

Katsumata, O., Hara-Yokoyama, M., Sautès-Fridman, C., Nagatsuka, Y., Katada, T., Hirabayashi, Y., Shimizu, K., Fujita-Yoshigaki, J., Sugiya, H. & Furuyama, S. (2001). Association of FcγRII with Low-density Detergent-resistant Membranes Is Important for Cross-linking-dependent Initiation of the Tyrosine Phosphorylation Pathway and Superoxide Generation, *J. Immunol.*, vol. 167 (10), pp. 5814–5823.

Knowles, G.C., McKeown, M., Sodek, J. & McCulloch, C.A. (1991). Mechanism of Collagen Phagocytosis by Human Gingival Fibroblasts: Importance of Collagen Structure in Cell Recognition and Internalization, *J. Cell Sci.*, vol. 98, pp. 551–558.

Kono, H., Suzuki, T., Yamamoto, K., Okada, M., Yamamoto, T. & Honda, Z. (2002). Spatial Raft Coalescence Represents an Initial Step in FcγR Signaling, *J.Immunol.*, vol. 169, pp. 193–203.

Koval, M., Preiterb, K., Adles, C., Stahl, P.D. & Steinberg, T.H. (1998). Size of IgG-opsonized Particles Determines Macrophage Response during Iternalization, *Exp. Cell Res.*, vol. 242, pp. 265–273.

Krigel, A., Felder-Schmittbuhl, M.P. & Hicks, D. (2010). Circadian-Clock Driven Cone-Like Photoreceptor Phagocytosis in the Neural Retina Leucine Zipper Gene Knockout Mouse, *Mol. Vis.*, vol. 16, pp. 2873–2881.

Kwiatkowska, K. & Sobota, A. (2001). The Clustered Fcγ Receptor II Is Recruited to Lyn-containing Membrane Domains and Undergoes Phosphorylation in a Cholesterol-dependent Manner, *Eur. J. Immunol.*, vol. 31 (4), pp. 989–998.

Kwiatkowska, K., Frey, J. & Sobota A. (2003). Phosphorylation of FcγRIIA Is Required for the Receptor-induced Actin Rearrangement and Capping: The Role of Membrane Rafts, *J. Cell Sci.*, vol. 116 (Pt 3), pp. 537–550.

Lee, W.L., Mason, D., Schreiber, A.D. & Grinstein S. (2007). Quantitative Analysis of Membrane Remodeling at the Phagocytic Cup. Molecular Biology of the Cell. Vol. 18, 2883–2892

Lekstrom-Himes, J.A. & Gallin, J.I. (2000). Immunodeficiency Diseases Caused by Defects in Phagocytes, *N. Engl. J. Med.*, vol. 343 (23), pp. 1703–1714.

Li, H. & Papadopoulos, V. (1998). Peripheral-type Benzodiazepine Receptor Function in Cholesterol Transport. Identification of a Putative Cholesterol Recognition/Interaction Amino Acid Sequence and Consensus Pattern, *Endocrinology*, vol. 139 (12), pp. 4991–4997.

Li, H., Yao, Z., Degenhardt, B., Teper, G. & Papadopoulos, V. (2001). Cholesterol Binding at the Cholesterol Recognition/Interaction Amino Acid Consensus (CRAC) of the

Peripheral-Type Benzodiazepine Receptor and Inhibition of Steroidogenesis by an HIV TAT-CRAC Peptide, *Proc. Natl. Acad. Sci. USA*, vol. 98, pp. 1267–1272.

Li, M.O., Sarkisian, M.R., Mehal, W.Z., Rakic, P. & Flavell R.A. (2003). Phosphatidylserine Receptor Is Required for Clearance of Apoptotic Cells, *Science*, vol. 302, pp. 1560–1563.

Lin, D., Lobell, S., Jewell, A. & Takemoto, D.J. (2004). Differential Phosphorylation of Connexin 46 and Connexin 50 by H_2O_2 Activation of Protein Kinase Cγ, *Mol. Vis.*, vol. 10, pp. 688–695.

Lin, D., Zhou, J., Zelenka, P.S. & Takemoto, D.J. (2003). Protein Kinase Cγ Regulation of Gap Junction Activity through Caveolin-1-containing Lipid Rafts, *Invest. Ophthalmol. Vis. Sci.*, vol. 44 (12), pp. 5259–5268.

Lin, E.Y,, Li, J.F., Gnatovskiy, L., Deng, Y., Zhu, L., Grzesik, D.A., Qian, H., Xue, X.N. & Pollard, J.W. (2006). Macrophages Regulate the Angiogenic Switch in a Mouse Model of Breast Cancer, *Cancer Res.*, vol. 66 (23), pp. 11238–11246.

Lingwood, D. & Simons, K. (2010). Lipid Rafts As a Membrane-Organizing Principle, *Science,* vol. 327, pp. 46–50.

Lucas, A.D. & Greaves, D.R. (2001). Atherosclerosis: Role of Chemokines and Macrophages, *Expert. Rev. Mol. Med.*, vol. 3 (25), pp. 1–18.

Marois, L., Paré, G., Vaillancourt, M., Rollet-Labelle, E. & Naccache, P.H. (2011) FcγRIIIb Triggers Raft-dependent Calcium Influx in IgG-mediated Responses in Human Neutrophils, *J. Biol. Chem.*, vol. 286 (5), pp. 3509–3519.

May, R.C. & Machesky, L.M. (2001). Phagocytosis and the Actin Cytoskeleton, *J. Cell Sci.*, vol. 114, pp. 1061–-1077.

Morrissette, N.S., Gold, E.S., Guo, J., Hamermann, J.A., Ozinsky, A., Bedian, V. & Aderem, A.A. (1999). Isolation and Characterization of Monoclonal Antibodies Directed against Novel Components of Macrophage Phagosomes, *J. Cell Science*, vol. 119, pp. 4705–4713.

Mosser, D.M. (1994). Receptors on Phagocytic Cells Involved in Microbial Recognition, *Immunol. Ser.*, vol. 60, pp. 99–114.

North, R.A. (2002). Molecular Physiology of P2X Receptors, *Physiol. Rev.*, vol. 82 (4), pp. 1013–1067.

Ofek, I., Goldhar, J. & Sharon, N. (1995). Nonopsonic Phagocytosis of Microorganisms, *Ann. Rev. Microbiol.*, vol. 49, pp. 239–276.

Pelkmans, L., Puentener, D. & Helenius, A. (2002). Local Actin Polymerization and Dynamin Recruitment in SV40-induced internalization of Caveolae, *Science*, vol. 296, pp. 535–539.

Peyron, P., Bordier, C., N'Diaye, E.-N. & Maridonneau-Parini, I. (2000). Nonopsonic Phagocytosis of *Mycobacterium kansasii* by Human Neutrophils Depends on Cholesterol and Is Mediated by CR3 Associated with Glycosylphosphatidylinositol-Anchored Proteins, *J. Immunol.*, vol. 165, pp. 5186–5191.

Pike, L.J. (2003). Lipid Rafts: Bringing Order to Chaos, *Journal of Lipid Research*, vol. 44 (4), pp. 655–667.

Pradhan, D., Krahling, S., Williamson, P. & Schlegel, R.A. (1997). Multiple Systems for Recognition of Apoptotic Lymphocytes by Macrophages, *Mol.Biol. Cell*, vol.8, pp. 767–778.

Pratten, M.K. & Lloyd, J.B. (1986). Pinocytosis and Phagocytosis: The Effect of Size of a Particulate Substrate on Its Mode of Capture by Rat Peritoneal Macrophages Cultured In Vitro, *Biochim. et biophys. Acta*, vol. 881, pp. 307–313.

Rabinovitch, M. (1995). Professional and Nonprofessional Phagocytes: An Introduction, *Trends Cell Biol.*, vol. 5, pp. 85–87.

Ross, G.D. & Větvicka, V. (1993). CR3 (CD11b, CD18): A Phagocyte and NK Cell Membrane Receptor with Multiple Ligand Specificities and Functions, *Clin. Exp. Immunol.*, vol. 92 (2), pp.181–184.

Saxena, R.K., Gilmour, M.I. & Hays, M.D. (2008). Isolation and Quantitative Estimation of Diesel Exhaust and Carbon Black Particles Ingested by Lung Epithelial Cells and Alveolar Macrophages In Vitro, *Biotechniques*, vol. 44 (6), pp. 799–805.

Schreiber, A.D., Parsons, J., Mcdermot, P. & Cooper R.A. (1975) Effect of Corticosteroids on the Human Monocyte IgG and Complement Receptors, J. Clin. Invest. Vol. 56, pp. 1189–1197.

Schubert, A.L, Schubert, W., Spray, D.C. & Lisanti, M.P. (2002). Connexin Family Members Target to Lipid Raft Domains and Interact with Cavcolin-1, *Biochemistry*, vol. 41 (18), pp. 5754–5764.

Simons, K. & Ikonen, E. (1997). Functional Rafts in Cell Membranes, *Nature*, vol. 387, pp. 569–572.

Takahashi, K., Takeya, M. & Sakashita, N. (2002). Multifunctional Roles of Macrophages in the Development and Progression of Atherosclerosis in Humans and Experimental Animals, *Med. Electron Microsc.*, vol. 35 (4), pp. 179–203.

Tidball, J.G. & Wehling-Henricks, M. (2007). Macrophages Promote Muscle Membrane Repair and Muscle Fibre Growth and Regeneration during Modified Muscle Loading in Mice In Vivo, *J. Physiol.*, vol. 578 (Pt 1), pp. 327–336.

Tjelle, T.E., Levdal, T. & Berg, T. (2000). Phagosome Dynamics and Function, *BioEssays*, vol. 22, pp. 255–263.

Torres, G.E., Egan, T.M. & Voigt, M.M. (1999). Hetero-oligomeric Assembly of P2X Receptor Subunits, *J. Biol. Chem.*, vol. 274, pp. 6653–6659.

Turner, R.A., Schumacher H. R. & Myers A.R. (1973). Phagocytic Function of Polymorphonuclear Leukocytes in Rheumatic Diseases, *J. Clin. Invest.*, vol. 52 (7), 1632–1635.

Vandivier, R.W., Fadok, V.A., Hoffman, P.R., Bratton, D.L., Penvari, C., Brown, K.K., Brain, J.D., Accurso, F.J. & Henson P.M. (2002). Elastase-Mediated Phosphatidylserine Receptor Cleavage Impairs Apoptotic Cell Clearance in Cystic Fibrosis and Bronchiectasis, *J. Clin. Invest.*, vol. 109, pp. 661–670.

Vishniakova Kh. S., Kireev I. I. & Dunina-Barkovskaya A.Ya. (2011). Effects of Cell Culture Density on Phagocytosis Parameters in IC-21 Macrophages, *Biochemistry (Moscow) Supplement Series A: Membrane and Cell Biology*, vol. 5 (4), pp. 355–363 [Original Russian Text in: *Biologicheskie membrany* (Rus.), vol. 28 (5), pp. 387–396].

Werb, Z. & Cohn, Z.A. (1972). Plasma Membrane Synthesis in the Macrophage Following Phagocytosis of Polystyrene Latex Particles, *J. Biol. Chem.*, vol. 247, pp. 2439–2446.

Yeung, T., Heit, B., Dubuisson, J.-F., Fairn, G.D., Chiu, B., Inman R., Kapus, A., Swanson, M., & Grinstein, S. (2006). Contribution of Phosphatidylserine to Membrane Surface Charge and Protein Targeting during Phagosome Maturation, *J. Cell Biol.*, vol. 185 (5), pp. 917–928.

Young J.D., Unkeless J.C., Cohn Z.A. (1985). Functional Ion Channel Formation by Mouse Macrophage IgG Fc Receptor Triggered by Specific Ligands. J. Cell Biochem., vol. 29, pp. 289–297.

Young J.D., Unkeless J.C., Young T.M., Mauro A., Cohn Z.A. (1983). Role for Mouse Macrophage IgG Fc Receptor as Ligand-Dependent Ion Channel, *Nature,* vol. 306 (5939), pp. 186–189.

Permissions

The contributors of this book come from diverse backgrounds, making this book a truly international effort. This book will bring forth new frontiers with its revolutionizing research information and detailed analysis of the nascent developments around the world.

We would like to thank Jianfeng Cai, for lending his expertise to make the book truly unique. He has played a crucial role in the development of this book. Without his invaluable contribution this book wouldn't have been possible. He has made vital efforts to compile up to date information on the varied aspects of this subject to make this book a valuable addition to the collection of many professionals and students.

This book was conceptualized with the vision of imparting up-to-date information and advanced data in this field. To ensure the same, a matchless editorial board was set up. Every individual on the board went through rigorous rounds of assessment to prove their worth. After which they invested a large part of their time researching and compiling the most relevant data for our readers. Conferences and sessions were held from time to time between the editorial board and the contributing authors to present the data in the most comprehensible form. The editorial team has worked tirelessly to provide valuable and valid information to help people across the globe.

Every chapter published in this book has been scrutinized by our experts. Their significance has been extensively debated. The topics covered herein carry significant findings which will fuel the growth of the discipline. They may even be implemented as practical applications or may be referred to as a beginning point for another development. Chapters in this book were first published by InTech; hereby published with permission under the Creative Commons Attribution License or equivalent.

The editorial board has been involved in producing this book since its inception. They have spent rigorous hours researching and exploring the diverse topics which have resulted in the successful publishing of this book. They have passed on their knowledge of decades through this book. To expedite this challenging task, the publisher supported the team at every step. A small team of assistant editors was also appointed to further simplify the editing procedure and attain best results for the readers.

Our editorial team has been hand-picked from every corner of the world. Their multi-ethnicity adds dynamic inputs to the discussions which result in innovative outcomes. These outcomes are then further discussed with the researchers and contributors who give their valuable feedback and opinion regarding the same. The feedback is then collaborated with the researches and they are edited in a comprehensive manner to aid the understanding of the subject.

Apart from the editorial board, the designing team has also invested a significant amount of their time in understanding the subject and creating the most relevant covers. They scrutinized every image to scout for the most suitable representation of the subject and create an appropriate cover for the book.

The publishing team has been involved in this book since its early stages. They were actively engaged in every process, be it collecting the data, connecting with the contributors or procuring relevant information. The team has been an ardent support to the editorial, designing and production team. Their endless efforts to recruit the best for this project, has resulted in the accomplishment of this book. They are a veteran in the field of academics and their pool of knowledge is as vast as their experience in printing. Their expertise and guidance has proved useful at every step. Their uncompromising quality standards have made this book an exceptional effort. Their encouragement from time to time has been an inspiration for everyone.

The publisher and the editorial board hope that this book will prove to be a valuable piece of knowledge for researchers, students, practitioners and scholars across the globe.

List of Contributors

N.V. Gorbunov, B.R. Garrison and M. Zhai
The Henry M. Jackson Foundation for the Advancement of Military Medicine, Inc.

D.P. McDaniel
The Department of Microbiology and Immunology, School of Medicine,

G.D. Ledney, T.B. Elliott and J.G. Kiang
Radiation Combined Injury Program, Armed Forces Radiobiology Research Institute, Uniformed Services University of the Health Sciences, Bethesda, Maryland, USA

Hong Yin and Jonathan Glass
Department of Medicine and the Feist-Weiller Cancer Center, LSU Health Sciences Center, Shreveport, LA

Kerry L. Blanchard
Eli Lilly & Company, Indianapolis, IN, USA

Youngshim Lee and Vadim Gaponenko
Department of Biochemistry and Molecular Genetics, University of Illinois at Chicago, Chicago, IL

Sherwin J. Abraham
Department of Molecular and Cellular Physiology, Stanford University, School of Medicine, Beckman Center, Stanford, CA, USA

Yifat Yanku and Amir Orian
Technion-Israel Institute of Technology, Israel

Natalie Zeytuni and Raz Zarivach
Department of Life Sciences, Ben-Gurion University of the Negev and the National Institute for Biotechnology in the Negev (NIBN), Beer sheva, Israel

Aditya Rao, Gopalakrishnan Bulusu, Rajgopal Srinivasan and Thomas Joseph
Life Sciences Division, TCS Innovation Labs, Tata Consultancy Services, Hyderabad India

Hiroaki Yokota
Institute for Integrated Cell-Material Sciences, Kyoto University, Japan

Verena Arndt
German Center for Neurodegenerative Diseases (DZNE), Bonn,

Ina Vorberg
Department of Neurology, Rheinische Friedrich-Wilhelms-University Bonn, Germany

Thomas L. Sims
Department of Biological Sciences, Northern Illinois University, USA

A.H.M. Nurun Nabi
Department of Biochemistry and Molecular Biology, University of Dhaka, Dhaka, Bangladesh

Fumiaki Suzuki
Laboratory of Animal Biochemistry, Faculty of Applied Biological Sciences, Gifu University, 1-1 Yanagido, Gifu, Japan

Antonina Dunina-Barkovskaya
Belozersky Institute of Physico-Chemical Biology at Moscow Lomonosov State University, Russia

Printed in the USA
CPSIA information can be obtained
at www.ICGtesting.com
JSHW011440221024
72173JS00004B/877